Molecular and Cell Biology of Human Gene Therapeutics

Molecular and Cell Biology of Human Diseases Series

Series Editors

D.J.M. WRIGHT MD FR CPath

Reader in Medical Microbiology,
Charing Cross and Westminster School, London, UK

L.C. ARCHARD PhD

Senior Lecturer in Biochemistry,
Charing Cross and Westminster Medical School, London, UK

The continuing developments in molecular biology have made possible a new approach to a whole range of different diseases. The books in this series each concentrate on a disease or group of diseases where real progress is being made in understanding the pathogenesis, diagnosis and management. Experts on aspects of each area provide a text accessible to scientists and clinicians in a form which records advances and points to the potential application of these advances in a clinical setting.

Other titles in this series

1 *Molecular and Cell Biology of Sexually Transmitted Diseases*
Edited by David Wright and Leonard Archard

2 *Molecular and Cell Biology of Opportunistic Infections in AIDS*
Edited by Steven Myint and Alan Cann

3 *Molecular and Cell Biology of Muscular Dystrophy*
Edited by Terence Partridge

4 *Molecular and Cell Biology of Neuropsychiatric Diseases*
Edited by Frank Owen and Ruth Itzhaki

Molecular and Cell Biology of Human Gene Therapeutics

EDITED BY

George Dickson

Department of Biochemistry,
Royal Holloway, University of London, UK

CHAPMAN & HALL

London · Glasgow · Weinheim · New York · Tokyo · Melbourne · Madras

Published by Chapman & Hall, 2–6 Boundary Row, London SE1 8HN, UK

Chapman & Hall, 2–6 Boundary Row, London SE1 8HN, UK

Blackie Academic & Professional, Wester Cleddens Road, Bishopbriggs, Glasgow G64 2NZ, UK

Chapman & Hall GmbH, Pappalallee 3, 69469 Weinheim, Germany

Chapman & Hall USA, 115 Fifth Avenue, New York NY10003, USA

Chapman & Hall Japan, ITP-Japan, Kyowa Building, 3F, 2–2–1 Hirakawacho, Chiyoda-ku, Tokyo 102, Japan

Chapman & Hall Australia, Thomas Nelson Australia, 102 Dodds Street, South Melbourne, Victoria 3205, Australia

Chapman & Hall India, R. Seshadri, 32 Second Main Road, CIT East, Madras 600 035, India

---

First edition 1995

Typeset in 10/12 Sabon by Florencetype Ltd, Stoodleigh, Devon
Printed in Great Britain by St Edmundsbury Press, Bury St Edmunds, Suffolk

ISBN 0 412 62550 4

A catalogue record for this book is available from the British Library

Library of Congress Cataloging-in-Publication data: 95–67601

∞ Printed on permanent acid-free text paper, manufactured in accordance with ANSI/NISO Z39.48-1992 and ANSI/NISO Z39.48–1984 (Permanence of Paper).

Contents

Contributors

Mary Ann Bender
University of Pittsburgh
School of Medicine
Department of Molecular Genetics and Biochemistry
E1246 Biomedical Science Tower
Pittsburgh
Pennsylvania 15261
USA

John Bridgewater
Chester Beatty Laboratories
Institute of Cancer Research
237 Fulham Road
London SW3 6JB
UK

Susan C. Brown
Department of Biochemistry
Royal Holloway
University of London
Egham
Surrey TW20 0EX
UK

Marinee K.L. Chuah
Clinical Gene Therapy Branch
National Center for Human Genome Research
National Institutes of Health
9000 Rockvill Pike
Bethesda
Maryland 20892; and
Genetic Therapy Inc
19 Firstfield Road
Gaithersburg
Maryland 20878
USA

Mary Collins
Chester Beatty Laboratories
Institute of Cancer Research
237 Fulham Road
London SW3 6JB
UK

Kenneth W. Culver
Human Gene Therapy Research Institute
1415 Woodland Avenue
Des Moines
Iowa 50309
USA

Olivier Danos
Laboratoire Rétrovirus et Transfert Génétique
Batiment SIDA/Rétrovirus
Institut Pasteur
28 rue du Dr Roux
75015 Paris
France

Heather L. Davis
Loeb Medical Research Institute
Ottawa Civic Hospital
1053 Carling Avenue
Ottawa, ON K1Y 4E9
Canada

Neal Deluca
University of Pittsburgh
School of Medicine
Department of Molecular Genetics and Biochemistry
E1246 Biomedical Science Tower
Pittsburgh
Pennsylvania 15261
USA

George Dickson
Department of Biochemistry
Royal Holloway
University of London
Egham
Surrey TW20 0EX
UK

Matthew G. Dunckley
Department of Biochemistry
Royal Holloway
University of London
Egham
Surrey TW20 0EX
UK

Christopher H. Evans
University of Pittsburgh School of Medicine
Departments of Orthopedic Surgery, Molecular Genetics and Biochemistry
986 Scaife Hall
Pittsburgh
PA 15261
USA

Stewart A. Fabb
Division of Medical and Molecular Genetics
UMDS Guy's Hospital
London Bridge
London SE1 9RT
UK

Ariberto Fassati
Department of Biochemistry
Royal Holloway
University of London
Egham
Surrey TW20 0EX
UK

David Fink
University of Pittsburgh
School of Medicine
Department of Molecular Genetics and Biochemistry
E1246 Biomedical Science Tower
Pittsburgh
Pennsylvania 15261
USA

Joseph Glorioso
University of Pittsburgh
School of Medicine
Department of Molecular Genetics and Biochemistry
E1246 Biomedical Science Tower
Pittsburgh
Pennsylvania 15261
USA

Jean-Michel Heard
Laboratoire Rétrovirus et Transfert Génétique
Batiment SIDA/Rétrovirus
Institut Pasteur
28 rue du Dr Roux
75015 Paris
France

Koji Inui
Department of Pediatrics
University of Osaka
Yamadaoka 2–2
Suita-Shi
Osaka 565
Japan

Yasufumi Kaneda
Institute for Molecular Biology
University of Osaka
Yamadaoka 2–2
Suita-Shi
Osaka 565
Japan

Christine Kinnon
Molecular Immunology Unit
Institute of Child Health
30 Guilford Street
London WC1 1EH
UK

Eric J. Kremer
Laboratoire de Génétique de Virus
Oncogenes
URA 1301 CNRS
PR2 Institut Gustave Roussy
94805 Villejuif
France

Martin G. Lee
Laboratoire de Génétique de Virus
Oncogenes
URA 1301 CNRS
PR2 Institut Gustave Roussy
94805 Villejuif
France

Pedro R. Lowenstein
Laboratory of Cellular and
Molecular Neurobiology
University of Wales College of
Cardiff
Museum Avenue
Cardiff CF? 1SS
UK

Donald W. Moorman
Human Gene Therapy Research
Institute
1415 Woodland Avenue
Des Moines
Iowa 50309
USA

Richard A. Morgan
Clinical Gene Therapy Branch
National Center for Human
Genome Research
Building 49
Room 2A–03
Bethesda
Maryland 20892

Shintaro Okada
Department of Pediatrics
University of Osaka
Yamadaoka 2–2
Suita-Shi
Osaka 565
Japan

Michel Perricaudet
Laboratoire de Génétique de Virus
Oncogenes
URA 1301 CNRS
PR2 Institut Gustave Roussy
94805 Villejuif
France

Howard Prentice
Department of Genetics and
Department of Medicine and
Therapeutics
University of Glasgow
Church Street
Glasgow G11 5JS
UK

Jiannis Ragoussis
Division of Medical and Molecular Genetics
UMDS Guy's Hospital
London Bridge
London SE1 9RT
UK

Paul D. Robbins
University of Pittsburgh School of Medicine
Departments of Molecular Genetics and Biochemistry
W1256 Biomedical Science Tower
Pittsburgh
PA 15261
USA

Alan E. Smith
Genzyme Corporation
PO Box 9322
One Mountain Road
Framingham
Massachusetts 01701–9322
USA

Clay Smith
Division of Hematology/Oncology
Duke University Medical Center
PO Box 3863
Durham
North Carolina 27710
USA

Bruce A. Sullenger
Howard Hughes Medical Institute
Box 215
Department of Biochemistry
University of Colorado
Boulder
Colorado 80309–0215
USA

Thierry vandenDriessche
Clinical Gene Therapy Branch
National Center for Human Genome Research
National Institutes of Health
9000 Rockvill Pike
Bethesda
Maryland 20892
USA

Samuel C. Wadsworth
Genzyme Corporation
PO Box 9322
One Mountain Road
Framingham
Massachusetts 01701–9322
USA

Keith A. Webster
Molecular Cardiology Laboratory
Life Sciences Division
SRI International
Menlo Park
California 94025
USA

Dominic J. Wells
Department of Veterinary Basic Sciences
The Royal Veterinary College
Royal College Street
London NW1 0TU
UK

Robert G. Whalen
Department of Molecular Biology
Institut Pasteur
25 rue du Dr Roux
75724 Paris cedex 15
France

Itaru Yanagihara
Department of Pediatrics
University of Osaka
Yamadaoka 2–2
Suita-Shi
Osaka 565
Japan

1

Retroviral vectors

ARIBERTO FASSATI, MATTHEW G. DUNCKLEY and GEORGE DICKSON

1.1 INTRODUCTION

The functional organization of retroviral genomes and the relative simplicity of their structure make retroviral-based systems attractive vectors for the introduction of foreign genes of interest into eukaryotic cells. Many features render modified retroviruses uniquely suitable for therapeutic gene transfer in humans. Most importantly, the retroviral genome is stably integrated into host cell DNA in a predictable orientation and is very well tolerated by the host cell in that normal cellular gene expression patterns are usually unaffected by the infection process. Virtually all human cells can be successfully infected by at least one type of retrovirus. Once integrated, retroviral promoters and enhancers can be used to drive high level expression of foreign genes in a wide range of cell types. Finally, in terms of vector construction, the relative simplicity of the retrovirus genome allows proteins necessary for all aspects of infection to be provided entirely *in trans*. Thus, it is possible to generate particles that are infectious but replication-defective and do not produce viral antigens, making them potentially very safe and controllable agents for gene therapy.

There are, however, limitations on the use of retroviruses as gene transfer reagents. Notably, retroviruses are small particles and can accommodate only relatively small genomes (approximately 9–10 kb). Therefore, transduction of large genes is not possible, especially if their regulatory sequences (promoters, introns, untranslated regions, etc.) are to be included. Most

Molecular and Cell Biology of Human Gene Therapeutics
Edited by George Dickson
Published in 1995 by Chapman & Hall. ISBN 0 412 62550 4

retroviruses can only infect and integrate into cells that are in, or close to, mitosis. From the safety angle, these agents are quite prone to recombination and rearrangements and great care is needed during the *in vitro* process of generating replication-defective viruses.

Another potential problem associated with the use of retroviral vectors is the danger of insertional mutagenesis and accidental oncogene activation due to the randomness of integration in the host genome. Nevertheless, although there are examples of retroviral-mediated cellular oncogene activation in mice, such an event has only been observed in a truly viremic situation in the presence of replication-competent virus. Despite these limitations, retroviral vectors are currently one of the most widely used gene transfer vehicles for mammalian cells and their clinical usefulness for human disease is witnessed by the proliferation of on-going clinical trials across the world accompanied by the development of novel therapeutic strategies.

In this chapter we shall outline the basic aspects of the retroviral genome structure and function that have been crucial for the understanding, use and development of retroviral vectors and describe a number of existing vectors and packaging cell systems.

1.2 THE RETROVIRAL GENOME

The retrovirus genome consists of two dimerized so-called plus (+) strands of RNA. A feature that characterizes all productive retroviral infections is that this RNA genome is 'reverse transcribed' by reverse transcriptase, an enzyme unique to this group of viruses, to form a double-stranded DNA provirus which is then integrated into the host genome (Varmus, 1982). Murine leukemia virus (MuLV)-related 'type C' retroviruses have been the most extensively used for gene transfer, especially the Moloney murine leukemia virus (MoMuLV) which has become archetypal in the development of retroviral vectors. Retroviruses belonging to this group share a very similar proviral structure (Weiss *et al.*, 1985) (Figure 1.1). The MoMuLV provirus is about 9 kb in length and is defined at both 5′ and 3′ ends by virtually identical untranslated regions resulting in direct repeat sequences, called long terminal repeats (LTRs) (Watson *et al.*, 1987). The provirus also contains three open reading frames encoding viral structural genes, namely *gag*, encoding viral core proteins, *pol*, encoding the enzyme reverse transcriptase, and related proteins, and *env*, encoding specific glycoproteins located at the viral envelope (Varmus, 1982). The LTR structures provide signals for initiation of transcription and polyadenylation and can be divided into three parts in the order 5′–U_3–R–U_5–3′. The U_3 region contains the enhancer and the viral promoter, including a TATA box, for initiation of transcription. The R region contains a Cap site that is predominantly active in the 5′LTR and a polyadenylation signal active in the U_5

region of the 3′ LTR (Watson *et al.*, 1987). At the 3′ boundary of the 5′LTR is the primer binding site (PBS), a sequence complementary to a cellular tRNA that is used to prime reverse transcription. Downstream from U_5 including a small portion of the *gag* coding sequence is the packaging signal (ψ) that allows complete viral RNAs to be encapsulated into new infectious particles (Watanabe and Temin, 1982; Mann *et al.*, 1983).

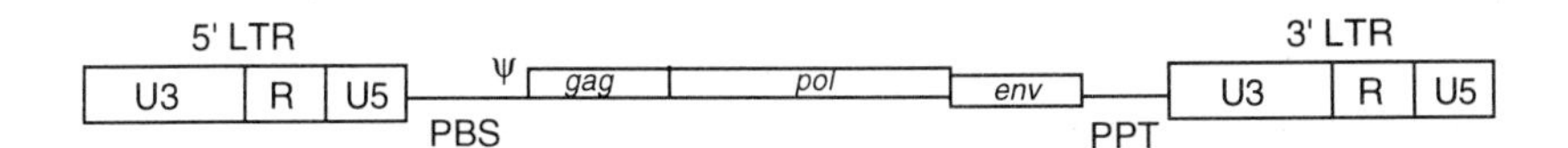

Figure 1.1 The proviral structure of MoMuLV. LTR, long terminal repeat; PBS, primer binding site; PPT, polypurine tract.

Transcription of the provirus produces several different mRNA species which are handled by the cell in distinct ways. The largest mRNA corresponding to virtually the full-length genome is mainly packaged into virions complexed with viral core proteins, but is also translated into the three structural proteins. The most abundant retroviral protein is polyprotein $Pr65^{gag}$, encoded by the *gag* gene, which is subsequently cleaved into a 30-kDa viral capsid protein as well as three other matrix and nucleocapsid proteins tightly associated with packaged retroviral RNA molecules (Weiss *et al.*, 1985). In addition, a glycosylated protein ($Pr85^{gag}$) from the *gag* region has also been detected, the function of which is not fully understood. In the MoMuLV genome, *gag* and *pol* are in the same reading frame, separated by a single amber termination codon (UAG), but spontaneous suppression of termination in about 10% of ribosomes leads to low levels of a large *gag–pol* polyprotein, that is eventually cleaved into low abundance *pol* proteins (Yoshinaka *et al.*, 1985). These include reverse transcriptase (RT), as well as protease and integrase enzymes vital for viral replication and integration into host cell DNA (Weiss *et al.*, 1985).

A proportion of retroviral RNA is spliced from 60 bp downstream of the 5′ LTR to about 300 bp 3′ to the end of the *pol* gene, yielding a smaller mRNA species corresponding to the *env* gene and which (lacking a ψ-site) cannot be packaged. Instead it is entirely used for production of the *env* gene products: a glycosylated protein spanning the lipid bilayer of the viral envelope (TM) and the outer surface glycoprotein (SU) which is responsible for virus–cell membrane interactions. As such, SU is the major determinant of the host range of the retrovirus as shown by interference assays (Rein, 1982). MuLVs can be of ecotropic, xenotropic, polytropic or amphotropic host range. Ecotropic MuLV can infect only rodent cells, xenotropic can only infect non-rodent cells, polytropic (or mink cell focus forming virus,

MuLV–MCF) and amphotropic can infect both rodent and non-rodent cells (Weiss, 1993).

1.3 THE RETROVIRAL LIFE CYCLE

The retroviral life cycle can be divided into several steps: after virus entry into the cell, reverse transcription of the RNA genome takes place in the cytoplasm of the cell giving a double-stranded proviral DNA. The provirus is transported to the nucleus as part of a nucleoprotein complex and eventually integrated into the host chromosome. The integrated provirus is then transcribed by host RNA polymerase and transcripts are in part translated to give viral proteins and in part packaged into new viral particles. Although many aspects of the retrovirus life cycle remain obscure, it is possible to describe it in some detail.

Mammalian retroviral infections mostly occur at neutral pH via receptor-mediated fusion of the viral envelope and the host cell membrane. Initially, glycoproteins at the surface of the retrovirus bind to a specific receptor, followed by endocytosis and uncoating at low pH by fusion of the viral envelope with the lysosomal membrane (McClure *et al.*, 1990). The ecotropic retrovirus receptor used by the Moloney murine leukemia virus has recently been characterized and found to be a cationic L-amino acid transporter (Wang *et al.*, 1991; Woo Kim *et al.*, 1991). In addition, the viral envelope of some retroviruses may also fuse directly with the cell membrane (Gilbert *et al.*, 1992).

Reverse transcription of the viral genome starts in the uncoated core structure once inside the cellular cytoplasm (Cann, 1993) and requires the presence of nucleocapsid proteins (Weiss *et al.*, 1985). Reverse transcription is initiated by the annealing of a specific cellular tRNA to the PBS (Figure 1.2). Next, reverse transcriptase (RT) synthesizes 100–180 bases of the first minus (−) strand of DNA, complementary to 5′ U_5 and R elements, while the RNAse H activity of RT degrades the RNA template. At this point the two complementary R sequences form a bridge by base pairing and the (−) strand of DNA is completed. The synthesis of the + DNA strand starts from a polypurine tract, an RNAse-resistant oligomer located at the 5′ of U_3 and continues to the end of the (−) strand where a second strand transfer allows the completion of transcription (Varmus, 1982; Allain *et al.*, 1994). Only one provirus is generated by reverse transcription of two RNA templates. A high rate of genetic rearrangement has been reported during reverse transcription, consisting of point mutations (estimated frequency $= 10^{-5}$), frameshifts (10^{-6}), deletions (2×10^{-6}), homologous recombination (2×10^{-4}) and non-homologous recombination (5×10^{-8}). Recombination occurs mainly during the synthesis of the (−) strand of DNA and the presence of different mechanisms involved in recombination and its high frequency indicate that the retroviral genome is organized to maximize variation (Temin, 1993).

Inside the cytoplasm the double-stranded viral DNA remains bound to viral nucleoproteins, forming a complex that carries all the activities necessary for integration. Moreover, a tight association of the viral nucleic acids with the viral replication machinery avoids any dilution of essential factors in the host cytoplasm or nucleus (Bowerman *et al.*, 1989). The size of the DNA–protein complex, however, precludes it from passing through the nuclear pores, so instead it must enter the nuclear compartment after breakdown of the nuclear membrane (Bukrinsky *et al.*,1993). Hence, integration of MuLV-type retroviruses requires the cell to be in mitosis (Miller *et al.*, 1990). As the viral DNA–protein complex reaches the nucleus, two base pairs are removed from the 3′ ends of both linear viral DNA strands by a viral integrase. Later, the 3′ ends of viral genome are joined to sites a few bases apart on the two strands of host DNA. Finally, the mismatches and gaps in this nicked intermediate are repared (Whitcomb and Hughes, 1992). In addition, four to six bases of cell DNA at the integration site are duplicated (Watson *et al.*, 1987) (Figure 1.3). It appears that the provirus integrates into post-replication host DNA during the S phase, so a maximum of 50% of the progeny of an infected cell can carry a copy (Hajihosseini *et al.*, 1993).

Once integrated, the provirus is transcribed by cellular RNA polymerase II and its expression is totally dependent upon cellular transcription machinery. Provirus–cellular interactions that regulate the expression of the viral genome are not fully understood. Expression of retroviral vectors in clonal hemopoietic stem cells is variable and often short-term (Apperley and Williams, 1990), and fibroblasts or myoblasts infected *in vitro* have been shown to gradually inactivate proviral expression if subsequently transplanted *in vivo* (Palmer *et al.*, 1991; Dai *et al.*, 1992). In addition, the MoMuLV LTR is inactive in pre-implantation embryos as well as in teratocarcinoma cells (Linney *et al.*, 1984; Gorman *et al.*, 1985). This may be because NF1, one of at least six different nuclear factors which bind to the MoMuLV enhancer, is absent in these cells (Speck and Baltimore, 1987). In addition, trans-acting factors present in undifferentiated cells can suppress transcription of the provirus by specifically acting on the viral enhancer (Gorman *et al.*, 1985). Chromosomal position also appears to be important in determining the extent of viral genes transcription, not least because proviral DNAs integrated in the proximity of inactive chromosomal sites are rapidly methylated and thus suppressed (Hoeben *et al.*, 1991), demonstrating that the regulatory elements in the LTR are not sufficient for substantial expression of the provirus in a position-independent manner.

A proportion of full-length viral RNAs are not translated but instead packaged to create new infectious particles. All the sequences necessary for packaging are located between the primer binding site and the *gag* coding

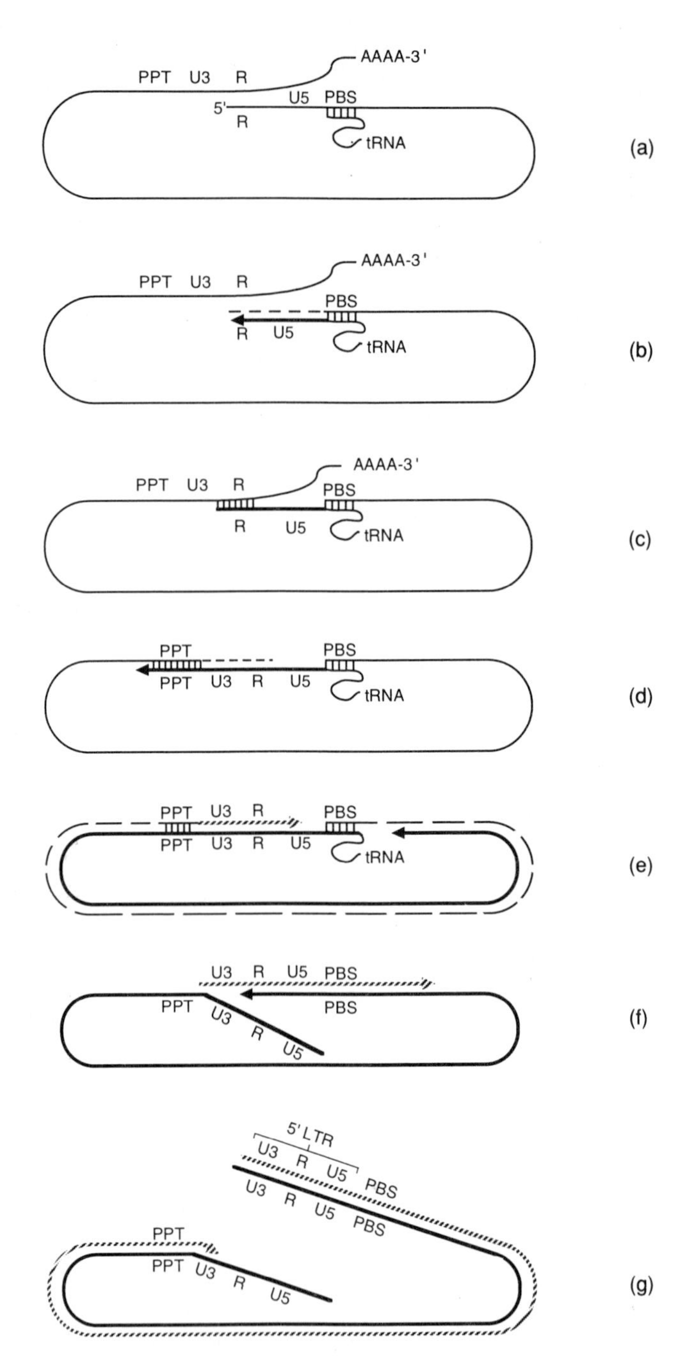

Figure 1.2 The process of reverse transcription. (a) Annealing of the tRNA to the PBS; (b) synthesis of the first 100–180 bp of the (−) strand of DNA and degradation of the RNA template by RNAse H; (c) a bridge is formed between the two complementary R sequences; (d) the 3′ end of RNA template is degraded; (e) the synthesis of the (−) strand of DNA is completed; (f) the synthesis of the (+) DNA strand starts at the polypurine track and a second strand transfer occurs; (g) the synthesis of the (+) DNA strand is completed.

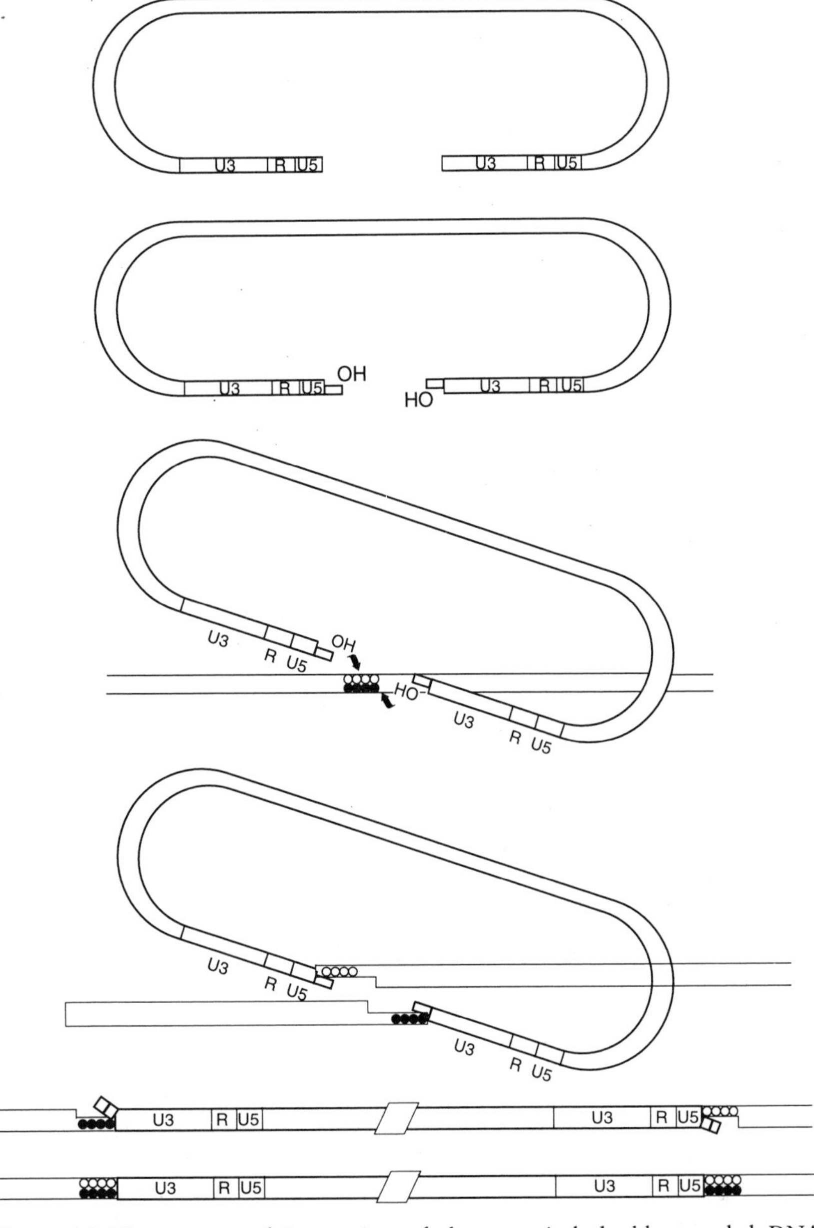

Figure 1.3 The process of integration of the retroviral double-stranded DNA genome: two base pairs are removed from the 3′ ends of both linear viral DNA strands by a viral integrase. Later, the 3′ ends of the viral genome are joined to sites a few bases apart on the two strands of host DNA. Finally, the mismatches and gaps in this nicked intermediate are repared and four to six bases of cell DNA at the integration site are duplicated.

region (Mann *et al.*, 1983) although in MoMuLV vectors the inclusion of limited portions of *gag* can enhance packaging efficiency (Armentano *et al.*, 1987). This encapsidation signal region also includes a dimerization signal and a recognition sequence/structure for nucleocapsid proteins (Prats *et al.*, 1990). Although the mechanisms responsible for viral RNA packaging are not fully understood, it appears that the process of dimerization is tightly linked to the process of encapsidation. The recognition of the packaging sequences appears to be largely dependent on the conformation of the two dimerized viral RNA molecules associated with nucleocapsid proteins (Yang and Temin, 1994).

1.4 THE DESIGN OF RETROVIRAL VECTORS

Acutely oncogenic retroviruses usually have defective genomes in which the coding region for retroviral proteins has been totally or partially replaced by an oncogene, rendering them incapable of independent replication. Analysis of these retroviruses has shown that, once a virus has entered the target cell, none of the viral coding regions is required for retroviral genomes to be packaged, reverse transcribed, or integrated into host DNA. Moreover, as long as all the appropriate retroviral proteins are present in the same cell, e.g. provided *in trans* by a concomitant wild-type virus infection, they may be expressed. These observations led to the conclusion that the only *cis*-acting sequences required for retroviral survival are the LTR, the packaging signal and the splicing sites (Weiss *et al.*, 1985), although efficient expression and high viral titers from non-splicing vectors have also been obtained (Morgernstern and Land, 1990). Thus, replication-defective retroviruses could be engineered to carry genes other than oncogenes and serve as retroviral vectors. In the process of designing retroviral vectors it is important to consider both the *cis*-acting sequences and the insert in order to achieve optimal gene expression and high titers of recombinant virus.

1.4.1 *Cis*-acting sequences

As previously explained, in certain cell types the MoMuLV LTR is inactive, of limited activity or switched off by cellular functions. The plasticity of the retroviral genome, however, has allowed the modification of the LTR to overcome transcriptional silencing and to increase the safety of retroviral vectors. For instance, in the vector pLNL6 a substitution of the 5′MoMuLV LTR with a Moloney murine sarcoma virus (MoMuSV) results, after reverse transcription, in a hybrid LTR containing the promoter and enhancer regions of the MoMuLV and R and U5 regions of the MoMuSV, which is very similar to MoMuLV but does not make the glycosylated $Pr85^{gag}$ (Miller and Rosman, 1989). Similarly, retroviral vectors with mutant or hybrid LTRs, in which the U3 enhancer has been replaced by a mutant polyoma virus enhancer have been successfully used to express recombinant genes in embryonic carcinoma cells and hemopoietic progenitor cells (Valerio *et al.*, 1989). The influence of the LTR has been reduced in so-called self-inactivating (SIN) vectors by deletion of the 3′ U3 region altogether. Following transfection of recombinant SIN vector DNA into packaging cells, normal levels of transcription are generated from the 5′ LTR and good viral titers can be achieved. However, after reverse transcription the deleted 3′ U3 region is copied to the 5′ end, inactivating the 5′ LTR as well (Yu *et al.*, 1986). The insertion of specific eukaryotic promoters in these vectors may allow tissue-specific expression of genes of interest without interference from the viral LTR (Cone *et al.*, 1987).

Titers of recombinant retroviruses generated by packaging cells are dependent on both the production levels of full-length genomic RNA and the efficiency of the packaging signal. Functional studies indicated that regions between nucleotides 215 and 539 of MoMuSV are sufficient for encapsidation of viral RNA but the titers obtained are severalfold lower than titers obtained using the region 215 to 1019 (Adam and Miller, 1988). Insertion of the packaging signal in inverse orientation results in very low titers, which is consistent with the requirements of secondary and tertiary structures of the viral RNA to be efficiently packaged (Yang and Temin, 1994). The addition of short 5′ sequences of the *gag* coding region to the packaging signal increases the efficiency of encapsidation severalfold and titers between 10^6 and 10^7 colony forming units (cfu)/ml can be generated (Armentano *et al.*, 1987). The location of the packaging signal near the 3′ LTR results in the encapsidation, reverse transcription and integration of both the unspliced and spliced mRNA that contains the *env* gene (Mann and Baltimore, 1985), once more demonstrating the extraordinary plasticity of retroviral genome. A virus-like sequence found in the rat (VL 30), when located next to the 5′ end of the Harvey mouse sarcoma virus genome, can drive encapsidation with a higher efficiency than the wild type packaging signal. This sequence is only 100 base pairs long and does not contain $Pr65^{gag}$ and $Pr85^{gag}$ coding sequences, rendering more unlikely the formation of replication competent virus via recombination of the viral RNA with the *gag–pol* RNA of the packaging cells. Therefore, its use should improve titers and safety of recombinant retroviruses (Torrent *et al.*, 1994).

It is often advantageous to eliminate the splicing donor site present in wild type MoMuLV to avoid artifacts due to the presence of cryptic splice acceptors within cDNA inserted in the retroviral vectors and to achieve the expression of high levels of full-length viral mRNA. Not all alterations of the splice donor site are compatible with the production of high viral titers: removal of a G residue at position 193 of MoMuLV genome results in low titers but a T to C mutation at position 194 that is naturally present in the non-splicing Harvey MoSV, leads to high titers production and very good expression of the gene of interest (Morgernstern and Land, 1990).

1.4.2 Inserted sequences: single gene vectors

Retroviral genomes can aquire many different types of oncogenes, showing that the genome is sufficiently plastic to tolerate a wide range of inserts. In the early retroviral vectors, a single gene coding for a selectable marker was inserted immediately downstream from the packaging element (Mann *et al.*, 1983). However, these early vectors yielded only moderate viral titers. To increase the titers, in later vectors the extended packaging signal was included and a mutation of the *gag* start codon into a stop codon reduced the possibilities of premature translation starts of the inserted gene and of any potential replication-competent virus arisen by recombination (Bender *et al.*, 1987). Some of these early vectors contained two polyadenylation signals, the first at the 3′ end of the inserted gene and the second in the U5 of the 3′ viral LTR. Two different mRNAs were detected after infection with these vectors: a full-length transcript and a shorter product resulting from the use of internal poly(A) signal. This transcriptional pattern yielded only low viral titers, due to the paucity of full-length transcript and possibly to competition for packaging (Miller *et al.*, 1983). Therefore, although it appears possible to use more than one poly(A) signal in retroviral vectors, it is a wise general strategy to remove those signals from the inserted gene (Weiss *et al.*, 1985). Single gene vectors carrying tissue-specific eukaryotic promoters were also developed: the thymidine kinase (*tk*) gene driven by its own promoter was inserted in a retroviral vector and it has been shown to efficiently express thymidine kinase, although it was not completely clear whether the expression was driven by the LTR or by the internal promoter (Tabin *et al.*, 1982). A vector carrying the human adenosine deaminase gene (ADA) and the phosphoglycerate kinase (*pgk*) promoter efficiently expressed ADA in murine hematopoietic stem cells, the transcription appearing to be dependent on *pgk* promoter (Lim *et al.*, 1987). Generally the orientation of the inserted promoter relative to the LTR does not matter greatly in the expression of the insert and the inserted promoter is able to correctly respond to the cellular transcriptional control signals (Weiss *et al.*, 1985). However, it is noteworthy to observe that the oncogenes naturally inserted in retroviral genome do not have promoters and the presence of eukaryotic promoters within the retroviral vectors does not always result in efficient and tissue-specific expression of the gene of interest, depending on the circumstances of individual cases.

1.4.3 Vectors expressing multiple genes

Vectors that express more than one gene are of particular usefulness, i.e. the expression of both a gene conferring a selectable advantage to cells and the gene of interest results in the possibility of easily cloning transfected cells or in the possibility to follow the fate of cells injected *in vivo* that express therapeutic proteins. Different strategies have been devised to design vectors that express more than one gene: the splicing vectors, the double promoter vectors and the internal ribosome entry site vectors. In the splicing vectors the first gene is located between the retroviral splicing signals while the second gene is located downstream of the splice acceptor. The vector transcript is expressed from the LTR promoter and it is spliced and translated to produce two proteins. However, it is generally difficult to predict the efficiency of splicing, and the production of the upstream protein is higher than the downstream protein (Boris-Lawrie and Temin, 1993). In the double promoter vectors one gene is transcribed from the LTR and the other gene from an internal promoter. The internal promoter can be either of viral origin, like SV40 or CMV (Price *et al.*, 1987; Osborne and Miller, 1988; Morgernstern and Land, 1990) or of cellular origin (Bowtell *et al.*, 1988). In addition, long-term expression *in vivo* of the gene of interest can be achieved by combination of an internal promoter of viral origin with a tissue-specific enhancer (Dai *et al.*, 1992). Although these vectors have been successfully used in many cases, the presence of promoters other than the LTR directing the transcription of two genes appears to be problematic. The 5′ gene can be suppressed when there is selection for the 3′ gene and vice versa, possibly due to epigenetic mechanisms (Emerman and Temin, 1984a, 1986a), although this effect is much stronger in retroviral vectors derived from spleen necrosis virus than in MoMuLV-based vectors (Emerman and Temin, 1986b). In addition, vectors containing two promoters are very susceptible to deletions and rearrangements (Emerman and Temin, 1984b), and their expression *in vivo* is not always satisfactory, once more depending on the circumstances of individual cases (Dai *et al.*, 1992).

A new generation of retroviral vectors which express multiple proteins from a single polycistronic mRNA have recently been developed. These vectors have been adapted from picornaviruses, a group of RNA viruses characterized by uncapped mRNA, (usually) long 5′ untranslated regions with multiple AUG codons, and a complex secondary structure. Ribosomes bind directly to these initiating AUG codons without scanning (Jang and Wimmer, 1990), therefore allowing translation of a polycistronic mRNA. The untranslated regions are designated 'internal ribosome entry sites' (IRES). IRES sequences cloned from poliovirus and encephalomyocarditis virus have been inserted in retroviral vectors between the genes of interest. The insertion of picornavirus IRES sequences between two coding regions in a single transcriptional unit resulted in translation of the downstream coding region at a very similar rate to the upstream region (Adam *et al.*, 1991). In addition, it has been shown that vectors containing three coding regions organized in a single polycistronic mRNA can be successfully expressed (Morgan *et al.*, 1992). In IRES vectors the selection for one gene should ensure expression of the second or third gene, thus allowing long-term cultures and high stability of the vectors themselves.

1.5 PACKAGING CELL LINES

In order to produce replication-defective retrovirus it is necessary to provide *in trans* all the proteins necessary for virus assembly, thus allowing the formation of new infectious virions that carry the defective genome. This has generally been achieved via genetically-engineered cultured strains of so-called packaging cells (Table 1.1). The early packaging cell lines were made by transfecting into mouse fibroblasts the helper virus genome from which the packaging signal was removed ($\Psi-$) (Mann *et al.*, 1983). However, these early packaging cells were prone to the generation of wild type virus because of low residual levels of encapsidation of $\Psi-$ genomes and subsequent recombination between co-packaged $\Psi-$ and $\Psi+$ genomes during the process of reverse transcription (Mann and Baltimore, 1985).

The deletion of the 5′ end of the 5′ LTR from the $\Psi-$ retroviral genome plus the deletion of the second strand initiation site, and of the 3′ LTR (that is substituted by the SV 40 poly (A) site), resulted in the amphotropic packaging cell line PA317. Helper virus is not generally detected in PA317 and high-titer retroviral stocks can be obtained (Miller and Buttimore, 1986). In this packaging cell line even if RNA coding for *gag–pol* and *env* is packaged, it cannot be reverse transcribed or integrated into the host chromosomes due to the lack of *cis*-acting sequences necessary for these processes to occur.

A different approach to generate high-titer, helper-free packaging cell lines is to split *gag–pol* and *env* coding regions into two separate transcriptional units (Watanabe and Temin, 1983). In ΨCRIP and ΨCRE packaging cell lines, well-characterized frameshift mutations in the *gag–pol* or *env* genes were transferred into $\Psi-$ constructs to generate functional *gag–pol*$^+$ *env*$^-$ and *gag–pol*$^-$ *env*$^+$ constructs. In addition, the 3′ LTR was deleted and substituted with the SV40 poly(A) site. Therefore, three recombination events are required for helper virus formation (Danos and Mulligan, 1988). Further improvement in this direction was achieved by separation, instead of inactivation, of the *gag–pol* and *env* genes on two different constructs, thus reducing the sequence overlap between the two plasmids and the chance of helper virus formation by recombination (Markowitz *et al.*, 1988b). However, since the *pol* gene ends after the start of the *env* gene it is difficult to completely eliminate overlaps between the two plasmids. To overcome this problem a *gag–pol* region from one virus and the *env* region from another have been used (Dougherty *et al.*, 1989), although titers obtained with this packaging cell lines are not very high. To reduce the homology among different retroviral plasmids it is also possible to delete both the 5′ and 3′ LTR and substitute them with the mouse metallothionein

Table 1.1 Retrovirus packaging cell lines

Name	*Type*	*Host range*	*Maximum titer*	*Drug resistance gene*	*Reference*
ψ-2	ψ-Deletion	Ecotropic	10^7	gpt	Mann *et al.* (1983)
ψ-AM	ψ-Deletion	Amphotropic	2×10^5	gpt	Cone and Mulligan (1984)
PA12	ψ-Deletion	Amphotropic	4×10^6	tk	Miller *et al.* (1985)
PA317	Multiple deletions	Amphotropic	4×10^7	tk	Miller and Buttimore (1986)
PE501	Multiple deletions	Ecotropic	10^7	tk	Miller and Rosman (1989)
clone 32	Split coding regions	Ecotropic	3×10^4	neo	Bosselman *et al.* (1987)
ψCRE	Split coding regions with deletions	Ecotropic	10^6	hph, gpt	Danos and Mulligan (1988)
ψCRIP	Split coding regions with deletions	Amphotropic	10^6	hph, gpt	Danos and Mulligan (1988)
GP+E-86	Separation of coding regions	Ecotropic	4×10^6	gpt	Markowitz *et al.* (1988b)
GP+envAm12	Separation of coding regions	Amphotropic	10^6	hph, gpt	Markowitz *et al.* (1988a)
DAN	Coding regions from different viruses, separated	Amphotropic	4×10^4	neo	Dougherty *et al.* (1989)
Ampli-GPE	Separation of coding regions, LTR deleted	Ecotropic	5×10^6	neo	Takahara *et al.* (1992)

I promoter and the human β-globin gene containing the poly(A) site respectively. In addition the bovine papillomavirus gene for gene amplification is inserted in the same plasmid to increase its copy number and to obtain high virus titers (Takahara *et al.*, 1992).

Retroviruses can expand or modify their host range by pseudotyping. For instance, co-infection of cells with HIV-l and xenotropic MuLV (XMuLV) results in broadening of the cellular tropism of HIV-l, due to the generation of pseudotypes containing mixed XMuLV and HIV-l envelope glycoproteins (Lusso *et al.*, 1990). The possibility of creating pseudotype retroviruses led to the construction of packaging cell lines with *gag–pol* gene from MoMuLV and *env* gene from vescicular stomatitis virus (VSV). The substitution of VSV for the MoMuLV envelope glycoproteins confers to the recombinant virus an increased host range and structural stability, thus allowing the preparation of very high-titer retroviral stocks by ultracentrifugation (up to 10^9 cfu/ml) (Burns *et al.*, 1993).

The isolation of high-titer retroviral vector producer cells is usually dependent on selection for stable integrations via single cell cloning, a process that takes 2–3 weeks. Recently, two different packaging cell lines have been described that allow transient production of high-titer retroviral stocks. The first packaging cell line is derived from COS-7 cells that contain the SV40 large T antigen. Transfection of the packaging genome and the retroviral vector plasmids that contain the SV40 origin of replication in COS cells results in replication of the plasmids and generation of high-titer retroviral stocks after 72 hours (Landau and Littman, 1992). The second packaging cell line is derived from 293T cells that are particularly efficiently transfected by the calcium phosphate precipitate method. Introduction of constructs that express retroviral packaging functions into 293T cells resulted in the ecotropic packaging cell line BOSC 23 that can transiently produce high-titer retrovirus stocks after calcium phosphate-mediated transfection with the retroviral plasmid (Pear *et al.*, 1993).

1.6 SAFETY CONSIDERATIONS

The technology of gene therapy has moved from the laboratory to clinical evaluation with considerable speed and the assessment of the risks and potential illness or injury to the patients deserves particular attention. Both the product and the intended study populations need to be considered: the treatment of patients with medium to long-term expectation of life obviously involves risk–benefits considerations that differ from those presented by the treatment of terminal patients. In terms of control of the product, three principles are central: control of the biologic source(s), control of the production process and control of the bulk of the final product (Kessler *et al.*, 1993). Gene therapy products derive from a variety of biologic sources: harvested autologous, allogeneic or xenogeneic cells as well as established cell lines and viral vectors. These sources must not be contaminated by pathogenic organisms and tumorigenicity of cell lines and genetically modified cells need to be investigated, particularly when the introduction of these cells into patients is involved (Culver *et al.*, 1992).

Minor changes in the conditions of cell cultures or in the materials used and conservation processes may significantly alter the characteristic of the final product; therefore the control of manufacturing process is a very important matter in ensuring its quality (Kessler *et al.*, 1993). The control of the final product is focused primarily, but not uniquely, on the absence of replication-competent virus. Recent findings indicate that replication-competent virus can be pathogenic and cause malignant T cell lymphomas in severely immunocompromised primates (Donahue *et al.*, 1992), underlining the necessity to use safe packaging cell lines and retroviral vectors and stringent tests to check for the presence of contaminant replicating virus. The presence of other antigens or contaminants that can cause acute severe immunoreactions should also be checked, as well as the effects of repeated administration of high doses of recombinant virus.

Other potential risks include activation of oncogenes as a consequence of random integration of the provirus and recombination between the vector and endogenous virus-like sequences, leading to the formation, *in vivo*, of replication-competent virus. Although these events could occur in theory, they are generally considered very unlikely possibilities and have not justified delay in the clinical trial of gene therapy (Temin, 1990; Cornetta *et al.*, 1991).

REFERENCES

Adam, M.A. and Miller, A.D. (1988) Identification of a signal in a murine retrovirus that is sufficient for packaging of nonretroviral RNA into virions. *J. Virol.*, **62**, 3802–6.

Adam, M.A., Ramesh, N., Miller, A.D. and Osborne, W.R.A. (1991) Internal initiation of translation in retroviral vectors carrying picornavirus 5′ nontranslated regions. *J. Virol.*, **65**, 4985–90.

Allain, B., Lapadat-Taplosky, M., Berlioz, C. and Darlix, J-L. (1994) Transactivation of the minus strand DNA transfer by nucleocapsid protein during reverse transcription of the retroviral genome. *EMBO Journal*, **13**, 973–81.

Apperley, J.F. and Williams, D.A. (1990) Gene therapy: current status and future directions. *Br. J. Haematol.*, **75**, 148–55.

Armentano, D., Yu, S-F., Kantoff, P.W., von Ruden, T., Anderson, W.F. and Gilboa, E. (1987) Effect of internal viral sequences on the utility of retroviral vectors. *J. Virol.*, **61**, 1647–50.

Bender, M.A., Palmer, T.D., Gelinas, R.E. and Miller, A.D. (1987) Evidence that the packaging signal of Moloney murine leukaemia virus extends into the gag region. *J. Virol.*, **61**, 1639–46.

Boris-Lawrie, K.A. and Temin, H.M. (1993) Recent advances in retrovirus vector technology. *Curr. Opin. Genet. Dev.*, **3**, 102–9.

Bosselman, R.A., Hsu, R-Y., Bruszewsky, J., Hu, F., Martin, F. and Nicolson, M. (1987) Replication defective chimeric helper proviruses and factors affecting generation of competent virus: expression of Moloney murine leukaemia virus structural genes via the metallothionein promoter. *Mol. Cell. Biol.*, **7**, 1797–806.

Bowerman, B., Brown, P.O., Bishop, J.M. and Varmus, H.E. (1989) A nucleoprotein complex mediates the integration of retroviral DNA. *Genes Dev.*, **3**, 469–78.

Bowtell, D.D.L., Cory, S., Johnson G.R. and Gonda, R.J. (1988) Comparison of expression in hemopoietic cells by retroviral vectors carrying two genes. *J. Virol.*, **62**, 2464–73.

Bukrinsky, M.I., Haggerty, S., Dempsey, P. *et al.* (1993) A nuclear localization signal within HIV-1 matrix protein that governs infection of non-dividing cells. *Nature*, **365**, 666–9.

Burns, J.C., Friedmann, T., Driever, W., Burrascano, M. and Yee, J.K. (1993) Vesicular stomatitis virus G glycoprotein pseudotyped retroviral vectors: concentration to very high titer and efficient gene transfer into mammalian and nonmammalian cells. *Proc. Natl Acad. Sci. USA*, **90**, 8033–7.

Cann, A.J. (1993) *Principles of Molecular Virology*, Academic Press, London.

Cone, R.D. and Mulligan, R.C. (1984) High efficient gene transfer into mammalian cells: generation of helper-free recombinant retrovirus with broad mammalian host range. *Proc. Natl Acad. Sci. USA*, **81**, 6349–53.

Cone, R.D., Reilly, E.B., Eisen, H.N. and Mulligan, R.C. (1987) Tissue-specific expression of functionally rearranged lambda 1g gene through a retrovirus vector. *Science*, **236**, 954–7.

Cornetta, K., Morgan, R. and Anderson, W.F. (1991) Safety issues related to retroviral-mediated gene transfer in humans. *Human Gene Ther.*, **2**, 5–14.

Culver, K.W, Ram, Z., Wallbridge, S., Ishii, H., Oldfield, E H. and Blease, R.M. (1992) *In vivo* gene transfer with retroviral vector producer cells for treatment of experimental brain tumors. *Science*, **256**, 1550–2.

Dai, Y., Roman, M., Naviaux, R.K. and Verma, I.M. (1992) Gene therapy via primary myoblasts: long-term expression of factor IX protein following transplantation *in vivo*. *Proc. Natl Acad. Sci. USA*, **89**, 10892–5.

Danos, O. and Mulligan, R.C. (1988) Safe and efficient generation of recombinant retroviruses with amphotripic and ecotropic host ranges. *Proc. Natl Acad. Sci. USA*, **85**, 6460–4.

Donahue, R.E., Kessler, S.W. and Bodine, D. (1992) Helper virus induced T cell lymphoma in nonhuman primates after retroviral mediated gene transfer. *J. Exp. Med.*, **176**, 1125–35.

Dougherty, J.P., Wisniewski, R., Yang, S., Rhode, B.W. and Temin, H.M. (1989) New retrovirus helper cells with almost no nucleotide sequence homology to retrovirus vectors. *J. Virol.*, **63**, 3209–12.

Emerman, M. and Temin, H.M. (1984a) High-frequency deletion in recovered retrovirus vectors containing exogenous DNA with promoters. *J. Virol.*, **50**, 42–9.

Emerman, M. and Temin, H.M. (1984b) Genes with promoters in retrovirus vectors can be independently suppressed by an epigenetic mechanism. *Cell*, **39**, 459–67.

Emerman, M. and Temin, H.M. (1986a). Quantitative analysis of gene suppression in integrated retrovirus vectors. *Mol. Cell. Biol.*, **6**, 792–800.

Emerman, M. and Temin, H.M. (1986b) Comparison of promoter suppression in avian and murine retrovirus vectors. *Nucleic Acids Res.*, **14**, 9381–96.

Gilbert, M.A., Cherreau, B., Vicart, P. and Nandi, P.K. (1992) Mechanism of entry of a xenotropic MMuLV-derived recombinant retrovirus into porcine cells using the expression of the reporter nlslacZ gene. *Arch. Virol.*, **124**, 57–67.

Gorman, C.M., Rigby, P.W.J. and Lane, D. (1985) Negative regulation of viral enhancers in undifferentiated embryonic stem cells. *Cell*, **42**, 519–26.

Hajihosseini, M., Iavachev, L. and Price, J. (1993) Evidence that retroviruses integrate into post-replication host DNA. *EMBO Journal*, **13**, 4969–74.

Hoeben, R.C., Mighielsen, A.J., van der Jagt, R., van Ormondt, H. and van der Eb, A. (1991) Inactivation of the Moloney murine leukaemia virus long terminal repeat in murine fibroblasts cell lines is associated with methylation and dependent on its chromosomal position. *J. Virol.*, **65**, 904–12.

Jang, S.K. and Wimmer, E. (1990) Cap-independent translation of encephalomyocarditis virus RNA: structural elements of the internal ribosome entry site and involvement of a cellular 57-kD RNA-binding protein. *Genes Dev.*, **4**, 1560–72.

Kessler, D.A., Siegel, J.P., Noguchi, P.D., Zoon, K.C., Feiden, K.L. and Woodcock, J. (1993) Regulation of somatic-cell therapy and gene therapy by the food and drug administration. *N. Engl. J. Med.*, **329**, 1169–73.

Landau, N.R. and Littman, D.R. (1992) Packaging system for rapid production of murine leukaemia virus vectors with variable tropism. *J. Virol.*, **66**, 5110–13.

Lim, B., Williams, D.A. and Orkin, S.H. (1987) Retrovirus mediated gene transfer of human adenosine deaminase: expression of functional enzyme in murine hematopoietic stem cells *in vivo*. *Mol. Cell. Biol.*, **7**, 3459–65.

Linney, E., Davis, B., Overhauser, J., Chao, E. and Fan, H. (1984) Non-function of a Moloney murine leukaemia virus regulatory sequence in F9 embryonal carcinoma cells. *Nature*, **308**, 470–2.

Lusso, P., Di Marzo Veronese, F., Ensoli, B. *et al.* (1990) Expanded HIV-1 cellular trophism by phenotypic mixing with murine endogenous retroviruses. *Science*, **247**, 848–51.

Mann, R. and Baltimore, D. (1985) Varying the position of a retrovirus packaging sequence results in the encapsidation of both unspliced and spliced RNAs. *J. Virol.*, **54**, 401–7.

Mann, R., Mulligan, R.C. and Baltimore, D. (1983) Construction of a retrovirus packaging mutant and its use to produce helper-free defective retrovirus. *Cell*, **33**, 153–9.

Markowitz, D., Goff, S. and Bank, A. (1988a) Construction and use of a safe and efficient amphotropic packaging cell line. *Virology*, **167**, 400–6.

Markowitz, D., Goff, S. and Bank, A. (1988b) A safe packaging line for gene transfer: separating viral genes on two different plasmids. *J. Virol.*, **62**, 1120–4.

McClure, M.O., Sommerfelt, M.A., Marsh, M. and Weiss, R.A. (1990) The pH independence of mammalian retrovirus infection. *J. Gen. Virol.*, **71**, 767–73.

Miller, A.D. and Buttimore, C. (1986) Redesign of retrovirus packaging cell lines to avoid recombination leading to helper virus production. *Mol. Cell. Biol.*, **6**,

2895–902.

Miller, A.D. and Rosman, G.J. (1989) Improved retroviral vectors for gene transfer and expression. *BioTechniques*, 7, 980–90.

Miller, A.D., Jolly, D.J., Friedmann, T. and Verma, I.M. (1983) A transmissible retrovirus expressing human hypoxanthine phosphoribosyltransferase (HPRT): gene transfer into cells obtained from humans deficient in HPRT. *Proc. Natl Acad. Sci. USA*, **80**, 4709–13.

Miller, A.D., Law, M.F. and Verma, I.M. (1985) Generation of helper free amphotropic retroviruses that transduce a dominant-acting, methotrexate-resistant dihydrofolate reductase gene. *Mol. Cell. Biol.*, **5**, 431–7.

Miller, D.G., Adam, M.A. and Miller, A.D. (1990) Gene transfer by retrovirus vectors occurs only in cells that are actively replicating at the time of infection. *Mol. Cell. Biol.*, **10**, 4239–42.

Morgan, R.A., Couture, L., Elroy-Stein, O., Ragheb, J., Moss, B. and Anderson, W.F. (1992) Retroviral vectors containing putative internal ribosome entry sites: development of a plycistronic gene transfer system and application to human gene therapy. *Nucleic Acids Res.*, **20**, 1293–8.

Morgernstern, J.P. and Land, H. (1990) Advanced mammalian gene transfer: high titre retroviral vectors with multiple drug selection markers and a complementary helper-free packaging cell line. *Nucleic Acids Res.*, **18**, 3587–96.

Osborne, W.R. and Miller, A.D. (1988) Design of vectors for efficient expression of human purine nucleoside phosphorylase in skin fibroblasts from enzyme-deficient humans. *Proc. Natl Acad. Sci. USA*, **85**, 6851–5.

Palmer, T.D., Rosman, G.J., Osborne, W.R. and Miller, A.D. (1991) Genetically modified skin fibroblasts persist long after transplantation but gradually inactivate introduced genes. *Proc. Natl Acad. Sci. USA*, **88**, 1330–4.

Pear, W.S., Nolan, G.P., Scott, M.L. and Baltimore, D. (1993) Production of high-titer helper-free retroviruses by transient transfection. *Proc. Natl Acad. Sci. USA*, **90**, 8392–6.

Prats, A-C., Roy, C., Wang, P. *et al.* (1990) *Cis* elements and *trans* acting factors involved in dimer formation of murine leukaemia virus RNA. *J. Virol.*, **64**, 774–83.

Price, J., Turner, D. and Cepko, C. (1987) Lineage analysis in the vertebrate nervous system by retrovirus-mediated gene transfer. *Proc. Natl Acad. Sci. USA*, **84**, 156–60.

Rein, A. (1982) Interference grouping of murine leukaemia viruses: a distinct receptor for the MCF-recombinant viruses in mouse cells. *Virology*, **120**, 251–7.

Speck, N.A. and Baltimore, D. (1987) Six distinct nuclear factors interact with the 75-base-pair repeat of the Moloney murine leukaemia virus enhancer. *Mol. Cell. Biol.*, **7**, 1101–10.

Tabin, D.J., Hoffman, J.W., Goff, S.P. and Weinberg, R.A. (1982) Adaptation of a retrovirus as a eukaryotic vector in transmiting the herpes simplex virus thymidine kinase gene. *Mol. Cell. Biol.*, **2**, 426–36.

Takahara, Y., Hamada, K. and Housman, D. (1992) A new retrovirus packaging cell for gene transfer constructed from amplified long terminal repeat-free chimeric proviral genes. *J. Virol.*, **66**, 3725–32.

Temin, H.M. (1990) Safety considerations in somatic gene therapy of human disease with retrovirus vectors. *Hum. Genet. Ther.*, **1**, 111–23.

Temin, H. (1993) Retroviral variation and reverse transcription: abnormal strand transfers result in retrovirus genetic variation. *Proc. Natl Acad. Sci. USA*, **90**, 6900–3.

Torrent, C., Gabus, C. and Darlix, J.L. (1994) A small and efficient dimerization/packaging signal for rat VL30 RNA and its use in murine leukaemia virus-VL30-derived vectors for gene transfer. *J. Virol.*, **68**, 661–7.

Valerio, D., Einerhand, M.P.W, Wamsley, P.M., Bakx, T.A., Li, C.L. and Verma, I.M. (1989) Retrovirus mediated gene transfer into embryonal carcinoma and hemopoietic stem cells: expression from a hybrid long terminal repeat. *Gene*, **84**, 419–27.

Varmus, H.E. (1982) Form and function of retroviral proviruses. *Science*, **216**, 812–20.

Wang, H., Kavanaugh, M.P., North, R.A. and Kabat, D. (1991) Cell surface receptor for ecotropic murine retroviruses is a basic amino-acid transporter. *Nature*, **352**, 729–31.

Watanabe, S. and Temin, H.M. (1982) Encapsidation sequences for spleen necrosis virus, an avian retrovirus, are between the 5′ long terminal repeat and the start of the *gag* gene. *Proc. Natl Acad. Sci. USA*, **79**, 5986–90.

Watanabe, S. and Temin, H.M. (1983) Construction of a helper cell line for avian reticuloendotheliosis virus cloning vectors. *Mol. Cell. Biol.*, **3**, 2241–9.

Watson, J.D., Hopkins, N.H., Roberts, J.W., Argetsinger Steitz, J. and Weiner, A.M. (eds) (1987) *Molecular Biology of the Gene*, 4th edn, Benjamin/Cummings, Menlo Park, CA, USA.

Weiss, R.A. (1993) Cellular receptors and viral glycoproteins involved in retrovirus entry, in *The Retroviridae*, (ed. J.A. Levy), vol. 2, Plenum Press, New York.

Weiss, R.A., Teich, N., Varmus, H.E. and Coffin, J.M. (eds) (1985) *Molecular Biology of Tumor Viruses: RNA Tumor Viruses*, 2nd edn, Cold Spring Harbor Laboratory, Cold Spring Harbor, NY.

Whitcomb, J.M. and Hughes, S.H. (1992) Retroviral reverse transcription and integration: progress and problems. *Annu. Rev. Cell. Biol.*, **8**, 275–306.

Woo Kim, J., Closs, E.I., Albritton, L.M. and Cunningham, M. (1991) Transport of cationic amino acids by the mouse ecotropic retrovirus receptor. *Nature*, **352**, 725–8.

Yang, S. and Temin, H.M. (1994) A double hairpin structure is necessary for the efficient encapsidation of spleen necrosis virus retroviral RNA. *EMBO Journal*, **13**, 713–26.

Yoshinaka, Y., Katoh, I., Copeland, T.D. and Oroszlan, S. (1985) Murine leukemia virus protease is encoded by the *gag–pol* gene and is synthesized through suppression of an amber termination codon. *Proc. Natl Acad. Sci. USA*, **82**, 1618–22.

Yu, S-F., von Ruder, T., Kantoff, P.W. *et al.* (1986) Self-inactivating retroviral vectors designed for transfer of whole genes into mammalian cells. *Proc. Natl Acad. Sci. USA*, **83**, 3194–8.

2

Adenoviral vectors

MARTIN G. LEE, ERIC J. KREMER and MICHEL PERRICAUDET

2.1 INTRODUCTION

The late Sir Peter Medawar described viruses as 'bad news wrapped in protein'. The goal of virally mediated gene therapy is to turn a piece of potentially dangerous bad news into positively good news wrapped in protein.

Viruses are nature's way of transferring and expressing foreign genes in a host's cells. As such, viruses are obvious candidates to modify in order to divert their current occupation of causing disease into the treatment of illnesses. The molecular biology revolution over the last 15 years has led to techniques where genes can be deleted from the viral genome, and exogenes, chosen for their therapeutic or purely research interest, can be substituted in their place. Adenoviruses (Ad) are being modified in order to be used as an alternative gene transfer vehicle. The resulting recombinant adenoviruses (R Ad) are able to infect cells in the same way as their wild type ancestors; however, they are now engineered to express an experimental exogene instead of their own viral gene(s). With the initial aim to treat congenital diseases, adenoviral gene therapy has extended its aim to treatment of acquired maladies. R Ad are currently being constructed and tested to treat a diverse range of diseases as described in this book.

Ad were first isolated in 1953 (Rowe *et al.*, 1953). Attempts at culturing excised adenoidal tissue were thwarted by the degeneration of the epithelia by a transmissible agent later named *adeno*-virus. This family of viruses, *Adenoviridae*, was quickly identified as the causative agent in some primarily respiratory and gastrointestinal illnesses. Ad have a number of properties that make them potentially useful candidates for gene therapy vectors. They

Molecular and Cell Biology of Human Gene Therapeutics
Edited by George Dickson
Published in 1995 by Chapman & Hall. ISBN 0 412 62550 4

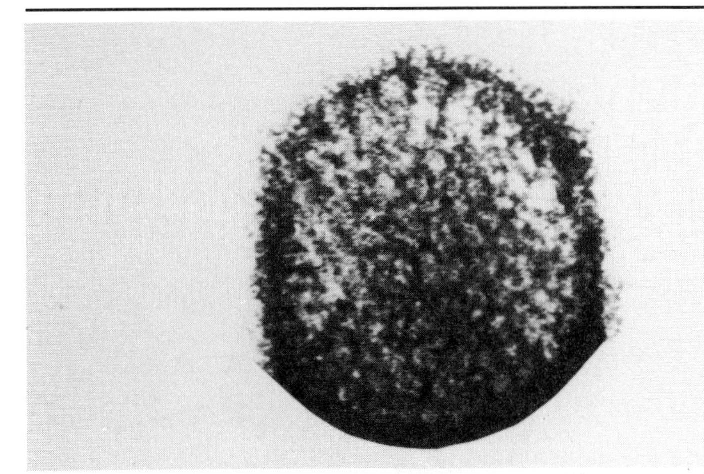

Figure 2.1 Electron micrograph of an adenovirus 5 wild-type particle.

have the ability to infect a wide range of dividing and quiescent cells, can be engineered to be replication-defective particles carrying 'large inserts', rarely integrate into the host genome, are capable of high stable titers and need no helper virus. This review will briefly describe the virus, its molecular biology, and why Ad could make promising vectors for gene therapy.

2.2 ADENOVIRUS STRUCTURE

Three loosely defined sets of protein exist in the mature virus: proteins that form the outer coat of the capsid; scaffolding proteins that hold the capsid together; and DNA binding proteins. The diameter of the icosahedral-shaped capsid varies from 65 to 80 nm depending on the serotype (Horne *et al.*, 1959). The capsid is composed of a total of 720 hexon and 60 penton subunit proteins, 360 monomers of polypeptide VI, 240 monomers of polypeptide IX, and 60 trimeric fiber proteins (Figures 2.1 and 2.2).

The capsid proteins are essentially the hexons, pentons and the fiber proteins. Bound to the penton subunits and protruding from the capsid is the fiber protein which mediates the initial attachment of the virus to a target cell (Londberg-Holm and Philipson, 1969). Polypeptides IX, IIIa and VI form the scaffolding which holds the capsid together. Polypeptide IX stabilizes the packing of adjacent hexons in the capsid, polypeptide IIIa spans the capsid to link hexons of adjacent faces, and polypeptide VI connects the structural proteins to the core which consists of DNA associated with polypeptides V, VII, μ and the terminal protein (Stewart *et al.*, 1993).

Ad contain double-stranded DNA as their genetic material. The percent-

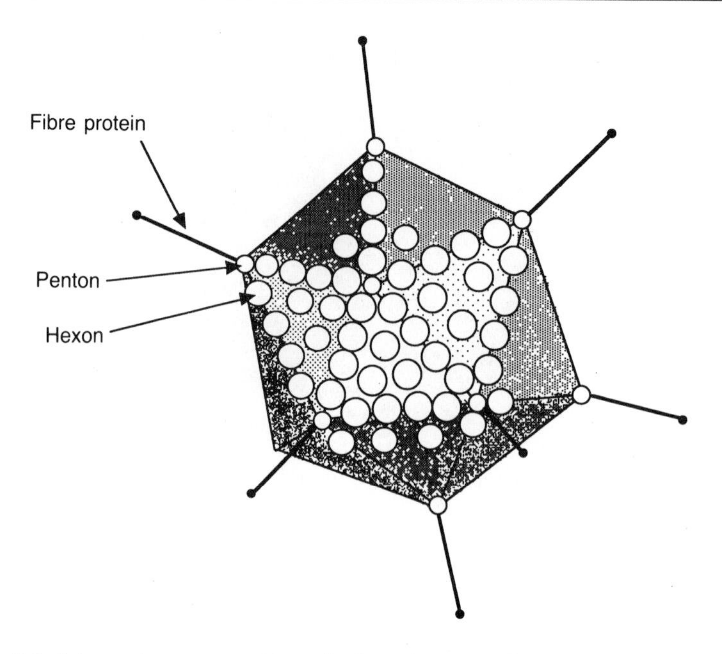

Figure 2.2 Schematic representation of adenovirus capsid.

age guanosine–cytosine (GC) content, the lengths of the viral genome (36 kb ≈ 11 μm) and inverted terminal repeats (100–140 bp) vary for the genome of the 47 characterized serotypes (Ad1–47) (Tooze, 1981). The genome is covalently linked at each 5′ end to individual 55-kDa terminal proteins. These polypeptides can associate with each other to circularize the DNA upon lysis of the virion (Robinson *et al.*, 1973). It is unknown if the DNA exists in a circular form inside the capsid; however the circular DNA form is an intermediate in viral DNA synthesis (Graham *et al.*, 1989).

2.3 MECHANISM OF VIRUS ENTRY AND DELIVERY TO THE NUCLEUS

The fiber protein initially attaches to an as yet unidentified cell surface receptor (Londberg-Holm and Philipson, 1969). A secondary cell surface receptor, the vitronectin binding integrins, binds the viral penton base and mediates virus–receptor endocytosis (Wickham *et al.*, 1993). The receptor and attached virus are pinocytosed via clatharin-coated pits to form an endosome (Pastan *et al.*, 1987). As the internal pH of the endosome falls to 5.5 a conformational change in the viral capsid is thought to take place (Wolfhart, 1988) which in turn ruptures the endosome at points of contact (Pastan *et al.*, 1987). The translocation from cell surface to cytoplasm takes

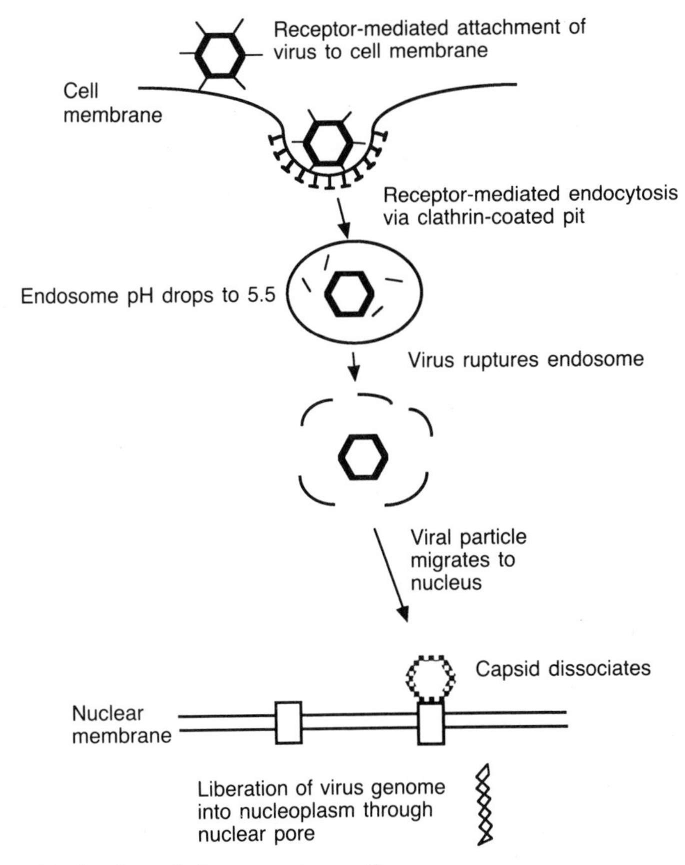

Figure 2.3 Mechanism of virus entry into cells.

approximately 15 minutes and is complete before the contents can be delivered to a lysosome (Svesson and Persson, 1984) (Figure 2.3).

This highly efficient adenoviral endolytic activity has been used as a means of entry of exogenous DNA into cells *in vitro*. Inactivated adenoviral particles, linked to plasmid DNA, can act as 'trojan horses' carrying DNA into the cell and delivering it to the cytoplasm (Cotten *et al.*, 1992; Wagner *et al.*, 1992).

Once in the cytoplasm, the virus migrates to the nucleus, binds to a nuclear pore complex, and releases the DNA into the nucleoplasm (Dales and Chardonnet, 1973; Greber *et al.*, 1993). This entire pathway is still poorly understood. What is clear is that the viral proteins are shed in a sequential manner as the virus migrates from the cell surface to the nucleus, and that the system is efficient. Based on a multiplicity of infection of one virus to one cell, approximately 40% of virus that attaches to the cell

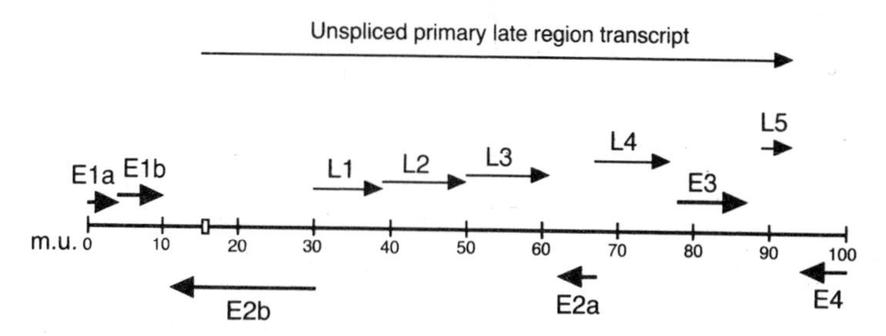

Figure 2.4 Schematic representation of 36-kb genome of Ad5. Heavy arrowed lines indicate unspliced transcripts from the early regions (E1a, E1b, E2a, E3 and E4). Major late promoter is indicated as an open box; single large pre-mRNA, from which the transcripts from the late regions (L1, L2, L3, L4 and L5) are subsequently spliced, are indicated as light arrowed lines.

surface succeed in delivering their DNA to the nucleoplasm (Greber *et al.*, 1993). The most complete description of virus entry and the shedding of the proteins to date has been reported by Greber *et al.* (1993).

2.4 MOLECULAR BIOLOGY OF ADENOVIRUS

The genome is functionally divided into two major overlapping regions, *early* and *late*, based on the time of transcription after infection. The switch from early to late gene expression takes places approximately 7 hours after infection. There are six distinct early regions; E1a, E1b, E2a, E2b, E3 and E4, each with individual promoters (except for the E2a–b region) and one late region, under the control of the major late promoter, with five well-characterized coding units (L1–L5) (Figure 2.4). There are also other minor intermediate and/or late transciptional regions less well characterized. Each region appears to contain a cassette of genes coding for proteins having related functions. The regions are transcribed initially as single RNAs which are then spliced into the various different mature mRNAs. More than 30 different mature RNA transcripts have been identified in Ad2, one of the most studied Ad serotypes.

Once the viral DNA is inside the nucleus, transcription is initiated from the viral E1a promoter. This is the only viral region that *must* be transcribed without the aid of viral coded transactivators. The E1a protein activates transcription of the other early regions and initiates viral gene expression leading to replication. The E1b protein in association with the E4–34 kDa protein quickly halts cellular protein synthesis. There are other regions that are also transcribed immediately after cell infection, suggesting that the E1 region is not the only region capable of being transcribed without viral transcription factors (Horowitz, 1990a). The E2 region codes for proteins involved directly in replication (e.g. the viral DNA polymerase, the terminal protein) and DNA binding proteins. In the E3 region, the nine predicted proteins are not required for Ad replication in cultured cells (Wold and Gooding, 1991). Of the six identified, four partially characterized proteins are involved in counteracting the immune system; gp19k prevents cytolysis by cytotoxic T lymphocytes (CTL), 14.7k and 10.4k/14.5k prevent tumor necrosis factor (TNF)-induced cytolysis. The E4 region appears to contain a cassette of genes whose products act to shut-down endogenous host gene expression and up-regulate transcription from other regions. Once viral DNA synthesis begins the late region genes are expressed, coding mainly for proteins involved in the structure and assembly of the virus particle.

2.5 ADVANTAGES OF ADENOVIRUS AS GENE THERAPY VECTORS

Ad satisfy several criteria which make them potential candidates for gene therapy. For example:

1. *Ad5 is able to accept 'large' DNA inserts.* Ad5 can safely package genome equivalents of 106 map units (6% larger than the wild type genome which represents 38 kb) without adversely altering their infectivity (Bett *et al.*, 1993). In order to provide more space within the genome it is necessary to delete regions. These lost gene functions can be supplied by a helper wild type virus such as Ad*wt*, or a cell line that complements these functions. Use of helper viruses causes problems not only during purification of the R Ad, but the Ad*wt* will replicate itself more efficiently than the R Ad. The 293 cell line is a human embryo kidney cell line stably transfected with a portion of the adenoviral E1 region (Graham *et al.*, 1977). This cell line can be used to propagate viruses deleted for map units 0–11 of the genome, representing 4 kb of DNA. A 2-kb portion of the E3 region, which is not required for viral replication, can also be easily deleted to permit a higher cloning capacity. The Ad5 d1324 mutant, deleted for the E1 region and part of the E3 region, has the theoretical capacity to accept 7 kb of insert and is frequently used as a base from which to construct recombinant vectors (Shenk and Williams, 1984).
2. *Ad are not host specific, and infect dividing as well as quiescent cells.* Human Ad*wt* can infect cells at any stage of the cell cycle and are not fastidious for a particular cell type or host. They primarily infect human epithelial cells from the eye, respiratory tract, gastrointestinal tract and urinary bladder (Kasel, 1980) but have also been reported to infect cells in the central nervous system, liver and lymphocytes (Kasel, 1980; Abken *et al.*, 1987). R Ad have been shown to be able to infect human macrophages (Haddada *et al.*, 1993), endothelial cells (Lemarchand *et al.*, 1992), fibroblasts, as well as murine muscle (Stratford-Perricaudet *et al.*, 1992), mouse liver (Stratford-Perricaudet *et al.*, 1990) and neural

cells (Akli *et al.*, 1993; Le Gal La Salle *et al.*, 1993). This lack of host specificity allows current animal models for a plethora of human diseases to be used to determine the feasibility of adenoviral-mediated gene therapy. The ability of human R Ad to infect murine cells has allowed the investigation of adenoviral-based gene therapy to the whole organism, as well as the investigation of physiological effects using transgenic mice disease models and therapeutic recombinant viruses.

Pathogenesis of R Ad administration can be followed in cotton rats (*Sigmodon hispidus*) and hamsters (*Cricetus aureatus*). These are the two non-human species identified as being permissive for human Ad, that allow viral replication as in human cells. In addition, intranasal inoculations result in respiratory tract disease with similar pulmonary histopathology to human adenoviral pulmonary disease (Pacini *et al.*, 1984; Hjorth *et al.*, 1988).

3. *Ad are stable and easy to prepare in high concentrations.* Ad, once purified, may be kept at −80°C for several months with only a slight decrease in the effective titer. Once thawed they are stable for many hours at 4°C but tend to lose their infectivity after 24 hours. R Ad replicate to high numbers in 293 cells, up to 1000 plaque forming units (pfu) can be collected from a single cell after a 30–40-hour cytolytic life cycle. Ad stocks are easily purified by two centrifugation steps on cesium chloride gradients, and titers as high as 1×10^{12} pfu/ml can be obtained.
4. *The expression of delivered genes is stable.* Expression of ornithine transcarbamylase from a R Ad injected intravenously in neonatal Spf-ash mice, a mutant strain that has less than 5% of the normal activity of this urea cycle enzyme, demonstrated that physiologically significant levels of expression of an exogene could last up to 15 months (Stratford-Perricaudet *et al.*, 1990). Also, intravenous injections of a R Ad containing the reporter gene β-galactosidase provided corroborating data. β-Galactosidase-positive nuclei were found in myocytes over 1 year after a single injection (Stratford-Perricaudet *et al.*, 1992). The post-mitotic state seems to offer favorable conditions for stability of transferred DNA.
5. *The most important consideration is safety.* Wild type adenoviral infections have not been shown to cause debilitating illness in immunocompetent adults and Ad vaccines have proven themselves safe in vaccination of more than 10 million military recruits over 20 years in North America (Rubin and Rorke, 1988). The rare cases of serious adenoviral illness are associated with specific adenoviral serotypes (e.g. serotypes 4, 7, 11, 21, 37) and are usually restricted to children or immunocompromised patients. Ad2 and 5, from which gene therapy vectors are based, cause little more than mild cold symptoms, although in some rare cases they can cause pneumonia. Indeed, many adenoviral infections do not result in symptoms and can be detected only by antibody formation and seroconversion (Horowitz, 1990a).

The viral genome rarely integrates into the host's chromosomes but exists as a stable extrachromosomal entity (Horowitz, 1990b). Indeed, regions of the genome of a R Ad have been reported to integrate with a frequency of 10^{-5} to 10^{-6} in rapidly dividing cultivated cells (Wang and Taylor, 1993). Thus, there is minimal risk that integration of the virus in the host's genome may alter expression of cellular genes or that viral promoters or enhancers may activate oncogenes. This does have the slight advantage/disadvantage in that the adenoviral vector DNA is not replicated in cell division. Thus the number of viral genomes per cell will diminish with each cell cycle, since the copies of the viral genome will be divided between the two daughter cells in a round of replication.

Ad, unlike retroviruses, have the regulatory advantage that their genome is not capable of being passed on to future generations. If germinal cells are infected then the many rounds of cell division required to produce a fully developed organism will result in the virus being diluted to an infinitesimally small fraction of the total body. This is assuming that the virus is maintained in the cell and not lost during a round of replication.

The observation that Ad 12 can induce tumors in hamsters led to a search for human cancers caused by Ad (Graham, 1984). As yet, no convincing evidence of causuality of human tumors by Ad have been shown (Horowitz, 1990a). The E1a transactivating protein was found to be the cause of the tumors in rodents. The E1 region is deleted to make the virus replication-defective in cells that do not complement E1a functions, reducing the risk of potential oncogenicity and to permit a greater cloning capacity. It is important to note that viral-induced illness is dependent on viral replication, therefore the E1 region deleted replication-deficient adenoviral vectors are unlikely themselves to be able to cause illness.

It is possible that co-infection of a patient treated with a replication-deficient R Ad, and an Ad*wt* may complement the deleted functions and lead to replication of the wild type virus *and* the recombinant, although this scenario has yet to be demonstrated *in vivo*.

2.6 ADENOVIRUS GENE THERAPY AND THE IMMUNE RESPONSE

The reported pneumonia in one patient in an American phase I adenoviral gene therapy trial for cystic fibrosis suggests that further tests of the vector are needed. The reasons behind this reaction have, at the time of writing, yet to be reported. It is known that E3 region-deleted Ad can cause inflammatory responses in cotton rats; however the pneumonia in this case may be completely unrelated to the treatment.

One of the major roles of the immune system is to destroy non-self entities. Circulating viral particles in the blood stimulate the production of specific antibodies which can prevent the virus from infecting cells and label the virus for destruction by macrophages and other phagocytic cells.

Cytotoxic T lymphocytes (CTLs) recognize cells infected by virus or cells expressing abnormally large amounts of a host protein and induce the cell's death.

This poses a problem for viral gene therapy. Repeat applications of a vector are likely to induce high titers of circulating, or secreted antibodies which may inactivate the virus before it can enter cells. Once the R Ad is inside a cell, some low level expression of the E2a DNA binding protein can occur. In 1% of the transfected cells in grafted tissue in nude mice a low level expression of the E2a binding protein was detected (Engelhardt *et al.*, 1993). It is suggested that these cells were undifferentiated when infected and produced E1a-like transactivating factor(s) which could induce viral transcription. These antigenic proteins may be presented in small amounts in association with class I major histocompatibility complex (MHC) molecules on the cell surface. It is this molecule that the CTLs probe to see if they recognize any abnormality in the expression of proteins within the cell.

The high expression of the exogene may pose the greatest problem. If the host does not express this gene, the host's immune system will recognize any cell presenting antigens of the resulting protein on the cell surface as foreign, thus provoking a CTL response to kill the altered cell (M.G.L. unpublished observation). An abnormally high expression of a host's gene in a cell may also provoke the immune system to destroy the abnormal cell. Both conditions are likely to be encountered in gene therapy to correct genetic abnormalities; either we wish a cell to produce a protein which it lacks the capacity to produce, or augment the production of a protein which is underproduced for the body's needs.

In the T cell deficient nude mouse model, expression of a virally encoded exogene lasted longer in transplanted human bronchial tissue than direct virus administration in the respiratory tract of immunocompetent animals, suggesting that it is indeed the T cell response which is involved in destroying the infected cells (Engelhardt *et al.*, 1993).

In order to circumvent this problem and increase the duration of expression of the exogene, the immune system may be supressed systemically by pharmacological means, as in organ transplants. However, this leads to numerous undesirable side effects. The same problem of evading the immune system has taxed viruses throughout their evolution. Ad*wt* are able to persist in normal human hosts for 24 months (Horowitz, 1990b). This feat is partly aided by the E3 region of the viral genome, which contains a cassette of genes coding for proteins that diminish immune recognition of infected cells (Wold and Gooding, 1991). TNF-induced lysis of the cell is unlikely to be a problem since the susceptibility to TNF is confered by the E1a protein (Duerksen-Hughes *et al.*, 1989; Gooding *et al.*, 1991). As already stated, this gene is deleted in current vectors. The most interesting of the E3 region gene products for the gene therapist is the protein gp19k. This glycoprotein has a high affinity for the MHC class I molecules and a strong signal for retention in the endoplasmic reticulum (ER). Thus gp19k and the associated MHC class I molecules rest in the ER, preventing the expression of the MHC complex on the cell surface. If the expression of MHC class I is diminished there is less presentation of antigens on the cell surface and less chance that the immune system will recognize the infected cell as unusual. Work is currently on-going to investigate the re-inclusion of this protein in vectors, since the E3 region being dispensible for viral replication is deleted in most current vectors.

Ad are not the only viruses to have developed methods of hiding from the immune system. Many different families of human viruses are known to produce factors that down-regulate the immune response. Human cytomegalovirus encodes a protein from the H301 gene that binds to β2-microglobulin to interfere with MHC class 1 maturation in the ER (Maudsley and Pound, 1991). Cowpox virus produces a serpin-like protein that interferes with interleukin (IL)-1β synthesis and a soluble glycoprotein with homology to the IL-1 receptor; these proteins block the action of lymphokines to prevent the immune effectors signaling the presence of an infection (Marrack and Kappler, 1994). Any of these or other genes could be integrated in a vector in order to down-regulate the immune response against the infected cells.

Another method of tricking the immune system to ignore altered cells is to induce tolerance. R Ad administered to neonatal mice results in greatly lengthened exogene expression in comparison with R Ad administration in adult mice (Stratford-Perricaudet *et al.*, 1990; Kass-Eisler *et al.*, 1993). Presumably the immature neonatal immune system learned to tolerate the virally altered cells as self early in development, thus suggesting a different, complementary strategy for avoiding the immune system in gene therapy. To use this system in human patients is far from easy. A newly born baby diagnosed *in utero* as being affected by a genetic disease must be treated as early as possible with the vector in order to induce tolerance. However, this may mean that the child will never reject a wild type adenovirus infection. There is much research to be undertaken and ethical questions to be addressed before this method can be used in human subjects.

There are certain circumstances where the stimulation of the immune system is the goal of the gene therapy, such as cancer therapy, discussed later in the book. In these cases deletion of the E3 region is necessary not only to free space but precisely to prevent escape from the watchful eyes of the immune system.

2.7 CONCLUSIONS

Adenoviruses are an effective and useful means by which to transfer exogenes to cells. They can infect many different cell types at all stages of the cell cycle via a widely expressed receptor. They are easy to prepare, have proven themselves to be safe as a vaccine, and can carry large DNA inserts in their genome. Since the first reported use of R Ad as a means of gene expression (Thummel *et al.*, 1981) and the first use as a therapeutic agent (Stratford-Perricaudet *et al.*, 1990) the field of research into adenoviral gene therapy has expanded at an incredible rate. Today, clinical protocols using adenoviral vectors are under evaluation for the treatment of cystic fibrosis, with protocols for other syndromes sure to follow. Hopefully in the near future Ad will be considered as positively good news for many different syndromes and no longer thought of as a potentially dangerous pathogen.

ACKNOWLEDGEMENTS

M. Lee gratefully thanks his wife and the IFSBM for their continuing support. Electron micrographs are provided by courtesy of Dr C. Dauguet, Service d'Oncologie Virale, Institut Pasteur, Paris, France.

REFERENCES

Abken, H., Butzler, C. and Willecke, K. (1987) Adenovirus type 5 persisting in human lymphocytes is unlikely to be involved in immortalization of lymphoid cells by fusion with cytoplasts or by transfection with DNA of mouse L cells. *Anticancer Res.*, 7, 553–8.

Akli, S., Caillaud, C., Vigne, E., Stratford-Perricaudet, L.D. *et al.* (1993) Transfer of a foreign gene into the brain using adenovirus vectors. *Nature Genetics*, **3**, 224–34.

Bett, A.J., Prevec, L. and Graham, F.L. (1993) Packaging capacity and stability of human adenovirus type 5 vectors. *J. Virol.*, **67**, 5911–21.

Cotten, M., Wagner, E., Zatloukal, K., Phillips, S., Curiel, D.T. and Birsteil, M.L. (1992) High-efficiency receptor-mediated delivery of small and large (48 kilobase) gene constructs using the endosome-disruption activity of defective or chemically inactivated adenovirus particles. *Proc. Natl Acad. Sci. USA*, **89**, 6094–8.

Dales, S. and Chardonnet, Y. (1973) Early events in the interaction of adenoviruses with HeLa cells IV. Association with microtubules and the nuclear pore complex during vectorial movement in the inoculum. *Virology*, **56**, 465–83.

Duerksen-Hughes, P., Wold, W.S.M. and Gooding, L.R. (1989) Adenovirus E1a renders infected cells sensitive to cytolysis by tumor necrosis factor. *J. Immunol.*, **143**, 4193–200.

Engelhardt, J.F., Yang, Y., Stratford-Perricaudet, L.D. *et al.* (1993) Direct gene transfer of human CFTR into human bronchial epithelia of xenografts with E1-deleted adenoviruses. *Nature Genetics*, **4**, 27–34.

Gooding, L R., Ranheim, T.S., Tollefson, A.E., Brady, H.A. and Wold, W.S.M. (1991) The 10,400- and 14,500-dalton proteins encoded by region E3 of adenovirus function together to protect many but not all mouse cell lines against lysis by tumor necrosis factor. *J. Virol.*, **65**, 4114–23.

Graham, F.L. (1984) Transformation and oncogenicity of human adenovirus, in *The Adenoviruses*, (ed. H. Ginsburg), Plenum Press, New York, pp. 339–98

Graham, F.L., Smiley, J., Russel, W.C. and Nairu, R. (1977) Characteristics of a human cell line transformed by the DNA from human adenovirus 5. *J. Gen. Virol.*, **36**, 59–72.

Graham, F.L., Rudy, J. and Brinkely, P. (1989) Infectious circular DNA of human adenovirus 5: regeneration of viral DNA termini from molecules lacking terminal sequences. *EMBO Journal*, **8**, 2077–85.

Greber, U.F., Willetts, M., Webster, P. and Helenius, A. (1993) Stepwise dismantling of adenovirus 2 during entry into cells. *Cell*, **75**, 477–86.

Haddada, H., Lopez, M., Martinache, C., Ragot, T., Abina, M.A. and Perricaudet, M. (1993) Efficient adenovirus-mediated gene transfer into human blood monocyte-derived macrophages. *Biochem. Biophys. Res. Commun.*, **195**, 1174–83.

Hjorth, R.N., Bonde, G.M., Pierzchala, W.A. *et al.* (1988) A new hamster model for adenoviral vaccination. *Arch. Virol.*, **100**, 279–83.

Horne, R.W., Bonner, S., Waterson A.P. and Wildy, P. (1959) The icosahedral form of an adenovirus. *J. Mol. Biol.*, **1**, 84–6.

Horowitz, M.S. (1990a) Adenoviridae and their replication, in *Virolgy*, 2nd edn, (eds B.N. Fields, D.M. Knipe, *et al.*), Raven Press, New York, pp. 1679–723.

Horowitz, M.S. (1990b) Adenoviruses, in *Virology*, (eds B.N. Fields, D.M. Knipe, *et al.*), Raven Press, New York, pp. 1723–41.

Kasel, J.A. (1980) Adenoviruses, in *Viral Rickettsial and Chlamydial Infections*, (eds E. Bennette and N. Schmidt), Public Health Association, New York, Chapter 9.

Kass-Eisler, A., Falck-Pedersen, E., Alvira, M. *et al.* (1993) Quantitative determination of adenovirus-mediated gene delivery to rat cardiac myocytes *in vitro* and *in vivo*. *Proc. Natl Acad. Sci. USA*, **90**, 11498–502.

Le Gal La Salle, G., Robert, J.J., Berrard, S. *et al.* (1993) An adenovirus vector for gene transfer into neurones and glial cells in the brain. *Science*, **259**, 988–90.

Lemarchand, P., Jaffe, H.A., Danel, C. *et al.* (1992) Adenovirus-mediated transfer of a recombinant human alpha1-antitrypsin cDNA to human endothelial cells. *Proc. Natl Acad. Sci. USA*, **89**, 6482–6.

Londberg-Holm, K. and Philipson, L. (1969) Early events of virus–cell interactions in an adenovirus system. *J. Virol.*, **4**, 323–38.

Marrack, P. and Kappler, J. (1994) Subversion of the immune system by pathogens. *Cell*, **76**, 323–32.

Maudsley, D.J. and Pound, J.D. (1991) Modulation of MHC antigen expression by viruses and oncogenes. *Immunol. Today*, **12**, 429–31.

Pacini, D.L., Dubovi, E.J. and Clyde, W.A. Jr (1984) A new animal model for human respiratory tract disease due to adenovirus. *J. Infect. Dis.*, **150**, 92–7.

Pastan, I., Seth, P., Fitzgerald, D. and Willingham, M. (1987) Adenovirus entry into cells: some new observations on an old problem, in *Concepts in Viral Pathogenesis*, (eds A. Notkins, M.B.A. Oldstone *et al.*), Springer-Verlag, New York, pp. 141–6

Robinson, A.J., Younghusband, H.B. and Bellet, A.J.D. (1973) A circular DNA protein complex from adenoviruses. *Virology*, **56**, 54–69.

Rowe, W.P., Huebner, R.J., Gilmore, L.K., Parrott, R.H. and Ward, T.G. (1953) Isolation of a cytopathogenic agent from human adenoids undergoing spontaneous degeneration in tissue culture. *Proc. Soc. Exp. Biol. Med.*, **84**, 570–3.

Rubin, S.A. and Rorke, L.B. (1988) Adenoviral vaccines, in *Vaccines*, (eds E.A. Mortimer, Jr and S.A. Plotkin), WB Saunders, Philadelphia, pp. 492–512.

Shenk, T. and Williams, J. (1984) Genetic analysis of adenoviruses. *Curr. Top. Microbiol. Immunol.*, **111**, 1–39.

Stewart, P.L., Fuller, S.D. and Burnett, R.M. (1993) Difference imaging of adenovirus: bridging the resolution gap between X-ray crystallography and electron microscopy. *EMBO Journal*, **12**, 2589–99.

Stratford-Perricaudet, L.D., Levrero, M., Chasse, J-F., Perricaudet, M. and Briand, P. (1990) Evaluation of the transfer and expression in mice of an enzyme encoding gene using a human adenovirus vector. *Hum. Gene Ther.*, **1**, 241–56.

Stratford-Perricaudet, L.D., Makeh, I., Perricaudet, M. and Briand, P. (1992) Widespread long-term gene transfer to mouse skeletal muscles and heart. *J. Clin. Invest.*, **90**, 626–30.

Svesson, U. and Persson, R. (1984) Entry of adenovirus 2 into HeLa cells. *J. Virol.*, **51**, 687–94.

Thummel, C., Tijan, R. and Grodzicker, T. (1981) Expression of SV40T antigen under the control of adenovirus promoters. *Cell*, **23**, 825–36.

Tooze, J. (1981) *DNA Tumor Viruses: Molecular Biology of Tumor Viruses*. Cold Spring Harbor Laboratory, New York, pp. 937–1036.

Wagner, E., Zatloukal, K., Cotten, M. *et al.* (1992) Coupling of adenovirus to transferrin–polylysine/DNA complexes greatly enhances receptor-mediated gene delivery and expression of transfected genes. *Proc. Natl Acad. Sci. USA*, **89**, 6099–103.

Wang, Q. and Taylor, M.T. (1993) Correction of a deletion mutant by gene targeting with an adenovirus vector. *Mol. Cell. Biol.*, **13**, 918–27.

Wickham, T.J., Mathais, P., Cheresh, D.A. and Nemerow, G.R. (1993) Integrins $\alpha_v\beta_3$ and $\alpha_v\beta_5$ promote adenovirus internalization but not virus attachment. *Cell*, **73**, 309–19.

Wold, W.S.M. and Gooding, L.R. (1991) Region E3 of adenovirus: a cassette of genes involved in host immunosurveillance and virus–cell interactions. *Virology*, **184**, 1–8.

Wolfhart, C. (1988) Neutralizing of adenovirus: kinetics, stoichiometry, and mechanisms. *J. Virol.*, **62**, 2321–8.

3

Herpes simplex virus vectors

JOSEPH GLORIOSO, MARY ANN BENDER, DAVID FINK
and NEAL DELUCA

3.1 INTRODUCTION

Advances in the methodology and practice of human gene therapy have been substantial in the past few years. In contrast to traditional palliative forms of therapy in which a therapeutic drug, protein or peptide product must be repeatedly administered to treat a disease, successful gene therapy would result in endogenous production of the therapeutic product *in situ*. Research on the design and application of gene transfer vehicles and genetically altered cells to the treatment of a variety of diseases has greatly expanded the dimension and scope of this field such that almost any genetic alteration of cells and tissues to effect a therapeutic outcome is now considered to employ the principles of gene therapy. Moreover, developments primarily in the use of novel viral vectors coupled with the rapid discovery of new disease-causing genes have driven the field with increasing vigor.

Yet, with few exceptions, progress in gene therapy is still impeded by several technical and practical problems. First, gene therapy methodology must be improved to avoid costly and laborious *ex vivo* approaches which rely on the *in vitro* transduction of autologous cells explanted from the patient. These cells are returned to the body after gene transfer, selection, expansion and characterization for gene expression. Aside from the risk of morbidity and possible cell transformation associated with these methods, it is difficult to grow many types of cells *in vitro*, and some tissues simply are not readily available or suitable for primary culture.

The difficulties associated with *ex vivo* approaches have arisen partly because the first and most widely used gene transfer vectors were retro-

Molecular and Cell Biology of Human Gene Therapeutics
Edited by George Dickson
Published in 1995 by Chapman & Hall. ISBN 0 412 62550 4

viruses. These vectors lack the ability to infect non-dividing cells, thus limiting their utility largely to cell transduction *in vitro*. Second, the maintenance and control of gene expression in vector-mediated gene transfer have proven to be a formidable problem. The frequent 'shut-off' in therapeutic gene expression following recombinant retrovirus integration into chromosomes is still poorly understood. Third, retroviral integration appears to be largely a random process creating the risk of oncogene activation by the retroviral promoter as a relatively rare, but potentially dangerous event. Moreover, viral stocks derived from producer cell lines occasionally contain active virus recombinants. Not all of these problems are unique to retroviral vectors but perhaps can be better controlled for using vectors which do not require producer lines or genome integration for stable gene expression. For gene transfer to rapidly dividing cells in the body, however, vectors which promote gene integration will remain the preferred method for gene transfer, as for example, the reconstitution of bone marrow using transduced stem cells.

In view of this discussion, it is desirable and often essential that methods for gene transfer be improved such that direct *in vivo* gene therapy using non-integrating viral vectors which readily infect non-dividing cells eventually becomes a practical reality. Adenovirus and herpesvirus vectors are representative of gene delivery systems usable for *in vivo* gene transfer because of the ability to efficiently infect non-dividing cells and have a broad host cell range. *In vivo* methods using these vectors should increase the versatility of gene therapy practice and clinical applications despite the fact that for most tissues, vector maintenance and expression will be transient due to cell turnover. Nevertheless, for the majority of applications, and depending on the tissue, transient expression will be sufficient and often desirable. These applications might include acute therapies for cancer, infectious disease, acute episodes of autoimmune disease, and some metabolic diseases. However, treatment of chronic diseases such as cystic fibrosis with transient gene transfer vectors will require repeated vector administration. This will no doubt require special attention to potential problems related to immune responses to vector proteins. For other applications involving certain non-dividing cell populations localized within well-differentiated tissues such as brain, non-integrating vectors should provide long-term benefits with only infrequent vector administration. Herpesvirus vectors should be ideal for gene transfer to the brain since this virus is neurotropic and has evolved in a manner to persist in neurons. Direct gene transfer to the brain is desirable for a number of reasons. First, the continuous production of a therapeutic substance from the transferred gene obviates the need for repeated administration of pharmaceutical agents. Second, because the gene product is released within the substance of the brain, the obstruction to systemic drug delivery imposed by the blood–brain barrier and the ependymal layer would be bypassed. Third, gene transfer has the potential to produce the therapeutic product within the target cell, especially for products whose activities require membrane localization or which can be released in a soluble form suitable for uptake by neighboring cells.

There is a wide variety of diseases which affect the nervous system. These include neuronal and glial cell malignancies, autoimmune disorders such as multiple sclerosis, a large number of metabolic deficiencies, and various forms of neurodegenerative diseases which are among the most common of any human afflictions. Determining the genetic basis of many of the neurodegenerative diseases has been the subject of intensive investigation in recent years. Some of the genes responsible for these pathological processes have now been cloned, and more global approaches to mapping and cloning of genes encoding neurotransmitters, their receptors, ion channels, and growth and differentiation factors are in progress. These methods include the production of antibodies to brain proteins, isolation of brain peptides, partial cloning of messenger RNAs expressed in mature brain, and physical mapping techniques designed to identify heritable disease genes. Notable examples of newly cloned genes include those responsible for amyotrophic lateral sclerosis (Rosen *et al.*, 1993) and Huntington's chorea (MacDonald *et al.*, 1993). On-going research is rapidly narrowing the number of genetic markers associated with common disorders such as Alzheimer's and Parkinson's disease which are likely to result from a number of different genetic defects affecting a common pathway to disease.

3.2 ENGINEERING HERPES SIMPLEX VIRUS VECTORS

This review will focus on herpes simplex virus (HSV) vectors for gene transfer to the nervous system. HSV-1 is a promising vector for central nervous system (CNS) applications since its natural biology involves long-term persistence in a latent state during which lytic genes are silenced, and only latency-specific intranuclear RNAs are detected in latently infected neuronal cell bodies. HSV-1 DNA has also been detected in normal brains of numerous individuals upon autopsy (Fraser *et al.*, 1981), suggesting that even wild type HSV can be harbored without apparent brain damage. Moreover, latency can be achieved by mutant viruses completely defective for replication and reactivation from the latent state thus demonstrating the feasibility of safe vector design.

HSV vectors have a number of additional advantages for *in vivo* gene therapy to brain and other tissues. These include: (i) their ability to efficiently infect a wide variety of both dividing and post-mitotic cell types; (ii) the ability of defective viruses to be replicated to exceedingly high levels (1000 pfu/cell) on available complementing cell lines (DeLucca *et al.*, 1985); (iii) their potential ability to be engineered to accommodate a very large amount of foreign genetic information (> 50 kb); and (iv) their ability to persist permanently in a latent state as neuronal and other post-mitotic cell types without viral protein synthesis, genome integration, or interference with host cell biology. This latter point is particularly relevant since neuronal cells harboring the viruses are not attacked by anti-viral immune responses.

3.2.1 Biological features of HSV-1 relevant to vector design

HSV-1 is a large neurotropic virus which infects a large majority of the human population, usually at an early age. The virus particle is composed of an envelope having at least 10 glycoprotein components, an icosahedral capsid containing seven different structural proteins and a matrix of proteins surrounding the capsids referred to as the tegument (Roizman and Furlong, 1974), which is composed of approximately 12 proteins (Roizman and Sears, 1993). The envelope proteins are required for virus attachment and entry into a variety of cell types by fusion of the virus envelope with the cell surface membrane (see Figure 3.3 b(i)). The capsid is released into the cytoplasm along with the tegument proteins and brought into juxtaposition with the nuclear membrane. The viral DNA is subsequently released through a viral penton (Newcomb and Brown, 1994) and transported into the nucleus through a nuclear pore.

The viral genome is a linear double-stranded DNA molecule 152 kb in length and is composed of a long and short unique segment each flanked by inverted repeats (Figure 3.1). The virus encodes at least 75 genes which are arranged in a linear and almost exclusively uninterrupted fashion along both DNA strands. This feature greatly simplifies the manipulation of the genome. Another important feature of the organization of the HSV genes is that the genes may be divided into two classes: approximately half of the genes are essential for virus growth in cell culture, while the other half encode functions which enhance its viability in the host but are non-essential for growth in cell culture. It is remarkable that these two classes of genes are distributed in a manner to make possible the removal of large segments of the viral genome without the need for complementation in order to produce virus. This feature should provide a unique opportunity to replace large segments of the viral genome with foreign DNA.

Some of the tegument proteins enter the nucleus and in particular one protein, the product of the UL48 gene (VP16, α TIF, Vmw65) plays a crucial role in the initiation and enhancement of the lytic cycle gene program (Figure 3.2). VP16 forms a complex with the cellular transcription factor, Octomer binding protein one (Oct-1) (Batterson and Roizman, 1983; Campbell *et al.*, 1984; Kristie and Roizman, 1987; McKnight *et al.*, 1987), and together these proteins recognize a specific core enhancer element having the consensus sequence TAATGARAT found in at least one copy in the promoters of all immediate early (IE) genes (Gaffney *et al.*, 1985). While the transcriptional enhancement activity of VP16 is not essential, it greatly increases the transcription of the five IE

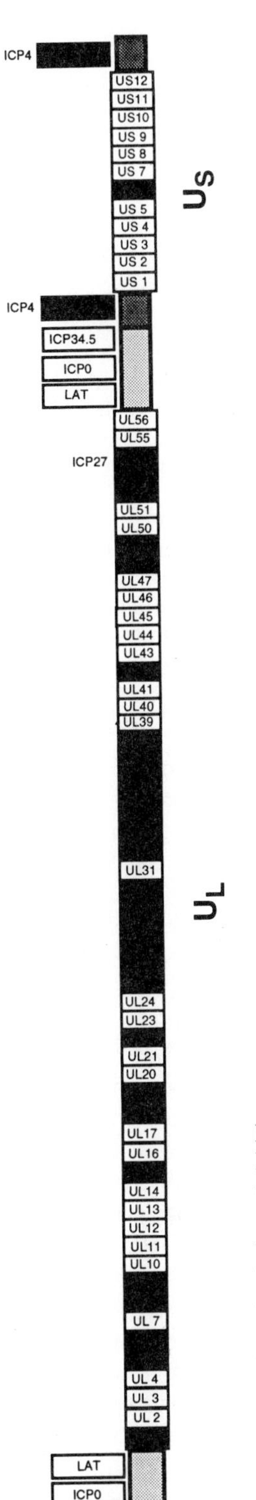

Figure 3.1 Map of the HSV genome. The 152-kb genome of HSV is organized into two regions, a unique long (U_L) and a unique short (U_S), each flanked by terminal repeat regions (shaded). The 75 genes of HSV are separated into two classes: those genes which are essential for growth in cell culture, and those which are not essential for growth *in vitro*, but may confer enhanced viability in the neuronal cell environment *in vivo*. The non-essential genes are listed above as clear boxes, while the essential viral genes are shown as black boxes. Those genes which map within the terminal repeats are shown above the shaded boxes of the repeat regions. It can be seen that much of the U_S region is not essential for virus growth in cell culture.

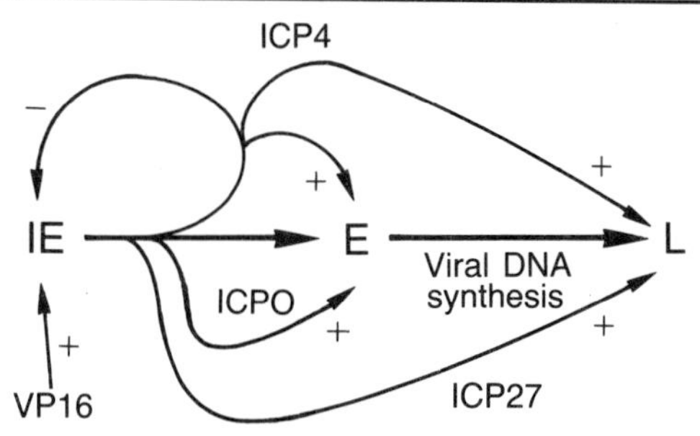

Figure 3.2 Transcriptional progression of HSV lytic cycle. The three classes of HSV genes expressed during the lytic cycle of growth are the immediate early (IE) or α, the early (E) or β, and the late (L) or γ genes. The IE genes are the first set of viral genes to be expressed, supplying transactivators which stimulate the expression of the E class of genes. The L genes are expressed after DNA replication has occurred. The viral gene products which regulate this cascade of transcription are shown, as well as the positive (+) or negative (−) effect they impose on transcription. The VP16 protein, present in the incoming virus' tegument layer, activates transcription of the IE genes. Three IE proteins, ICP4, ICP0 and ICP27 activate expression of both the E and L genes. ICP4 also inhibits the expression of the IE genes, once transcription of the E class of genes is initiated.

genes referred to as infected cell polypeptide (ICP), 0, 4, 22, 27 and 47. Only two of the IE gene products, ICP4 and ICP27, are essential for virus replication (Watson and Clements, 1980; Preston, 1979; Dixon and Schaffer, 1980; Sacks *et al.*, 1985). ICP4 is required for early (β) and late (γ) gene expression while ICP27 is required for efficient late gene expression (McCarthy *et al.*, 1989). ICP4 can also down-regulate its own transcription. Late gene transcription also strictly requires viral DNA synthesis (Holland *et al.*, 1980). The other IE genes are non-essential but express products important to the virus life cycle *in vivo*. ICP22 is required for late gene expression in certain cell types (Sears *et al.*, 1985) and ICP47 has recently been shown to down-regulate the expression of class I histocompatibility molecules on the cell surface, rendering the infected cell less susceptible to immune attack (York *et al.*, 1994). ICP0 is a promiscuous transactivator which is involved with ICP4 and ICP27 in regulating HSV gene expression (McMahan and Schaffer, 1990). In addition, ICP6, which encodes the large subunit of the viral ribonucleotide reductase, is induced at the IE time and is up-regulated by the ICP0 gene product as an early (β) gene (Desai *et al.*, 1993).

Figure 3.3a illustrates diagrammatically the virus life cycle in natural human infections, while Figure 3.3b demonstrates ultrastructural and biochemical features corresponding to the diagram. The virus replicates primarily in skin or mucosal epithelial cells before entering regional axon terminals. Viral capsids subsequently undergo retrograde transport to nerve cell bodies of sensory ganglia, and the viral DNA enters the nucleus where either the lytic or latency pathway ensues (Roizman and Sears, 1990). While some cells may be destroyed, the lytic cycle is aborted in the majority of infections with the accompanying establishment of latency. Latent virus can be reactivated by a variety of stimuli followed by anterograde transport and productive infection at or near the primary site. During latency the viral genomes become partially methylated and sequestered into episomal minichromosome-like structures bound by nucleosomes (Dressler *et al.*, 1987; Deshmane and Fraser, 1989). Some nuclei may contain numerous copies of the genome, and recent studies demonstrated that the vast majority of susceptible nuclei harbor latent virus (Ramakrishnan *et al.*, 1994). In animal models, infection of the peripheral sensory ganglia and latency can be achieved through inoculation of the skin (Cook and Stevens, 1973), cornea (Seiler and Schwab, 1984), or the olfactory bulb after intranasal inoculation (Stroop and Schaefer, 1987). Focal infection and latency can be achieved within specific brain regions by stereotactic inoculation of small volumes of virus into targeted brain regions (Bak *et al.*, 1977; McFarland *et al.*, 1986).

Latently infected neurons show a highly restricted and characteristic pattern of viral gene expression (Hill, 1985). The only transcriptionally active region of the viral genome maps to a segment of the inverted repeat sequences of the unique long (UL) region of the viral genome just downstream of the structural gene for ICP0 (Figure 3.4a). The most abundant latency-associated transcript (LAT) is 2 kb in size, although several less abundant transcripts (1.5 and 1.45 kb) can also be detected by Northern analysis of RNA extracted from latently infected ganglia (Spivack and Fraser, 1987, 1988; Wagner *et al.*, 1988b; Stevens, 1989; Farrell *et al.*, 1991). The 2-kb LAT is also readily detected late in lytic infections in cell culture while the smaller LATs have been detected only in latently infected neurons (Devi-Rao *et al.*, 1991). The smaller LATs are thought to be processed from the 2-kb LAT. LAT appears not to be capped or polyadenylated (Wagner *et al.*, 1988a) and may be a stable intron spliced from a large 8.77-kb poly A+ RNA (Farrell *et al.*, 1991). This large RNA appears to be very unstable and thus difficult to detect in latency. In cell culture experiments, some LAT molecules have been found associated with polyribosomes, suggesting that LAT could play a role in regulating mRNA translation (N. Fraser, personal communication). Moreover, LAT has been predicted to function as an antisense molecule capable of inhibiting ICP0 mRNA translation. Recent cell culture experiments where LAT is overexpressed by an IE promoter during infection suggest this is the case, and perhaps LAT even destabilizes the ICP0 mRNA (Karpinski and DeLuca, in preparation). The LAT RNA contains several open reading frames but so far no LAT-related viral proteins have been detected in latency (Doerig *et al.*, 1991; Spivack *et al.*, 1991; Wechsler *et al.*, 1989). During latency the LATs are largely localized in cell nuclei as determined by *in situ* hybridiz-

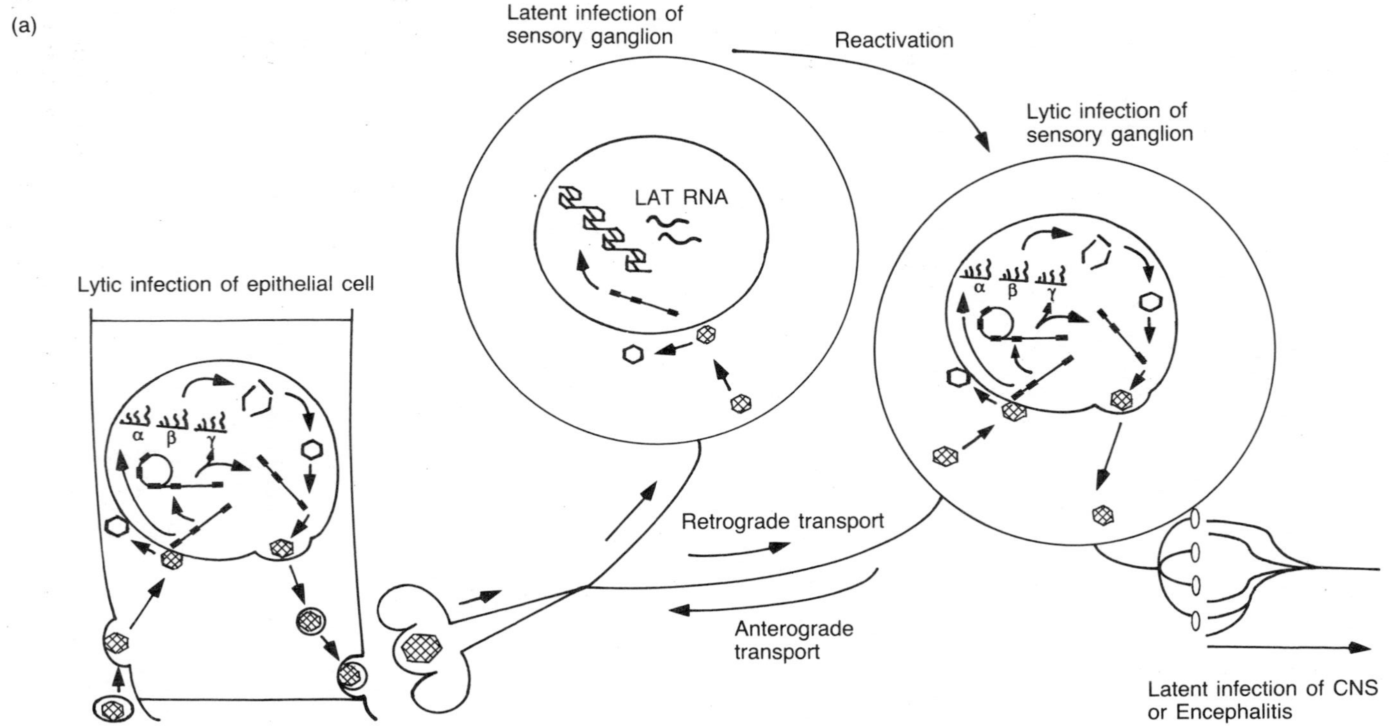

Figure 3.3 (a) *See p. 42 for caption.*

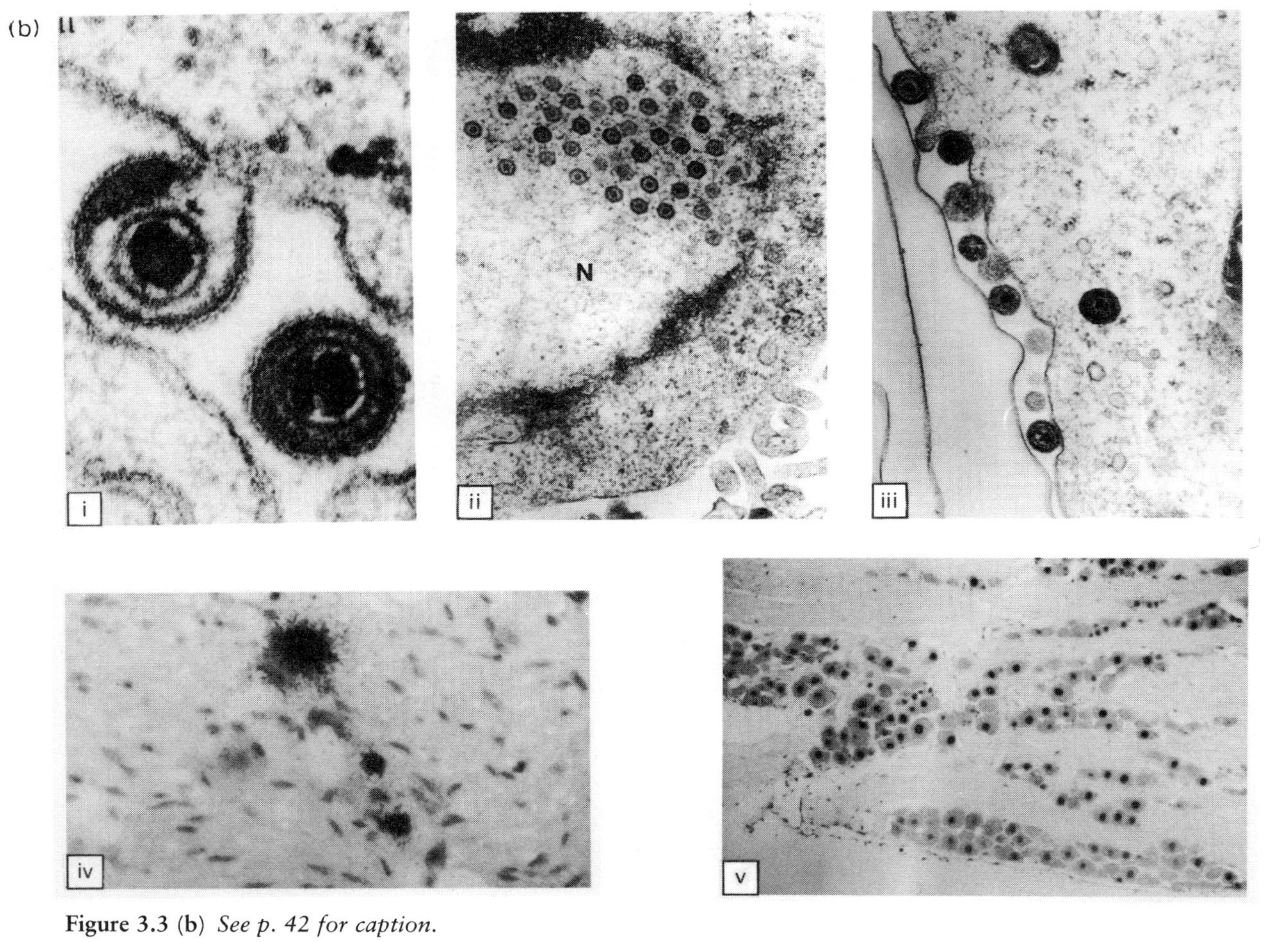

Figure 3.3 (**b**) *See p. 42 for caption.*

Figure 3.3 (a) Life cycle of herpes simplex virus (HSV-1). A lytic HSV infection is initiated by the fusion of the viral envelope with the cell membrane of either an epithelial or a mucosal cell at the site of primary infection. The virus is then transported to the nucleus of the cell where the viral DNA is released and is replicated via a rolling circle mechanism. (Steps in virus life cycle proceed clockwise.) Both before and after the initiation of viral DNA replication, there is a regulated order of viral transcription which provides for the expression of proteins required for replication as well as for the formation of new viral capsids. Newly synthesized viral DNA is packaged by unit-lengths into the new empty capsids. The full viral capsids emerge from the nucleus, acquiring an envelope, and are released from the cell, with accompanying cell death. Foci of infected cells permit contact of the virus with nerve terminals innervating the area. Upon entry of the virus into a nerve, the virus is transported by rapid retrograde flow to the nucleus of the neuron. There, the virus sets up one of two types of infection. In a latent infection, the viral DNA becomes associated with chromatin and does not replicate. During latency, a set of latency-associated transcripts (LATs) is the only viral RNA expressed by the virus, with no viral protein produced. Conversely, the virus may establish a lytic infection with the production of viral progeny. Latency may last for the lifetime of the host, but upon stimulation by stress, drugs or other agents, the lytic cycle may be activated. The progeny viruses are then transported either: (1) in an anterograde manner back to the site of the initial infection; or (2) to CNS neurons where either a productive encephalitis occurs or latency is re-established. (b) Features of an HSV infection. (i) Ultrastructure of the HSV virus. The dark inner core of the virus containing the viral DNA and core proteins is seen centrally within each virion. The core is surrounded by the symmetric icosahedral capsid. An amorphous tegument region separates the capsid from the outer envelope of the virus. The virus envelop fuses with the cell surface membrane to initiate infection. (ii) Electronmicrograph of progeny virions accumulating within the nucleus of an infected cell. Each capsid contains a inner core of DNA and protein. As the viruses leave the nucleus, they will pick up an outer envelope containing viral glycoproteins which have become embedded within the nuclear membrane. (iii) Electronmicrograph of progeny virus at the outer cellular membrane of a productively infected cell. The virions are now enveloped and can enter axon terminals. (iv) Detection of LAT synthesis during latency. Peripheral infection of mouse cornea with wild type HSV-1 results in virus entry into the trigeminal ganglion neurons where it establishes latency and expresses the intranuclear LAT RNA. The LATs are detected in approximately 5% of neurons at 30 days post-infection by *in situ* hybridization. Those neurons which are positive for LAT are marked by the deposition of the darkly-staining granules. (v) Detection of latently-infected neurons within peripheral nervous tissue. At 30 days following wild type infection, latent viral DNA can be detected in the majority of the trigeminal ganglion neurons by *in situ* polymerase chain reaction (PCR). The viral sequences which code for the glycoprotein B (gB) protein were amplified by the PCR reaction, and neurons positive for the viral gB sequences exhibit dark staining. This extensive labelling demonstrates the virus' capacity to establish latency within a large number of neurons within the sensory ganglion.

ation (Rock *et al.*, 1987; Stevens *et al.*, 1987; Deatly *et al.*, 1988). It is unclear whether all neurons harboring viral DNA express LAT. Our experiments in the CNS suggest LAT is unstable at least over several weeks, requiring continued synthesis for its detection. The functional role of LAT *in vivo* is still unclear, since studies with LAT deletion mutants have yielded mixed results. Several studies with animal model systems have demonstrated that LAT^- viruses are capable of establishing latency within neurons (Leib *et al.*, 1989; Sedarati *et al.*, 1989; Hill *et al.*, 1990). However, these viruses exhibit differing abilities to reactivate, displaying either wild-type (Javier *et al.*, 1988; Sedarati *et al.*, 1989; Block *et al.*, 1990; Natarajan *et al.*, 1991) or delayed reactivation patterns (Steiner *et al.*, 1989; Hill *et al.*, 1990; Sawtell and Thompson, 1992). These differences could depend on the actual number of viral genomes which establish latency within nerve tissue as well as the relationship of the host cell factors leading to reactivation.

The origin and function of the LATs are further complicated by the promoter/regulatory elements which control LAT expression. Two latency active promoters (LAPs), LAP1 (Javier *et al.*, 1988; Dobson *et al.*, 1989; Leib *et al.*, 1989; Steiner *et al.*, 1989; Zwaagstra *et al.*, 1989, 1990, 1991; Batchelor and O'Hare, 1990, 1992; Devi-Rao *et al.*, 1991; Sawtell and Thompson, 1992; Goins *et al.*, 1994) and LAP2 (Nicosia *et al.*, 1993; Goins *et al.*, 1994) (Figure 3.4b), have been identified upstream of the LAT coding sequence. LAP1 contains a TATA box with recognizable upstream control elements such as CAAT, USFI, CRE and Sp1 sites. In addition, there is an initiator (INR) element at the start of transcription which also contains an ICP4 binding site (Soares *et al.*, unpublished). The former promotes transcription, while the latter down-regulates LAT expression (Rivera-Gonzalez *et al.*, 1994). LAP2 is quite different, lacking a TATA box but containing elements found in a wide variety of housekeeping gene promoters. LAP2 is down-regulated by ICP4 which might be expected for a latency promoter since the role of this viral product is to activate lytic gene expression. LAP2 contains a GC-rich and a C/T-rich sequence separated by a long uninterrupted sequence of 23 thymidines. All three elements are required for LAP2 activity in transient gene expression assays (French *et al.*, unpublished). The C/T-rich sequence forms triplex DNA (French *et al.*, unpublished) and binds to a family of related transcription factors (Postel *et al.*, 1989; Bossone *et al.*, 1992; Kolluri *et al.*, 1992; Pyrc *et al.*, 1992). LAP2 is over 75% homologous between −1 and −100 with the TATAless promoter for brain derived neurotropic factor (BDNF) which is expressed in specific regions of the CNS (Timmusk *et al.*, 1993). Recent experiments suggest that LAP2 can express LAT as a primary transcript but at levels approximately 50–100-fold lower that LAP1 (Chen and Glorioso, unpublished results).

The expression of LAT *in vitro* requires viral replication while LAT is readily detected following infection of the nervous system by non-replicating viral mutants. The ICP4 deletion mutant (d120), for example, does not

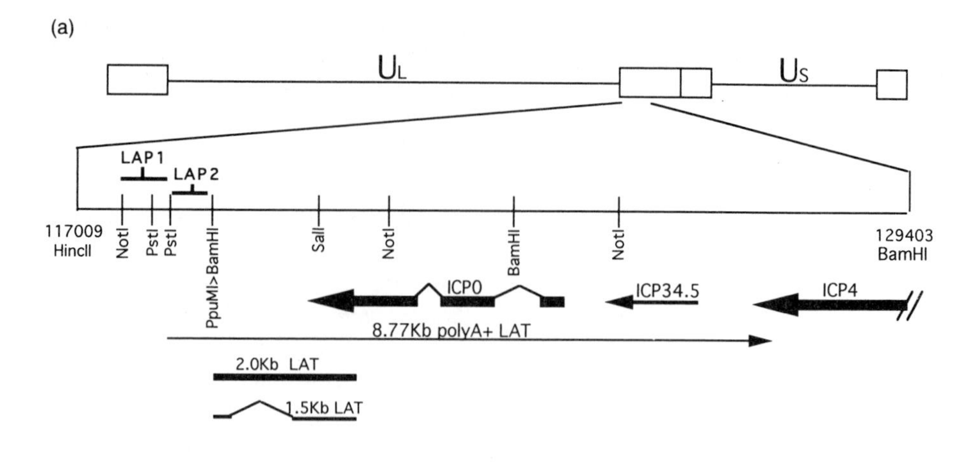

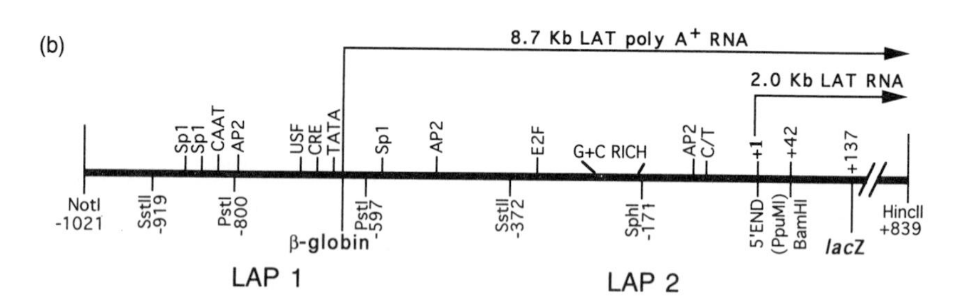

Figure 3.4 Transcription within the HSV-1 LAT region. (a) The genomic location and restriction map of the LAT region with the position of relevant restriction sites numbered according to the nucleotide position from the left-hand end of the prototype genome. Below the genomic map is the transcription pattern of the LAT region. The 8.77-kb polyadenylated LAT transcript, present during productive infection, is depicted. The stable 2-kb LAT transcript, also detected during productive infection, is the major transcript during latency, and the smaller 1.5-kb transcript is a spliced variant found exclusively during latent infection. (b) An expansion of the LAT promoter regions displaying the *cis*-acting sites within LAP1 and LAP2. LAP1 contains a TATA box, several Sp1 sites, a CRE, a USF-1 site and is located over 600 bp upstream from the 5′ end of the stable 2-kb LAT. The site at which β-globin was inserted downstream of LAP1 in the LAP1-β-globin expression vector (Dobson *et al.*, 1989) is depicted. LAP2, located just upstream of the 5′ end of the 2-kb stable LAT (Goins *et al.*, 1994), lacks a TATA box consensus yet displays sequence similarity to several eukaryotic RNA polIII housekeeping gene promoters. A C/T-rich element which is present within these promoters is also located within LAP2. The location of the *lacZ* insertion downstream of LAP2 in the RH142 recombinant virus (Ho and Mocarski, 1989) is denoted.

express LAT in Vero or B103 rat neuroma cells as detected by Northern blotting (Figure 3.5a). However, d120 readily expresses LAT *in vivo* as detected using *in situ* hybridization methods (Figure 3.5b). Infection of E5 complementing cells which allow the virus to replicate restored LAT expression. Thus, LAT expression *in vivo* during latency must rely on a neuronal enhancer which is active in at least some neuronal cell types *in vivo*. This is a fortuitous result for several reasons. First, evidence that tissue-specific enhancers are capable of functioning in a non-replicating genome during latency is at hand, suggesting the possibility of tissue targeted gene expression using the appropriate *cis*-acting transcriptional regulatory elements. Second, it may be possible to introduce this enhancer into stronger promoters providing a mechanism to up-regulate gene expression during latency.

3.2.2 Problems and solutions to HSV vector design

An ideal HSV vector would require: (i) the removal of genes which allow the virus to replicate and cause cytopathology; and (ii) the design of promoter systems able to produce persistent and appropriate levels of therapeutic gene expression from the latent viral genome. Experimental approaches to solving these two key problems are listed in Table 3.1 and our results thus far are discussed below.

3.3 REMOVAL OF CYTOTOXIC GENES

Evidence that the HSV particle itself does not carry cytotoxic gene products derives from early experiments in which ultraviolet-irradiated virus preparations were not found to be cytotoxic for cells in culture and have been used to biochemically transform cells (Leiden *et al.*, 1980). The infecting virus does carry into the cell, as part of the tegument, the product of the UL41 gene. This product referred to as the virion host shut-off (vhs) function interferes with the synthesis of cellular proteins (Kwong and Frenkel, 1987) through the indiscriminate degradation of mRNA (Oroskar and Read, 1989). The *vhs* gene can be deleted without compromising virus replication (Read and Frenkel, 1983), and such mutants fail to shut down host cell protein production. Removal of U_L41 will improve vector utility. It has also been well documented that deletion of either essential IE gene encoding ICP4 (both copies) or ICP27 compromises virus replication but does not block viral killing on infection. Deletion of both of these essential genes in addition to the *vhs* function and the ICP6 gene (Figure 3.6), however, renders the multiply mutated virus much less toxic for cells in culture (Wu *et al.*, unpublished). These multiple mutants are propagated on a specifically engineered complementing cell line carrying the required essential viral genes which can be induced by VP16 on infection. Recombinant wild type virus rescuants have not been isolated from concen-

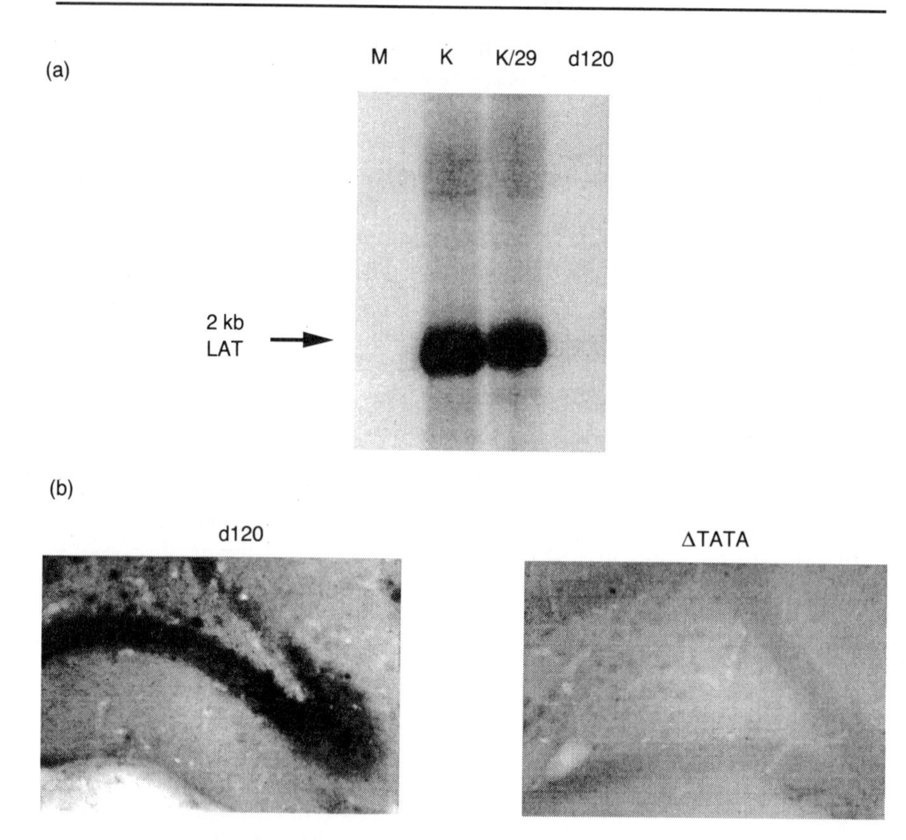

Figure 3.5 Expression of the latency-associated transcripts (LATs). (a) Northern analysis of the expression of LAT RNA during infection *in vitro*. Vero cells were infected with virus particles of either wild type KOS virus, KOS/29 virus, or d120 virus. KOS/29 contains a PstI–PstI deletion within the LAP1 promoter, removing the TATA box and much of the regulatory sequences of that promoter (Dobson *et al.*, 1989) (see Figure 3.4b). d120 is a non-replicating virus which contains a deletion of both copies of the immediate early gene, ICP4, a major viral transcriptional activator (DeLuca *et al.*, 1985). Total cell RNA was isolated 20 hours post-infection and was separated on a formaldehyde gel, transferred to Nytran paper, and hybridized to a DNA probe specific for LAT RNA sequences (PstI (bp #118866) to Kpn I (bp #119980) containing LAP2 and the +1 to +516 bp of the 2-kb LAT). LAT-specific RNAs are detected in the KOS and the KOS/29 infections, but not in the d120 infection. (b) Production of LAT in brain after infection with viral mutants d120 and d120 ΔTATA LAP1. A 1μl suspension of virus particles (2×10^6 pfu) was stereotactically injected into the hippocampal region of rats. Seven days after injection, the brain tissue was analyzed for the production of LAT by *in situ* hybridization using a digoxigenin-labeled LAT riboprobe. Intense staining for LAT was observed after an infection with the d120 ($ICP4^-$) virus, but no staining was seen after an infection with a d120 virus which contains a deletion of the TATA box of both copies of the LAP1 promoter (see Figure 3.4b). This deletion defines the LAP promoter during latency in brain.

Table 3.1 Engineering HSV vectors

Problems	*Solutions*	
1. Cytotoxicity	(i)	Systematically delete non-essential and essential IE genes
	(ii)	Delete virus encoded host shut-off gene
	(iii)	Block leakage of delayed early gene expression using transdominant negative IE genes
2. Gene expression		
Stability	(i)	Exploit viral latency active promoters
	(ii)	Employ autoregulatory genes
Regulation	(i)	Exploit drug-inducible promoters
	(ii)	Design latency active promoter with regulatable cellular enhancers

trated stocks of this complex deletion mutant, and remarkably the ICP6 promoter has retained the ability to express the *lacZ* gene for up to 3 weeks in standard cell cultures. We are also developing additional multiple mutant viruses which should also fail to express ICP0. If successful, these multiple mutants will provide a non-cytotoxic vector backbone into which multiple foreign genes can be introduced and expressed in the absence of viral lytic functions. Moreover, these vectors will retain the ability to express ICP47 which should decrease their likelihood for immune recognition and rejection of the infected cell.

3.4 EXPRESSION OF THERAPEUTIC GENES

To employ HSV as an effective gene transfer vector, the virus should be able to provide appropriate levels of the therapeutic gene product(s). For some diseases, a high level of gene expression over a short period of time may be sufficient, while in other cases, a more constitutive rate of expression, either high or low, may be required. The ability to regulate or induce gene expression by drug or hormone administration could provide an additional level of control over the expression of the delivered gene. All of these cases require different tactics in promoter design which can accommodate these requirements. In the past 10 years, foreign genes have been placed in a variety of positions within the viral backbone under the direction of both HSV and non-viral promoters and tested for their ability to drive transgene expression.

The first evidence of the efficacy of HSV as a gene delivery tool was demonstrated in cultured cells with the production of the hepatitis B surface antigen (Shih *et al.*, 1984), the α and β globins (Smiley *et al.*, 1987; Panning and Smiley, 1989) and the enzyme hypoxanthine phosphoribosyltransferase

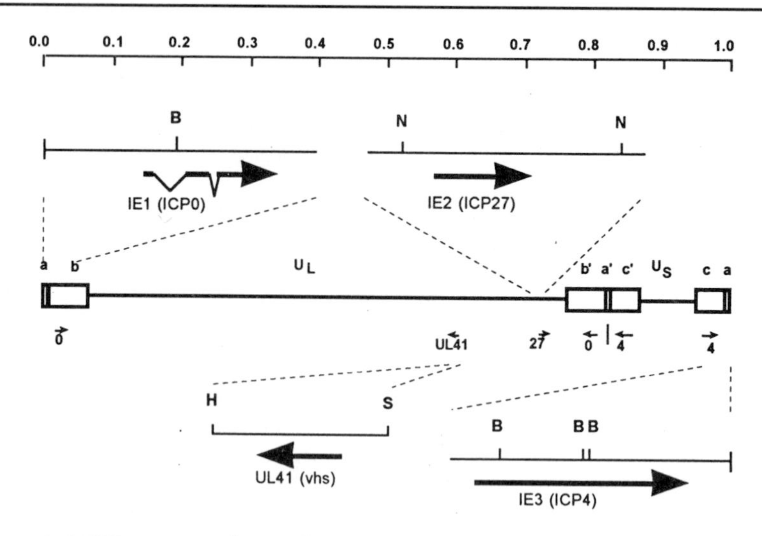

Figure 3.6 HSV gene products affecting gene expression and cytotoxicity. The HSV genome consists of a long and short component. The long component is composed of the unique long (U_L) sequence bracketed by inverted copies of the 'ab' sequences. The short component is composed of the unique short (U_S) sequence bracketed by inverted copies of the 'ca' sequences. The 'a' sequences are directly repeated at the genomic termini. The genes for ICP4 (IE3), ICP0 (IE1), ICP27 (IE2) and UL41 (vhs) are shown to scale relative to their map locations on the HSV-1 genome. Note that ICP0 and ICP4 are present in two copies per genome due to their location in the 'b' and 'c' sequences, respectively. The regions encoding these genes are expanded showing the location of the sequences encoding the indicated mRNAs, and also giving some restriction enzyme cleavage sites as landmarks. The ICP0 mRNA is spliced in two places as shown in the figure. The sites are: Bam HI (B), Hind III (H), Nru I (N), Sal I (S).

(HPRT) (Palella *et al.*, 1988). The time course of foreign gene expression *in vitro* followed HSV early gene kinetics, rising to a peak just a few hours after infection. Similar kinetics and short-term expression were also observed *in vivo* in the CNS. Upon injection of a replicating HPRT-expression vector into the brains of mice, HPRT-specific mRNA was detected 3–5 days after injection (Palella *et al.*, 1989), but the animals died from viral infection.

Our initial experiments to develop an HSV gene transfer vector for the CNS were aimed at determining whether a non-replicating virus blocked at a very early stage in the lytic cycle was capable of efficiently infecting and establishing latency in different brain regions following stereotactic inoculation of virus. The use of non-replicating defective vectors is essential since the expression of viral lytic genes is toxic to cells and would stimulate an immune response, particularly since the majority of people to whom these vectors might ultimately be administered are infected with this virus and possess immunologic memory. Most of our studies have been carried out with the replication defective mutant virus (d120) deleted for both copies of the immediate early (IE) gene ICP4 (DeLuca *et al.*, 1985). d120 can be propagated only on a complementing Vero cell line designated E5. Because ICP4 is essential to the expression of delayed early and late viral genes, d120 is blocked at a very early stage in the virus growth cycle. This mutant was further modified to contain a P1 phage recombination site, *lox*P, engineered into the viral thymidine kinase (tk) locus rendering the mutant TK deficient and providing a target site for recombining plasmids into the genome using the phage *lox*-specific recombinase *Cre* in a cell-free system (Gage *et al.*, 1992). To report the location of infected cells in brain, a reporter gene cassette composed of the HCMV IE promoter driving expression of the *lac* Z gene was introduced into the d120 mutant (Figure 3.7a).

Stereotactic inoculation of this recombinant virus into rat hippocampus revealed the following. First, inoculation of very high doses (e.g. 10^7 pfu/μl) of the vector into this and other brain regions did not cause encephalitis or show any obvious changes in brain architecture or neuronal cell structure on evaluation by histochemical methods. Moreover, the virus very efficiently infected the neuronal cells, expressed high levels of the reporter gene as determined by X-gal staining for the β-galactosidase gene product (Figure 3.7b) and indicated that the virus did not spread from the injected site. However, expression was transient, and the signal disappeared within 7 days.

To determine whether the loss of gene expression was due to the loss of the viral genome or the shut-down of the HCMV promoter, a quantitative polymerase chain reaction (PCR) technique was used to determine the number of viral genome equivalents retained in the brain at 2, 7 and 56 days post-infection with the vector (Ramakrishnan *et al.*, 1994). The results demonstrated that a large number of genomes were retained (approximately 10% of the inoculated viral DNA) which remained stable for months. Moreover, the LAT RNA was readily detected in the nuclei of infected neurons for several weeks by *in situ* hybridization, and the LAT RNA could be detected long-term from d120 infected neurons by RT–PCR indicating that although LAT expression declined, it clearly continued to be expressed from the latent defective viral genomes. A number of other neuronal-specific cellular promoters (e.g. nerve-specific enolase, neurofilament, and nerve cell-specific sodium channel), viral promoters (RSV LTR, SV40) and several polIII promoters including the adenovirus VA1 gene gave similar transient expression in brain to that observed with the HCMV promoter. Only the LAP remained active long term.

Results from other laboratories have also reported this short-lived pattern of transgene expression in the brain. Breakefield and co-workers have employed various replication-defective HSV vectors with the *lac*Z gene under the control of different neuronal promoters to demonstrate that expression increases within specific regions of the brain over the first 3 days

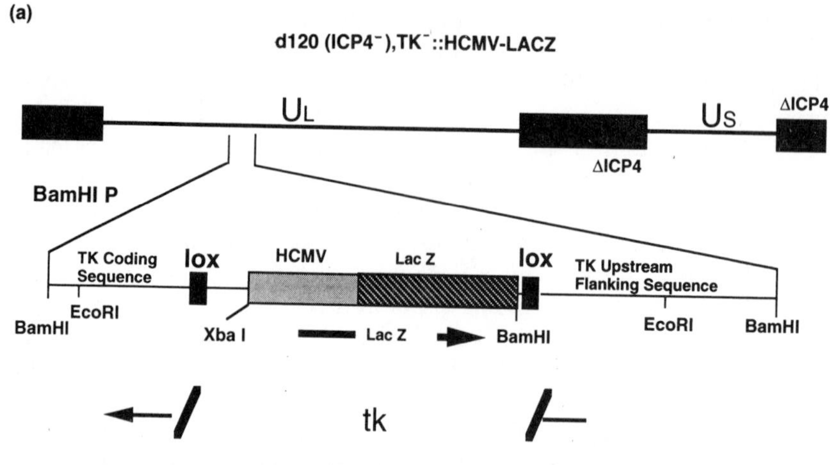

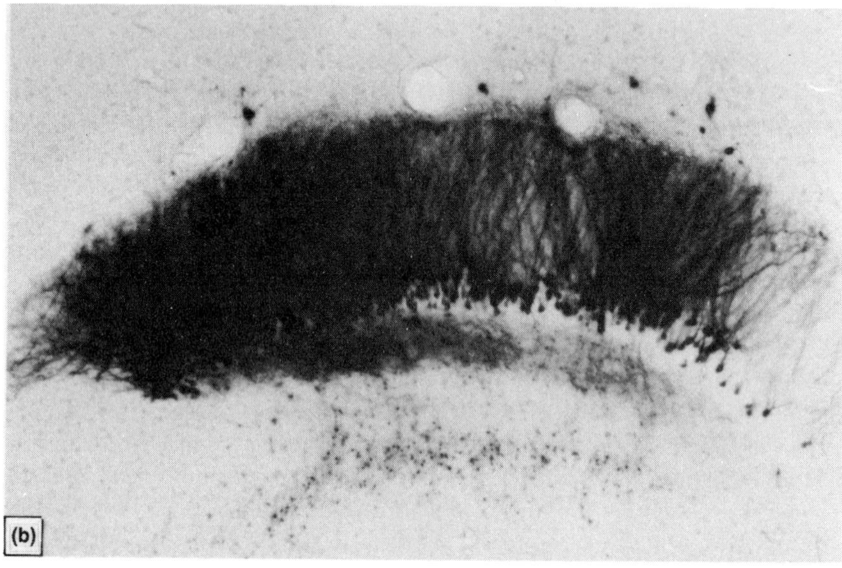

Figure 3.7 *In vivo* expression of *lacZ* from the HCMV IE promoter in rat CNS following stereotactic inoculation of mutant defective in the essential gene ICP4. (a) The HCMV IE promoter–*lacZ* reporter gene cassette was recombined into the tk locus of the ICP4 defective virus ICP4$^-$,tk::lox [a derivative of d120 (DeLuca *et al.*, 1985)] using cre–lox recombination (Gage *et al.*, 1992) to produce the recombinant ICP4$^-$,TK$^-$::HCMV IEp-*lacZ*. (b) Stereotactic inoculation of 5 μl containing 5 × 10^7 pfu of this recombinant into rat hippocampus resulted in intense staining of granule cell neurons of the dentate gyrus at 2 days post-inoculation. Although expression of *lacZ* was transient, disappearing by day 7, *lacZ* expression could not be detected in either the contralateral hippocampus or other brain regions with projections to the hippocampal region, confirming the inability of this recombinant to spread. In addition, we have been unable to detect any significant pathology resulting from this mutant, even in long-term animals.

post-injection, but regresses by day 14 (Chiocca *et al.*, 1990; Andersen *et al.*, 1992). Similarly, only a few isolated cells expressing the *lacZ* gene product have been detected at 1 to 4 months after inoculation, using the mammalian nerve-specific enolase promoter in the striatal region of the brain (Andersen *et al.*, 1992).

There are several potential applications of a viral vector which is able to support foreign gene expression for short time periods in the brain. For example, treatment of cancerous growth in the brain might be accomplished by inoculation of the tumor with an HSV vector that produces a toxic product within the tumor mass or activates an anti-cancer pro-drug. These vectors can also be engineered to co-express immunomodulatory genes which recruit and amplify specific immune cells. Such a vector could be injected within a tumor site in the brain to initiate an immunological response against the tumor at that site as well as other sites of metastasis in other brain regions.

3.4.1 Long-term expression using the latency-active promoters

In order to obtain long-term expression of foreign genes from the latent HSV genome and to identify the viral sequences responsible for latency-specific expression, several laboratories have engineered a variety of viral recombinants in which foreign or reporter genes are driven by either LAP1 or LAP2 in the native LAT or ectopic sites within the viral genome.

LAP1

A recombinant with LAP1 driving expression of the rabbit β-globin genomic clone in the native LAT locus expressed β-globin-specific mRNA at 3 weeks in latently infected mouse lumbosacral ganglia neurons as determined by *in situ* hybridization (Dobson *et al.*, 1989) (see Figure 3.4b). However, the intensity of the hybridization signal and the number of neurons expressing the transgene decreased 20-fold during latency (Margolis *et al.*, 1993). We have engineered a similar construct in which the murine interferon (α-IFN) cDNA was inserted at the same site 14 bp downstream of the LAP1 TATA box in the native LAT locus of the ICP4 deletion mutant (d120). This recombinant produced α-IFN-specific mRNA and protein during productive infection of the ICP4 complementing E5 cell line. However, in contrast to the results of Dobson *et al.* (1989) using the β-globin gene, no α-IFN-specific message was detected by *in situ* hybridization following stereotactic inoculation of rat brain (Fink *et al.*, unpublished). The reason for this difference is unknown, but LAP1-driven β-globin expression in the Dobson virus recombinant may be due to the presence of an enhancer/regulatory region within the β-globin genomic clone, perhaps within an intron. More recent results indicate that sequences downstream of the site of α-IFN insertion are important to LAP1 activity and could account for the absence of α-IFN expression in latency. This could either be due to the insertion of the gene 5′ to the LAP1 initiator and/or to downstream sequences 5′ to the 2-kb LAT which significantly affect LAP1 activity. John Wolfe and colleagues have found that a wild type virus recombinant in which the β-glucuronidase cDNA was inserted into the LAT locus downstream of LAP1 expressed β-glucuronidase enzymatic activity in a number of cells in the trigeminal ganglion and brain stem following infection by corneal scarification. Enzyme activity was detected during acute and latent (2–18 weeks) times (Wolfe *et al.*, 1992). Again, however, the number of cells expressing the transgene in the brain decreased, resulting in insufficient levels to obtain accurate measurements of enzymatic activity. In other studies, viruses containing the LAP1 regulatory region driving β-galactosidase in the native LAT (Margolis *et al.*, 1993) expressed *lacZ* during acute infection in mouse dorsal root ganglia, yet latently infected neurons failed to express the reporter gene as determined by *in situ* hybridization with a *lacZ*-specific probe (Margolis *et al.*, 1993). A similar recombinant containing LAP1 controlling the murine β-nerve growth factor cDNA (β-NGF) expressed NGF message and protein but only during acute infection (Margolis *et al.*, 1993).

LAP2

All of the above recombinants lacked sequences that encompass the second latency promoter (LAP2) region, raising the possibility they may be required for long-term expression in peripheral nervous system (PNS) neurons. A recombinant virus in which the *lacZ* reporter gene was inserted 137 bp downstream of the 5′ end of the major 2-kb LAT (i.e. immediately downstream of LAP2) expressed the reporter gene in 0.1–1.0% of mouse trigeminal ganglion neurons at 8 weeks post-infection by corneal scarification (Ho and Mocarski, 1989) (Figure 3.4b). A characteristic punctuate X-gal staining pattern was detected in these neurons, but not in corneal epithelial cells at the site of primary infection. The 5′ end of the chimeric LAT–*lacZ* mRNA mapped to a site at or near that of the 2-kb LAT, suggesting that LAP2 was responsible for driving expression of the reporter gene. In addition, the size of the *lacZ*-specific message detected by Northern blot analysis was in agreement with initiation of the *lacZ* mRNA from LAP2 and not LAP1. We have constructed recombinant viruses in which the *lacZ* reporter gene was inserted downstream of LAP2 in either the native LAT (Chen *et al.*, unpublished results) or ectopic gC locus (Goins *et al.*, 1994). These viruses were capable of expressing the reporter gene within neurons in culture and in trigeminal ganglia neurons at times consistent with latent infection (42 to 300 days post-infection) as detected by X-gal staining or immunohistochemistry for β-galactosidase. In addition, RT–PCR assays have detected *lacZ*-specific RNA during latency in trigeminal ganglia from animals infected with these recombinants, thereby confirming the latency-specific activity of this promoter. We have recently shown that a replication defective vector containing the LAP2–*lacZ* gene cassette in the gC locus was capable of expressing the reporter gene at 7 and 21 days following stereotactic inoculation of the vector into the hippocampus of rats using RT–PCR, even though we could not detect β-gal in expressing neurons by X-gal staining several weeks post-inoculation. These recent data suggest that LAP2 is capable of driving long-term expression of a foreign gene product from a defective HSV vector in neurons of the CNS. However, the LAP2 region requires modification to express the transgene at higher levels.

One avenue to bolster LAP2's promoter function could involve the introduction of additional *cis*-acting transcriptional elements into the promoter sequence. For example, insertion of enhancer elements found in polyoma virus (Bryan and Folk, 1986; Pikaart *et al.*, 1992) or other sequences which may provide for methylation resistance, DNAseI hypersensitivity and the ability to bind to transcription factors. Such a strategy was reported by the group working with Inder Verma (Dai *et al.*, 1992) who introduced the muscle creatine kinease (MCK) enhancer element into the HCMV promoter, which allowed it to remain active long-term in differentiated myoblasts *in vivo*. Another approach may involve a juxtaposition of the LAP region to another stronger promoter such as a retroviral LTR. This recombinant promoter construct might confer the ability of the LTR to remain active during latency. Once higher levels of activity are achieved, then a variety of neurodegenerative diseases may be treated with the constitutive levels of expression of the therapeutic gene from the modified LAP promoter. For example, sufferers of Parkinson's disease may benefit from transfer of the gene for tyrosine hydroxylase, the rate-limiting enzyme in the production of the dopamine precursor L-DOPA. Production of the dopamine neurotransmitter in undamaged regions of the brain could supply dopamine to areas containing dopamine-deprived neurons which have efferent connections to the site of gene transfer.

3.4.2 Inducible vector gene expression

In the latent state, the HSV DNA has been shown to be associated with nucleosomes in a chromatin-like structure (Deshmane and Fraser, 1989). Nucleosome positioning over promoter regions suppresses transcription initiation and elongation, probably by prohibiting access to the TATA box/initiator region by transcription factors. This may partially contribute to the absence of lytic gene expression and the low level of LAT transcription observed during HSV latency. The observation that the C/T-rich region of the LAP2 promoter (Figure 3.4b) contains S1 nuclease-sensitive sites, corresponding to sequences known to exclude nucleosomes (French *et al.*, unpublished) suggests that this LAP may employ such elements. Upon transcriptional activation of genes packaged within chromatin structures, there is an accompanying rearrangement of the nucleosomal positioning over the promoter (Eissenberg *et al.*, 1985; Gross and Garrard, 1987). This displacement of nucleosomes can be mediated by sequence-specific DNA-binding transcription factors which physically exclude the nucleosomes from the promoter region.

Our laboratories are attempting to develop alternate strategies to achieve sustained gene expression from the latent HSV vector which overcome potential promoter shut-off due to chromatin formation. These include the design and use of recombinant genes whose products are uniquely capable of transactivating their own promoters, as well as similar promoters driving expression of a therapeutic gene. Using such an autogene system may be capable of maintaining vigorous gene expression, despite the formation of chromatin. Autogene transactivation might be most effective if they can shift the position of nucleosomes in a manner to 'open up' the promoter for access to other transcription factors. Once such potent transcriptional activator is a fusion product of the DNA-binding region of the yeast activator GAL4 and the strong acidic activating domain of the HSV VP16 transactivator (Sadowski *et al.*, 1988). Not only is GAL4:VP16 able to stimulate heterologous promoters which contain multimers of the 17-mer GAL4 binding site in transient assays (Sadowski *et al.*, 1988; Chasman *et al.*, 1989; Carey *et al.*, 1990), but it can release nucleosome repression of chromatin-bound promoters during transcriptional activation in frog oocytes and yeast (Axelrod *et al.*, 1993; Xu *et al.*, 1993).

We have employed the GAL4:VP16 transactivation model for the delivery of foreign genes by HSV in an attempt to support continued expression of the foreign gene during viral latency. The experimental paradigm was to construct a self-propelled transcription system through the continuous self-activation of the GAL4:VP16 protein which then transactivates expression of the foreign gene positioned at a secondary locus within the virus or on a second virus (Figure 3.8). In this way, the GAL4:VP16 product would be capable of maintaining not only its own transcription, but also that of a second transgene. The strong HCMV IE promoter was modified by insertion of five tandem GAL4 binding sites and juxtaposed to the GAL4:VP16 gene (Bender *et al.*, in preparation). In *in vitro* transient transfection/infection assays, the GALA:VP16 virus elicited a 12–35-fold stimulation of expression of the chloramphenicol acetyl transferase (CAT) reporter gene from a minimal heterologous promoter which had been modified with GAL4 binding sites. The gene cassette was then inserted into the tk locus of

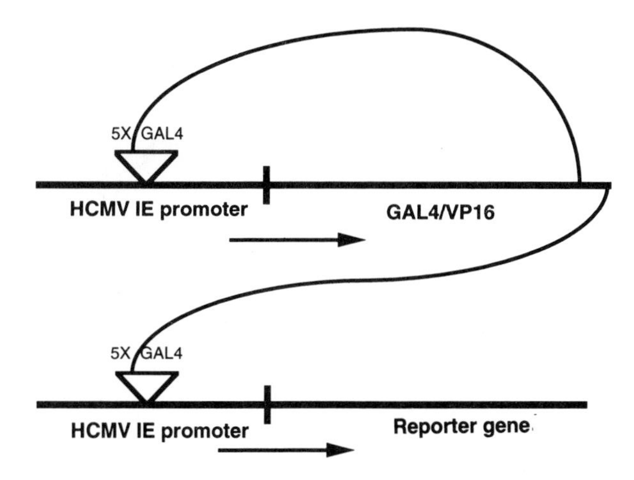

Figure 3.8 Stimulation of modified HCMV IE promoters by GAL4:VP16 transactivation. To set up an autogene promoter system, the HCMV IE promoter is modified to contain five tandem GAL4 binding sites by insertion of the binding sites at the Sna BI site at position −161 of the promoter. This modified HCMV IE promoter is then placed upstream of both the GAL4:VP16 transactivator protein and a reporter (or therapeutic) gene. The two constructs may be placed within a single virus at two distinct locations, or they may be inserted within two separate viruses and co-infected into cells. The HCMV IE promoter will drive the expression of the GAL4:VP16 protein which, in turn, will bind to its own promoter, creating an autogene capable of stimulating its own activity. Likewise, the GAL4:VP16 protein will also activate the second HCMV IE promoter upstream of the reporter gene. The paradigm proposes that with the continuous auto-stimulation and production of the GAL4:VP16 activator, there will also be sustained production of the reporter gene.

the d120 virus using the *in vitro* recombination system involving the *Cre* recombinase enzyme (Figure 3.9a).

The GAL4:VP16 virus was tested for its ability to stimulate a susceptible promoter placed within a cellular chromosome. A Vero cell line which did not support d120 viral replication was transfected with a plasmid which contained a GAL4-sensitive minimal promoter driving the CAT reporter gene. After integration of the CAT gene cassette within the cellular chromosome, these Vero cells were infected with the GAL4:VP16 virus. Expression of the virally-encoded transactivator gene resulted in a 99.9% activation of the CAT gene (Figure 3.9b). Finally, in co-infection experiments with a similar d120 virus carrying the GAL4-sensitive HCMV IE promoter driving the rat cDNA for tyrosine hydroxylase (TH), the GAL4:VP16 virus was able to stimulate virally-based TH synthesis (Figure 3.9c). In the brain, however, the synthesis of the GAL4:VP16 mRNA within infected rat hippocampal brain displayed the same early gene kinetics that had been observed with other promoters, disappearing by day 10 post-infection. This result

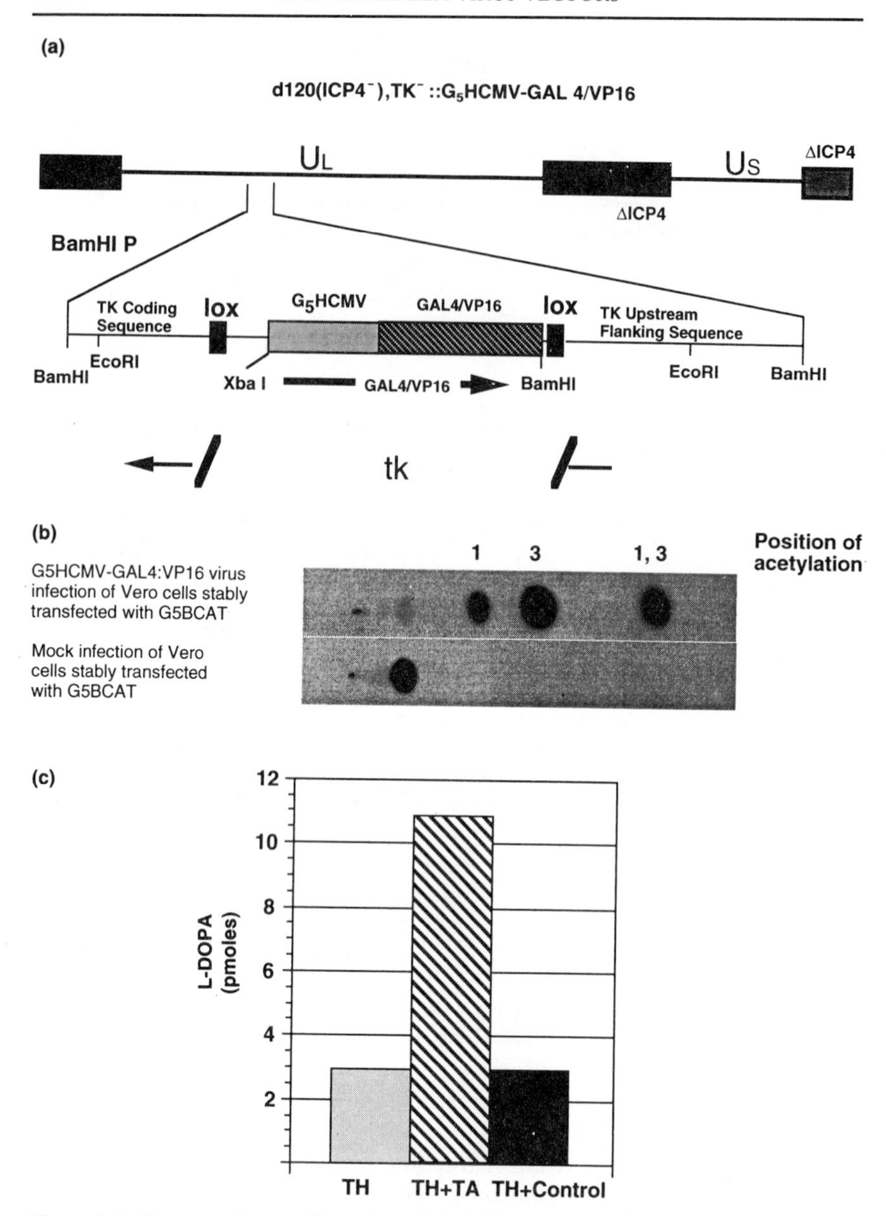

Figure 3.9 Construction and function of the HSV vector producing GAL4:VP16. (a) The modified HCMV IE promoter with the insertion of five GAL4 binding sites was placed upstream of the GAL4:VP16 transactivator sequences. This cassette was then inserted into the d120 viral backbone at the tk locus through the cell-free *cre*–lox recombination system. The recombinant virus was detected through dot-blot hybridization for GAL4:VP16 sequences. (b) Vero cells were stably transfected with a plasmid containing a minimal TATA box promoter and five GAL4 binding sites upstream of the chloramphenicol acetyl transferase (CAT) reporter gene. Five

suggested that the strong transcriptional activation by GAL4:VP16 was unable to overcome the repression of viral promoters observed during the establishment of latency in the brain. Experiments are currently underway to determine whether transcriptional factor squelching by GAL4:VP16 plays a role in the loss of function in the brain over time.

3.4.3 Future research

One possible strategy for effecting long-term and possibly regulated gene expression during latency is the application of a drug-inducible gene therapy approach. We are employing a system which renders the GAL4:VP16 activation system sensitive to the action of the RU486 progesterone analog. O'Malley and co-workers have fused the GAL4:VP16 sequences to a truncated hormone-binding domain (HBD) of the progesterone receptor which no longer binds to progesterone. However, this modified HBD will be recognized by the drug RU486 (Vegeto *et al.*, 1992) which is small enough to pass through the blood–brain barrier. Only upon the intracellular binding of the HBD portion of the chimeric protein to the RU486 steroid does the GAL4:VP16 transactivator become functional, thus setting up a drug-inducible transcription system (Wang *et al.*, 1994). Such an approach could be applied to foreign gene delivery to any tissue in the body, affording control of the transgene expression with small amounts of a drug. This

Figure 3.9 *Continued:* micrograms of this plasmid and a plasmid coding for the neomycin-resistance gene were co-transfected into Vero cells, and neomycin-resistant clones were isolated under G418 selection. Resistant clones that were positive for the CAT sequences were isolated by dot-blot hybridization. One such clone was infected with the d120, (ICP4$^-$), TK$^-$::G_5HCMV-GAL4:VP16 virus at a multiplicity of infection (MOI) of 3, and at 24 hours post-infection, the activity of the integrated CAT gene was assayed. Mock-infected transformed cells gave no CAT activity. Only after viral infection was the minimal promoter stimulated to express the CAT gene to over 99.9% activity. (c) Rat B103 neuroma cells which are negative for the production of tyrosine hydroxylase (TH) were co-infected at an MOI of 0.01 with the d120(ICP4$^-$), TK$^-$::GAL4:VP16 virus (TA, transactivator virus) and a similarly constructed d120(ICP4$^-$), TK$^-$::G_5HCMV-TH virus (TH virus) which contained the modified HCMV promoter driving the TH gene. After 8 days post-infection (dpi), both the media and lysates of the infected cells were analyzed for the synthesis of L-DOPA, the immediate product of TH activity, by HPLC. L-DOPA production was detected in both media and cell lysates at both time points, as was the production of dopamine by these cells (data not shown). Shown here are the results from the B103 cell lysates at 8 dpi. TH represents an infection with the TH virus alone; TH+TA represents a co-infection of the TA and TH viruses and shows stimulation of TH expression. TH+control represents a co-infection of the TH virus with a similar virus producing *lacZ* driven from the HCMV promoter.

alternate strategy may be useful in overcoming the difficulties related to the achievement of persistent gene expression from the latent HSV genome.

3.5 CONCLUSIONS

HSV-1 is a logical choice for development as a versatile and effective gene transfer system for brain and other tissues. The virus has many features which make it suitable for gene transfer. Key among these is its ability to remain in cell nuclei for long periods without cytotoxicity, raising the possibility that it will be effective for long-term gene expression. Still, more work has yet to be done to achieve this goal. Specifically, this involves developing greater insight into gene expression during latency. This knowledge should provide useful information for the design of more vigorous and tissue-specific promoter systems. Research on the design of promoters responsive to drugs administered to patients further provides the exciting possibility that appropriate and therapeutic levels of gene product can be achieved in tissues. In the near future, efforts to develop these methods will be intensified, as will efforts to use currently available vectors for gene therapy of certain human cancers.

REFERENCES

Anderson J.K., Garber D.A., Meaney C.A. and Breakefield X.O. (1992) Gene transfer into mammalian central nervous system using herpesvirus vectors: extended expression of bacterial *lacZ* in neurons using the neuron-specific enolase promoter. *Human Gene Ther.*, **3**, 487–99.

Axelrod, J.D., Reagan, M.S. and Majors, J. (1993) GAL4 disrupts a repressing nucleosome during activation of GAL 1 transcription *in vivo*. *Genes Dev.*, **7**, 857–69.

Bak, I.J., Markhan, C.H. and Cook, M.L. (1977) Intra-axonal transport of herpes simplex virus in the rat central nervous system. *Brain Res.*, **136**, 415–29.

Batchelor, A.H. and O'Hare, P.O. (1990) Regulation and cell-type-specific activity of a promoter located upstream of the latency-associated transcript of herpes simplex virus type 1. *J. Virol.*, **64**, 3269–79.

Batchelor, A.H. and O'Hare, P.O. (1992) Localization of cis-acting sequence requirements in the promoter of the latency-associated transcript of herpes simplex virus type 1 required for cell-type-specific activity. *J. Virol.*, **66**, 3573–82.

Batterson, W. and Roizman, B. (1983) Characterization of the herpes simplex virion-associated factor responsible for the induction of α-genes. *J. Virol.*, **46**, 371–7.

Block, T.M., Spivack, J.G., Steiner, I., *et al.* (1990) A herpes simplex virus type 1 latency-associated transcript mutant reactivates with normal kinetics from latent infection. *J. Virol.*, **64**, 3417–26.

Bossone, S.A., Asselin, C., Patel, A.J. and Marcu, K.B. (1992) MAZ a zinc finger protein binds to c-MYC and C2 gene sequences regulating transcriptional initiation and termination. *Proc. Natl Acad. Sci. USA*, **89**, 7452–6.

Bryan, P.N. and Folk. W.R. (1986) Enhancer sequences responsible for DNase I hypersensitivity in polyomavirus chromatin. *Mol. Cell. Biol.*, **6**, 2249–52.

Campbell, M.E.M., Palfreyman, J.W. and Preston, C.M. (1984) Identification of herpes simplex virus DNA sequences which encode a trans-acting polypeptide responsible for stimulation of immediate early transcription. *J. Mol. Biol.*, **180**, 1–19.

Carey, M., Leatherwood, J. and Ptashne, M. (1990) A potent GAL4 derivative activates transcription at a distance *in vitro*. *Science*, **247**, 710–12.

Chasman, D.I., Leatherwood, M., Carey, M., Ptashne, M. and Kornberg, R.D. (1989) Activation of yeast polymerase II transcription by herpesvirus VP16 and GAL4 derivative *in vitro*. *Mol. Cell. Biol.*, **9**, 4746–9.

Chiocca, A.E., Choi, B.B., Cai, W. *et al.* (1990) Transfer and expression of the *lacZ* gene in rat brain neurons by herpes simplex virus mutants. *New Biol.*, **2**, 739–46.

Cook, M.L. and Stevens, J.G. (1973) Pathogenesis of herpetic neuritis and ganglionitis in mice: evidence of intra-axonal transport of infection. *Infect. Immun.*, **7**, 272–88.

Dai, Y., Roman, M., Naviaux, R.K. and Verma, I.M. (1992) Gene therapy via primary myoblasts: long-term expression of factor IX protein following transplantation *in vivo*. *Proc. Natl Acad. Sci. USA*, **89**, 10892–5.

Deatly, A.M., Spivack, J.G., Lavi, E. *et al.* (1988) Latent herpes simplex virus type 1 transcripts in peripheral and central nervous systems tissues of mice map to similar regions of the viral genome. *J. Virol.*, **62**, 749–56.

DeLuca, N.A., McCarthy, A.M. and Schaffer, P.A. (1985) Isolation and characterization of deletion mutants of herpes simplex virus type 1 in the gene encoding immediate-early regulatory protein ICP4. *J. Virol.*, **56**, 558–70.

Desai, P., Ramakrishnan, R., Lin, Z.W. *et al.* (1993) The RRI gene of herpes simplex virus type 1 is uniquely *trans* activated by ICP0 during infection. *J. Virol.*, **67**, 6125–35.

Deshmane, S.L. and Fraser, N.W. (1989) During latency herpes simplex virus type 1 DNA is associated with nucleosomes in a chromatin structure. *J. Virol.*, **63**, 943–7.

Devi-Rao, G.B., Goddart, S.A., Hecht, L.M. *et al.* (1991) Relationship between polyadenylated and nonpolyadenylated, HSV type 1 latency-associated transcripts. *J. Gen. Virol.*, **65**, 2179–90.

Dixon, R.A.F. and Schaffer, P.A. (1980) Fine-structure mapping and functional analysis of temperature-sensitive mutants in the gene encoding the herpes simplex virus type 1 immediate early protein VP175. *J. Virol.*, **36**, 189–203.

Dobson, A.T., Sederati, F., Devi-Rao, G. *et al.* (1989) Identification of the latency-associated transcript promoter by expression of rabbit β-globin mRNA in mouse sensory nerve ganglia latently infected with a recombinant herpes simplex virus. *J. Virol.*, **63**, 3844–51.

Doerig, C., Pizer, L.I. and Wilcox, C.L. (1991) An antigen encoded by the latency-associated transcripts in neuronal cell cultures latently infected with herpes simplex virus type 1. *J. Virol.*, **65**, 2724–7.

Dressler, G.R., Rock, D.L. and Fraser, N.W. (1987) Latent herpes simplex virus type 1 DNA is not extensively methylated *in vivo*. *J. Gen. Virol.*, **68**, 1761–5.

Eissenberg, J.C., Cartwright, I.L. Thomas, G.H. and Elgin, S.C.R. (1985) Selected topics in chromatin structure. *Annu. Rev. Genet.*, **19**, 485–536.

Farrell, M.J., Dobson, A.T. and Feldman, L.T. (1991) Herpes simplex virus latency-associated transcript is a stable intron. *Proc. Natl Acad. Sci. USA*, **88**, 790–4.

Fraser, N.W., Lawrence, W.C., Wroblewska, Z., Gilden, D.H. and Koprowski, H. (1981) Herpes simplex virus type 1 DNA in human brain tissue. *Proc. Natl Acad. Sci. USA*, **78**, 6461–5.

Gaffney, D.F., McLaughlin, J., Whitton, J.L. and Clements, J.B. (1985) A molecular system for the assay of transcription regulatory signals: the sequence TAATGARAT is required for herpes simplex virus immediate early gene activation. *Nucleic Acids Res.*, **13**, 7847–63.

Gage, P.J., Sauer, B., Levine, M. and Glorioso, J.C. (1992) A cell-free recombination system for site-specific integration of multigenic shuttle plasmids into the herpes simplex virus type 1 genome. *J. Virol.*, **66**, 5509–15.

Goins, W.F., Sternberg, L.R., Croen, K.D. *et al.* (1994) A novel latency-active promoter is contained within the herpes simplex virus type 1 UL flanking repeats. *J. Virol.*, **68**, 2239–52.

Gross, D.S. and Garrard, W.T. (1987) Poising chromatin for transcription. *Trends Biochem. Sci.*, **12**, 293–7.

Hill, J.M., Sedarati, F., Javier, R.T., Wagner, E.K. and Stevens, J.G. (1990) Herpes simplex virus latent phase transcription facilitates *in vivo* reactivation. *Virology*, **174**, 117–25.

Hill, T.J. (1985) Herpes simplex virus latency, in *The Herpesviruses*, (ed. B. Roizman), Plenum Press, New York, pp. 175–240.

Ho, D.Y. and Mocarski, E.S. (1989) Herpes simplex virus latent RNA (LAT) is not required for latent infection in the mouse. *Proc. Natl Acad. Sci. USA*, **86**, 7596–600.

Holland, L.E., Anderson, K.P., Shipman, C. and Wagner, E.K. (1980) Viral DNA synthesis is required for the effective expression of specific herpes virus type 1 mRNA species. *Virology*, **101**, 10–24.

Javier, R.T., Stevens, E.S., Dissene, V.B. and Wagner, E.K. (1988) A herpes simplex virus transcript abundant in latently infected neurons is dispensible for establishment of the latent state. *Virology*, **166**, 254–7.

Kolluri, R., Torrey, T.A. and Kinniburgh, A.J. (1992) A CT promoter element binding protein: definition of a double-strand and a novel single-strand DNA binding motif. *Nucleic Acids Res.*, **20**, 111–16.

Kristie, T.M. and Roizman, B. (1987) Host cell proteins bind to the cis-acting site for virion-mediated induction of herpes simplex virus 1α-genes. *Proc. Natl Acad. Sci. USA*, **84**, 71–5.

Kwong, A.D. and Frenkel, N. (1987) Herpes simplex virus-infected cells contain a function(s) that destabilizes both host and viral mRNAs. *Proc. Natl Acad. Sci. USA*, **84**, 1926–30.

Leib, D.A., Bogard, C.L., Kosz-Vnenchak, M. *et al.* (1989) A deletion mutant of the latency-associated transcript of herpes simplex virus type 1 reactivates from the latent infection. *J. Virol.*, **63**, 2893–900.

Leiden, J.M., Frenkel, N. and Rapp, F. (1980) Identification of the herpes simplex virus DNA sequences present in six herpes simplex virus thymidine kinase-transformed mouse cell lines. *J. Virol.*, **33**, 272–85.

Margolis, T.P., Bloom, D.C., Dobson, A.T., Feldman, L.T. and Stevens, J.G. (1993) Decreased reporter gene expression during latent infection with HSV LAT promoter constructs. *Virology*, **197**, 585–92.

McCarthy, A.M., McMahan, L. and Schaffer, P.A. (1989) Herpes simplex virus type 1 ICP27 deletion mutants exhibit altered patterns of transcription and are DNA deficent. *J. Virol.*, **63**, 18–27.

MacDonald, M.E., Ambrose, C.M., Duyao, M.P. *et al.* (1993) A novel gene containing a trinucleotide repeat that is expanded and unstable on Huntington's disease chromosomes. *Cell*, **72**, 971–83.

McFarland, D.J., Sikora, E. and Hotchkin, J. (1986) The production of focal herpes encephalitis in mice by stereotaxic inoculation of virus. Anatomical and behavioral effects. *J. Neurol. Sci.*, **72**, 307–18.

McKnight, J.L.C., Kristie, T.M. and Roizman, B. (1987) Binding of the virion protein mediating α gene induction in herpes simplex virus 1-infected cells to its cis-site requires cellular proteins. *Proc. Natl Acad. Sci. USA*, **84**, 7061–5.

McMahan, L. and Schaffer, P.A. (1990) The repressing and enhancing functions of the herpes simplex virus regulatory protein ICP27 map to C-terminal regions and are required to modulate viral gene expression very early in infection. *J. Virol.*, **64**, 3471–85.

Natarajan, R., Deshmane, S., Valyr-Nagy, T., Everett, R. and Fraser, N.W. (1991) A herpes simplex virus type 1 mutant lacking the ICP0 introns reactivates with normal efficiency. *J. Virol.*, **65**, 5569–73.

Newcomb, W.W. and Brown, J.C. (1994) Induced extrusion of DNA from the capsid of herpes simplex virus type 1. *J. Virol.*, **68**, 433–40.

Nicosia, M., Deshmane, S.L., Zabolotny, J.M., Valyi-Nagy, T. and Fraser, N.W. (1993) Herpes simplex virus type 1 latency-associated transcript (LAT) promoter deletion mutants can express a 2-kilobase transcript mapping to the LAT region. *J. Virol.*, **67**, 7276–83.

Oroskar, A.A. and Read, G.S. (1989) Control of mRNA stability by the virion host shut-off function of herpes simplex virus. *J. Virol.*, **63**, 1897–906.

Palella, T.D., Silverman, L.J., Schroll, C.T. *et al.* (1988) Herpes simplex virus-mediated human hypoxanthine-guanine phosphoribosyl-transferase gene transfer into neuronal cells. *Mol. Cell. Biol.*, **8**, 457–460.

Palella T.D., Hidaka, Y., Silverman, L.J. *et al.* (1989) Expression of human HPRT mRNA in brains of mice infected with a recombinant herpes simplex virus type 1 vector. *Gene*, **80**, 137–44.

Panning, B. and Smiley, J.R. (1989) Regulation of cellular genes transduced by herpes simplex virus. *J. Virol.*, **63**, 1929–37.

Pikaart, M., Feng, J. and Villeponteau, B. (1992) The polyomavirus enhancer activates chromatin accessibility on integration into the HPRT gene. *Mol. Cell. Biol.*, **12**, 5785–92.

Postel, E.H., Mango, S.E. and Flint, S.J. (1989) A nuclease-hypersensitive element of the human c-myc promoter interacts with a transcription initiation. *Mol. Cell. Biol.*, **9**, 5123–33.

Preston, C.M. (1979) Control of herpes simplex virus type 1 mRNA synthesis in cells infected with wild-type virus or the temperature-sensitive mutant tsK. *J. Virol.*, **29**, 275–84.

Pyrc, J.J., Moberg, K.H. and Hall, D.J. (1992) Isolation of a novel cDNA encoding a zinc-finger protein that binds to two sites within the c-myc promoter. *Biochemistry*, **31**, 4102–10.

Ramakrishnan, R., Fink, D.J., Guihua, J., Desai, P., Glorioso, J.C. and Levine, M. (1994) Competitive quantitative polymerase chain reaction (PCR) analysis of herpes simplex virus type 1 DNA and LAT RNA in latently infected cells of brain. *J. Virol.*, **68**, 1864–70.

Read, G.S. and Frenkel, N. (1983) Herpes simplex virus mutants defective in the virion-associated shutoff of host polypeptide synthesis and abnormal synthesis of a (immediate-early) viral polypeptides. *J. Virol.*, **46**, 498–512.

Rivera-Gonzalez, R., Imbalzano, D.J., Gu, B. and DeLuca, N.A. (1994) The role of ICP4 repressor activity in the temporal regulation of the IE-3 and LAT promoter during HSV-1 infection. *Virology*, **202**, 550–640.

Rock, D.L., Nesburn, A.B., Ghiasi, H. *et al.* (1987) Detection of latency-related viral RNAs in trigeminal ganglia of rabbits infect with herpes simplex virus type 1. *J. Virol.*, **61**, 3820–6.

Roizman, B. and Furlong, D. (1974) The replication of herpesviruses, in *Comprehensive Virology*, Vol. 3, (eds H. Fraenkel-Conrat and R.R. Wagner), Plenum Press, New York, pp. 229–403.

Roizman, B. and Sears, A.E. (1990) Herpes simplex viruses and their replication, in *Virology*, 2nd edn, (eds B.N. Fields, D.N. Knipe *et al.*), Raven Press, New York, pp. 1795–841.

Roizman, B. and Sears, A.E. (1993) Herpes simplex viruses and their replication, in *The Human Herpesviruses*, (eds B. Roizman, R.J. Whitley and C. Lopez), Raven Press, New York, pp. 11–68.

Rosen, D.R., Siddique, T., Patterson, D. *et al.* (1993) Mutations in Cu/Zn superoxide dismutase gene are associated with familial amyotrophic lateral sclerosis. *Nature*, **362**, 59–62.

Sacks, W.R., Greene, C.C., Aschman, D.A. and Schaffer, P.A. (1985) Herpes simplex virus type 1 ICP27 is an essential regulatory protein. *J. Virol.*, **55**, 796–805.

Sadowski, I., Ma, J., Triezenberg, S. and Ptashne, M. (1988) GAL4:VP16 is an unusually potent transcriptional activator. *Nature*, **335**, 563–4.

Sawtell, N.M. and Thompson, R.L. (1992) Herpes simplex virus type 1 latency-associated transcription unit promotes anatomical site-dependent establishment and reactivation from latency. *J. Virol.*, **66**, 2157–69.

Sears, A.E., Halliburton, I.W,. Meignier, B. *et al.* (1985) Herpes simplex virus 1 mutant deleted in the α22 gene: growth and gene expression in permissive and restrictive cells and establishment of latency in mice. *J. Virol.*, **55**, 338–6.

Sedarati, F., Izumi, K.M., Wagner, E.K. and Stevens, J.G. (1989) Herpes simplex virus type 1 latency-associated transcript plays no role in establishment or maintenance of a latent infection in murine sensory neurons. *J. Virol.*, **63**, 4455–8.

Seiler, M. and Schwab, M.E. (1984) Specific retrograde transport of nerve growth factor (NGF) from neocortex to nucleus basalis in the rat. *Brain Res.*, **300**, 33–9.

Shih, M.-F., Arsenakis, M., Tiollais, P. and Roizman, B. (1984) Expression of hepatitis B virus S gene by herpes simplex virus type 1 vectors carrying a- and β-regulated gene chimeras. *Proc. Natl Acad. Sci. USA*, **81**, 5867–70.

Smiley, J.R., Smibert, C. and Everett, R.D. (1987) Expression of a cellular gene cloned in herpes simplex virus: rabbit *b*-globin is regulated as an early viral gene in infected fibroblasts. *J. Virol.*, **61**, 2368–77.

Spivack, J.G. and Fraser, N.W. (1987) Detection of herpes simplex virus type 1 transcripts during latent infection in mice. *J. Virol.*, **61**, 3841–7.

Spivack, J.G. and Fraser, N.W. (1988) Expression of herpes simplex virus type 1 latency-associated transcripts in the trigeminal ganglia of mice during acute infection and reactivation of latent infection. *J. Virol.*, **62**, 1479–85.

Spivack, J.G., Woods, G.M. and Fraser, N.W. (1991) Identification of a novel latency-specific splice donor signal within HSV type 1 2.0-kilobase latency-associated transcript (LAT): translation inhibition of LAT open reading frames by the intron within the 2.0-kilobase LAT. *J. Virol.*, **65**, 6800–10.

Steiner, I., Spivack, J.G., Lirette, R.P. *et al.* (1989) Herpes simplex virus type 1 latency-associated transcripts are evidently not essential for latent infection. *EMBO Journal*, **8**, 505–11.

Stevens, J.G. (1989) Human herpesviruses: a consideration of the latent state. *Microbiol. Rev.*, **53**, 318–32.

Stevens, J.G., Wagner, E.K., Devi-Rao, G.B., Cook, M.L. and Feldman, L.T. (1987) RNA complementary to a herpesviruses a gene mRNA is prominent in latently infected neurons. *Science*, **255**, 1056–9.

Stroop, W.G. and Schaefer, D.C. (1987) Herpes simplex virus, type 1 invasion of the rabbit and mouse nervous systems revealed by *in situ* hybridization. *Acta Neuropathol.*, **74**, 124–32.

Timmusk, T., Palm, K., Metsis, M. *et al.* (1993) Multiple promoters direct tissue specific expression of the rat BDNF gene. *Neuron*, **10**, 475–89.

Vegeto, E., Allan, G.F., Schrader, W.T. *et al.* (1992) The mechanism of RU486 antagonism is dependent on the conformation of the carboxy-terminal tail of the human progesterone receptor. *Cell*, **69**, 703–13.

Wagner, E.K., Devi-Rao, G., Feldman, L.T. *et al.* (1988a) Physical characterization of the herpes simplex virus latency-associated transcript in neurons. *J. Virol.*, **63**, 1194–2002.

Wagner, E.K., Flanagan, W.M. and Devi-Rao, G.B. (1988b) The herpes simplex virus latency-associated transcript is spliced during the latent phase of infection. *J. Virol.*, **62**, 4577–85.

Wang, Y., O'Malley, Jr, B.W., Tsai, S.Y. and O'Malley, B.W. (1994) A novel regulatory system for gene transfer. *Proc. Natl Acad. Sci. USA*, **91**, 8180–4.

Watson, R.J. and Clements, J.B. (1980) A herpes simplex virus type 1 function continuously required for early and late virus RNA synthesis. *Nature*, **285**, 329–30.

Wechsler, S.L., Nesburn, J., Zwaagstra, N. and Ghiasi, H. (1989) Sequence of the latency related gene of herpes simplex virus type 1. *Virology*, **168**, 168–72.

Wolfe, J.H., Deshmane, S.L. and Fraser, N.W. (1992) Herpes virus vector gene transfer and expression of β-glucuronidase in the central nervous system of MPS VII mice. *Nature Genetics*, **1**, 379–84.

Xu, L., Schaffner, W. and Rungger, D. (1993) Transcription activation by recombinant GAL4/VP16 in the *Xenopus* oocyte. *Nucleic Acids Res.*, **21**, 2775.

York, I.A., Roop, C., Andrews, D.W. *et al.* (1994) A cytosolic herpes simplex virus protein inhibits antigen presentation to $CD8^+$ T lymphocytes. *Cell*, 77, 525–55.

Zwaagstra, J., Ghiasi, H., Nesburn, A.B. and Wechsler, S.L. (1989) In vitro promoter activity associated with the latency-associated transcript gene of herpes simplex virus type 1. *J. Gen. Virol.*, **70**, 2163–9.

Zwaagstra, J.C., Ghiasi, H., Slanina, S.M. *et al.* (1990) Activity of herpes simplex virus type 1 latency-associated transcript (LAT) promoter in neuron-derived cells: evidence for neuron specificity and for a large LAT transcript. *J. Virol.*, **64**, 5019–28.

Zwaagstra, J., Ghiasi, H., Nesburn, A.B. and Wechsler, S.L. (1991) Identification of a major regulatory sequence in the latency-associated transcript (LAT) promoter of herpes simplex virus type 1 (HSV-1). *Virology*, **182**, 287–97.

4

Liposome-mediated gene transfer

ITARU YANAGIHARA, YASUFUMI KANEDA, KOJI INUI
and SHINTARO OKADA

4.1 INTRODUCTION

Liposomes are artificial spherical vesicles which contain a small volume of aqueous solution enclosed in a lipid bilayer. Liposomes consist mainly of phospholipids and cholesterol, major components of biological membranes. From the mid-1960s, numerous studies on liposomes have been presented as models of biological membranes to examine physical and chemical characteristics, membrane permeability, membrane leakage, cell-to-cell membrane fusion and cell–membrane interaction. Clinically, from the early 1970s, liposomes have been studied as a drug delivery system (DDS) using techniques of entrapment of ionized or low-molecular weight substances into liposome. Furthermore, integration of antigenic proteins into liposome membranes is being used for vaccination against virus infection. Now, genetic advances are heralding a new application to the use of liposomes, namely liposome-mediated gene delivery.

Viral vectors, such as retroviral or adenoviral vectors, are now widely used in clinical trials of gene therapy because of the stability and high efficacy of transgene expression; however, liposome-mediated gene transfer deserves attention as a safe and non-invasive delivery method. For example, liposomes consist only of biological lipids and could be applicable for repeated gene therapy, having less antigenicity, and no concerns about proliferation of recombinant virus in the host. Although we have not as yet achieved higher efficacy and stable expression in liposomal methods than with viral vectors, safety and easy handling of liposomes is a great benefit in human gene therapy. In the meantime, immunotherapy by liposome-

Molecular and Cell Biology of Human Gene Therapeutics
Edited by George Dickson
Published in 1995 by Chapman & Hall. ISBN 0 412 62550 4

mediated gene transfer has started against progressive melanoma (Nabel *et al.*, 1993), and correcting the abnormal ion transport in the animal model of cystic fibrosis has been reported by instillation (Hyde *et al.*, 1993) and nebulization (Alton *et al.*, 1993).

In this chapter, we overview the biological aspects of liposomes, discuss their current applications and describe our newly developed HVJ–liposome gene transfer methods.

4.2 BASIC ASPECTS OF LIPOSOMES

4.2.1 Cell membrane and liposome structure

The cell membrane is composed mainly of lipids and proteins, and cell membrane structure is thought to reflect the fluid mosaic model of Singer and Nicolson (1972). According to this model, membranous proteins are buried in the phospholipid bilayer, and lipid molecules in the bilayer are weakly associated with each other by hydrophobic bonds, but maintain some fluidity. In 1965, Bangham *et al.* found that egg-yolk phosphatidylcholine formed vesicles in water, and that his enclosed vesicle could entrap cations and anions inside. Later, this vesicle was called a liposome. In aqueous solution, hydrophobic fatty acid tails of phospholipid self-associate within the lipid bilayer to exclude water, while hydrophilic heads interact with the liquid, resulting in the formation of a continuous membrane of lipid bilayer (Figure 4.1).

Liposomes are classified into three subgroups by their morphological features: multilamellar vesicles (MLV); large unilamellar vesicles (LUV); and small unilamellar vesicles (SUV) (Figure 4.2). The structure of MLVs is strikingly different from SUVs and LUVs, because MLVs have a multilamellar lipid bilayer, like an onion bulb. The diameters of MLVs range from 100 nm to several microns. MLVs can be easily formed and are superior in stability to LUVs and SUVs. However, it is difficult to consistently form the same size of MLV and to place an entrapped transgene into the target cells due to the multilamellar barrier. The diameters of LUVs range from 100 to 1000 nm. LUVs can entrap high-molecular-weight substances, but are inferior for intravenous gene therapy because they are easily entrapped by the reticuloendothelial system. The diameters of SUVs are less than 100 nm. SUVs cannot entrap as large molecules as MLVs or LUVs; however, the size of SUVs is highly uniform and the single lipid bilayer is superior in transferring genes into target cells.

4.2.2 Liposome–cell membrane interaction

Liposome and cell membrane interactions have been studied as a model of biological membrane interactions. There are four basic types of inter-

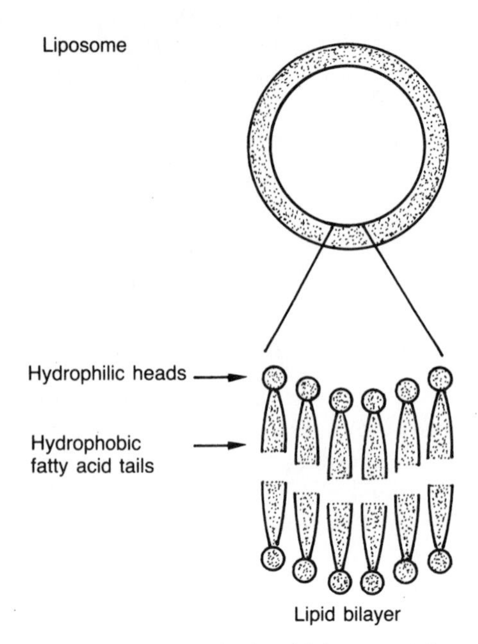

Figure 4.1 Structure of the liposome. The lipid bilayer mainly consists of phospholipids, the hydrophobic fatty acid tails of which face each other.

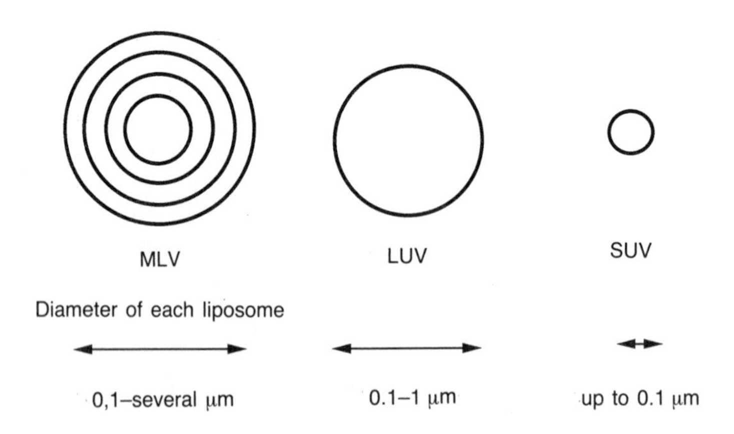

Figure 4.2 Schema of the three subgroups of liposomes.

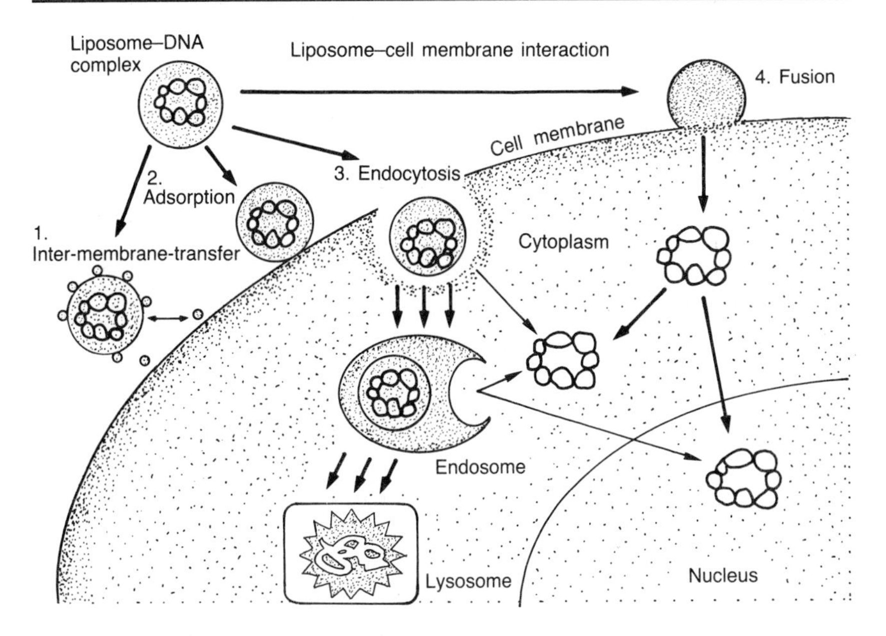

Figure 4.3 Schema of liposome and cell membrane interactions. 1. Inter-membrane transfer mediated lipid transfer between liposome and cell membrane. 2. Liposomal adsorption can be enhanced by modification of lipid bilayer, with antibodies against cell surface antigens, various receptors, or giving electric charge to liposome. 3. Some of the liposomes fuse with endosomal membrane to release the entrapped substances directly into the cytoplasm, such as low pH sensitive liposome. 4. Through the fusion process, entrapped materials are directly released into the cytoplasm and some of the plasmids are transferred into the nucleus.

action between liposomes and cell membranes which have been described (Figure 4.3).

Inter-membrane transfer

It has been observed that lipids or proteins can exchange randomly between liposomes and cell membranes. Inter-membrane transfer differs from membrane fusion in that it usually occurs very slowly, and is insufficient to translocate phospholipids. However, phospholipids can transfer more easily into the cell membrane with lipid-carrier proteins, which often consist of lipoproteins. Because of the similarity of the structure between liposomes and a phospholipid monolayer-containing lipoproteins, lipoproteins also can exchange with the liposome. Using fluorescence-labeled lipid, inter-membrane lipid transfer from liposomes to Golgi membranes, cell plasma membranes and to newly prepared liposomes has been demonstrated (Lipsky and Pagano, 1985).

Liposome adsorption to cell membranes

Some glycolipids and proteins (including antibodies) have the ability to bind to cell membranes. Such molecules, when inserted into the lipid bilayer of a liposome, can be recognized by specific cell membrane components. For example, liposomes containing sialyl Lewis^x (Le^x, known to be a ligand for L-selectin, one of the cell adhesion molecules), show efficient adsorption to F9 teratocarcinoma cells (Eggens *et al.*, 1989). This adsorption does not however always result in intracellular transfer of the materials entrapped in the liposome. Cationic liposome also adsorbs to negatively charged cell membranes by electrostatic bonds.

Endocytosis of liposomes into the cell

After adsorption to the cell membrane, liposomes which associate with coated pits are internalized into the coated or uncoated vesicle. Liposomes are sorted into secondary lysosomes via the endocytosis pathway and then exposed to lysosomal enzymes, resulting in the destruction of liposome membrane and the entrapped materials. However, development of pH-sensitive liposomes has enabled transfer of entrapped materials to cytoplasm from the endosome compartment (Connor *et al.*, 1984).

Liposomes and cell membrane fusion

Fusion between liposomes and the cell membrane is the most ideal model for gene transfer, because the fusion process enables internalization of foreign genetic material directly into the cell cytoplasm without lysosomal degradation. Although it is well known that fusion between liposomes occurs easily, fusion between liposomes and cell membranes rarely happens. For example, phosphatidylcholine liposomes are known to have the ability to fuse to cell membranes, but this requires several days to occur. This slow fusion has great disadvantages for *in vivo* gene transfer, because free liposomes are readily captured by the reticuloendothelial system. From this point, developing rapid fusion systems is necessary for liposome-mediated gene therapy.

Several viruses are known to exhibit membrane fusion ability mediated by glycoproteins of the viral envelope. There are two main fusion pathways between viral envelopes and cell membranes: endosomal membrane fusion, and cell membrane fusion. Some single-stranded RNA viruses (influenza viruses, alphaviruses, rhabdoviruses and flaviviruses) are introduced into cells by receptor-mediated endocytosis. After exposure to acid pH in the endosome, the viral envelope fuses with the endosomal membrane, resulting in the viral genome being released into the cytoplasm. On the other hand, paramyxoviruses, coronaviruses and retroviruses fuse directly with cell membrane via viral envelope glycoproteins (Lamb, 1993). For gene therapy, it is important to internalize genes of interest into cytoplasm without degradation in the endosome–lysosome pathway.

4.2.3 Toxicity and antigenicity

Although liposomes consist mainly of biodegradable lipids, liposome toxicity and antigenicity must always be a concern in the development of clinical products. Cholesterol, phosphatidylcholine, phosphatidic acid and phosphatidylglycerol have no effect on the synthesis of the DNA in the cells, whereas, stearylamine (SA), phosphatidylserine and dicetyl phosphate (DCP) reduce DNA synthesis. DCP and SA have also been reported as toxic lipids when injected intracerebrally into mouse brain. Furthermore, it is known that while cholesterol hemisuccinate is a benign agent for normal rats, it is a lethal agent for hypophysectomized rats (Weiner, 1989). As a result, we must pay attention in clinical liposome trials when patients have lipid metabolism disorders or disturbances. In one recent experiment 1000-fold higher concentrations of a particular DNA and lipid formulation (dimyristyloxypropyl-3-dimethyl-hydroxyethyl ammonium (DMRIE) and dioleoylphosphatidylethanolamine (DOPE)) than utilized in human gene therapy protocols was shown not to cause any pathogenic change (San *et al.*, 1993). Generally, hydrophobic proteins exposed at the surface of the outer membrane of liposome are highly antigenic, whereas fully entrapped proteins exhibit no antigenicity.

4.3 CONVENTIONAL LIPOSOME SYSTEMS

4.3.1 Simple liposomes

Negatively charged liposomes, consisting of cholesterol and phosphatidylserine, have been used for *in vitro* gene delivery. The thymidine kinase (TK) gene was introduced into mouse L cells which has no TK activity. Transient expression of TK gene was observed in 10% of total L cells, and 0.02% of total L cells were selected as stable transformants on day 14 after transfection. This simple liposome was three times more efficient for transient expression than the standard calcium phosphate gene transfer method, and had about half the efficacy for stable expression compared with calcium phosphate method (Schaefer-Ridder *et al.*, 1982).

4.3.2 Cationic liposomes

Cationic liposomes consist of positively charged lipids and are more efficient in transferring nucleic acids than neutral or anionic liposomes, because they bind directly with negatively charged DNA and RNA, and the

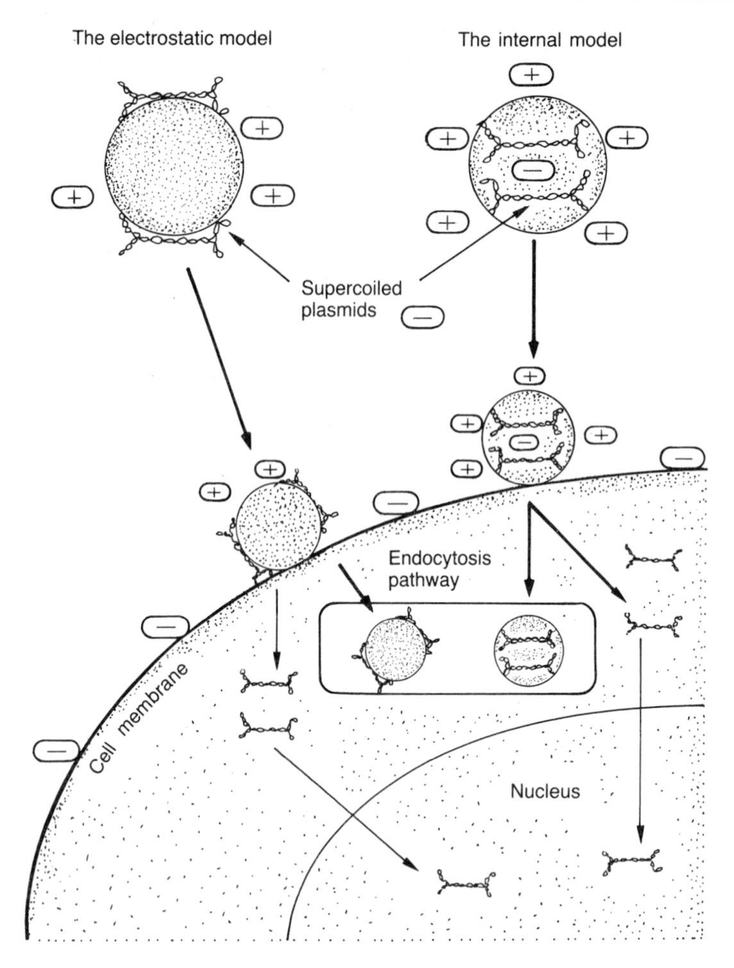

Figure 4.4 The cationic liposome gene delivery system.

cell membrane (Figure 4.4). In addition, cationic liposomes overcome some of the difficulties of conventional liposomes in stability, preparation at uniform size and poor transfection efficacy. In relation to the structure of liposome and nucleic acids, two models of cationic liposomes have been developed.

The electrostatic model

Lipofectin (Lipofectin®, GibcoBRL) (Felgner *et al.*, 1987) uses aqueous cationic liposomes, which are composed of *N*-[1-(2,3-dioleyloxy)propyl]-*N*,*N*,*N*-trimethylammonium chloride (DOTMA) and DOPE in an equal ratio (w/w). In water, DOTMA is known to form lipid bilayers alone or in combination with other phospholipids. It is also known that DOTMA can complex with nucleic acids (DNA and RNA) following simple mixing. On the subject of gene expression efficiency, the ratio of liposome and DNA is very important, because while the liposomes are adsorbed to the cell membrane electrostatically, these liposome–DNA complexes must remain positively charged as a whole in order to be adsorbed onto negatively charged cell membranes and not excluded by excess negatively charged DNA. An additional drawback is that serum sulfated proteoglycans and other negatively charged serum substances inhibit the lipofectin system.

Felgner and Ringold (1989) have hypothesized that linear DNA is surrounded by four cationic liposomes. On the other hand, plasmid can form a supercoil that reduces its size by 41% of total DNA length and has four branches from a central core branch. Therefore supercoiled DNA–liposome complexes are thought to exist in a more compact form than linear DNA–liposome complexes; moreover, these compact complexes may be aggregated with each other (Smith *et al.*, 1993).

In a fluorescent DNA–liposome study, lipofectin was found to fuse efficiently with target cell membranes with over 99% of the cells having plasmid. However, reporter gene (β-*gal*) expression occurred only in 25% of the cells. From studies of plasmid formed in the nuclear fraction per cell, intact plasmid was found to exist below 1% (Felgner and Ringold, 1989).

The Lipofectin® system has proved very useful for gene transfection because of its ease of use, wide host cell range *in vitro*, and capacity for large DNA. For instance, a 150 kb YAC clone containing the α1 (I) collagen locus was introduced by lipofectin and intact RNA was expressed in murine fibroblasts (Strauss and Jaenisch, 1992). However, it has proved difficult to obtain stable transformants and to avoid lysosomal pathway degradation with lipofectin-mediated gene transfer.

The lipofection gene transfer system seems also to have some advantages to express foreign gene directly *in vivo* especially in lung and artery. Cationic liposome-mediated gene therapy for the cystic fibrosis mouse model has been carried out by instillation (Hyde *et al.*, 1993) and nebulization (Alton *et al.*, 1993). Partial correction of the ion transport defect was reported following transfer of the human cystic fibrosis transmembrane regulator gene by both routes. Surprisingly, about 50% of the deficit was restored by nebulization therapy. By aerosol and intravenous transfection, human α_1-antitrypsin gene was expressed in New Zealand White rabbits (Canonico *et al.*,1994), with expression being detected up to 7 days following a single administration. The human growth hormone gene was also expressed via rabbit ear artery yielding levels up to 3.8 ng/ml after 20 days (Losordo *et al.*, 1994).

Clinically, liposomes have been tested for therapy of malignant melanoma. HLA-B7, a major histocompatibility complex (HMC) protein, was introduced into five HLA-B7-negative patients at the stage IV melanoma. This therapy is designed to raise an immune response to attack the foreign antigen with cytotoxic T cells stimulated by the foreign MHC (class I) signal; as a result, tumor regression was reported in one patient (Nabel *et al.*, 1993).

The internal model

Genetransfer™ (Koshizaka *et al.*, 1989) is a cationic liposome which has the ability to entrap DNA inside the lipid bilayer. Genetransfer™ is composed of *N*-(alpha-trimethylacetyl)-didodecyl-D-glutamate chloride (TMAG), dilauroylphosphatidylcholine (DLPC) and DOPE in molar ratio of 1 : 2 : 2. With this reagent entrapment of DNA is a simple procedure involving straightforward mixing and vortexing, with the entrapment of negatively charged DNA not affecting the cationic charge on the outer lipid bilayer. Hence, this system can be used under conditions where serum is added to the medium *in vitro*. The human β-interferon gene has been transfected into cultured glioma cell by the Genetransfer™ liposome system, with significant inhibition of cell proliferation resulting (Mizuno *et al.*, 1990).

4.4 HEMAGGLUTINATING VIRUS OF JAPAN (HVJ OR SENDAI VIRUS)–LIPOSOMES

4.4.1 Principles

There are two major difficulties for liposome-mediated gene transfer. Firstly, it is necessary to place DNA directly into the cytoplasm to avoid degradation. To this end, we have utilized HVJ to fuse liposomes with cell membranes, which leads to direct introduction of DNA into the cytoplasm (Kaneda *et al.*, 1987). Secondly, it is important to develop efficient delivery of DNA into the nucleus. For this purpose, co-introduction of DNA with so-called high mobility group-1 protein (HMG-1) facilitates DNA migration into the nucleus and enhances its expression (Kaneda *et al.*, 1989a).

The HVJ component

HVJ is a member of the paramyxovirus family, which has an HN glycoprotein and an F glycoprotein in its viral envelope (Figure 4.5). The HN glycoprotein has neuraminidase activity and mediates surface receptor binding via recognition of glycoproteins and glycolipids containing sialic acid.

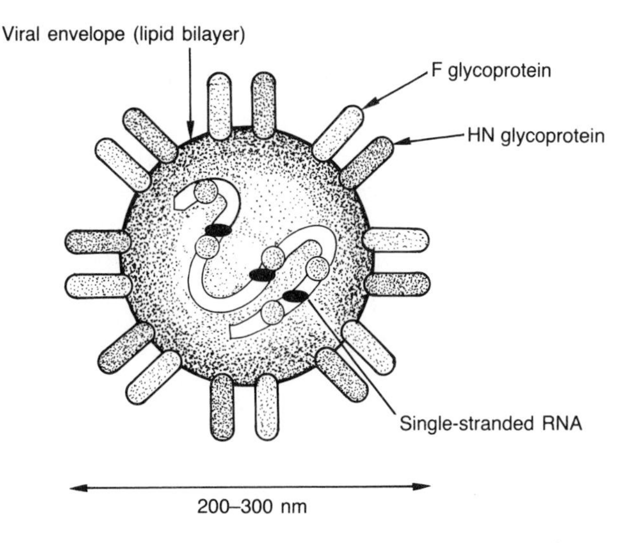

Figure 4.5 Schema of hemagglutinating virus of Japan (HVJ). HN and F proteins are buried in the lipid layer of the viral envelope, and are responsible for viral attachment and fusion between the cell membrane and virus.

The F (or fusion) glycoprotein has fusion activity both at acidic and at neutral pH. The exact mechanism of virus-to-cell fusion is still unclear, but it is thought that after HN glycoprotein recognition and binding to the cell membrane, conformational changes occur to the F glycoprotein to release hydrophobic fusion peptides into the target cell membrane (Lamb, 1993).

As far as safety is concerned in potential clinical applications, HVJ is non-pathogenic in humans and can be completely inactivated by appropriate ultra-violet irradiation without loss of fusion competence.

The HMG-1 component

Recently, many DNA-binding transcriptional factors have been reported. HMG-1 is a member of the family of HMG proteins known to regulate DNA replication, nucleosome assembly, and nucleolar and mitochondrial RNA polymerase transcription. It is a non-histone chromosomal protein of 28 kDa and is localized at the linker regions between nucleosomes. HMG-1, which might be involved in DNA recombination, preferentially binds with poly(CA)·poly(TG)-associated four-stranded DNA structures (Gaillard and Strauss, 1994), but does not bind to linear duplex or single-stranded DNA (Bianchi *et al.*, 1989). Once introduced into the cell cytoplasm, HMG-1 rapidly migrates into the nucleus (Tsuneoka *et al.*, 1986).

4.4.2 Gene (cDNA) delivery system

Preparation of liposomes

Non-cationic liposomes are prepared which consist of bovine brain phosphatidylserine-sodium salt (PS), egg-yolk phosphatidylcholine (PC) and cholesterol. PS (10 mg) is dissolved in tetrahydrofuran (THF, 0.9 ml) and 0.01M K_2PO_4, pH 6.8 (0.1 ml). PC (48 mg) and cholesterol (20 mg) were each dissolved in 1 ml THF. PS, PC and cholesterol are then mixed in a weight ratio of 1 : 4.8 : 2 and 0.9 ml of THF is added. The lipid mixture is divided into glass tubes (0.5 ml, 10 mg each). To avoid oxidation, the reaction vessels must be filled with nitrogen gas and stored at −20°C. Organic solvent evaporation is performed under vacuum (initially at 200 mmHg) using a rotary evaporator at least for 10 minutes, whereupon a thin layer of lipids is formed inside the glass tube.

Harvesting HVJ

A 0.1 ml portion of HVJ seed (suspended in 1% w/v polypeptone, 0.2% w/v NaCl, pH 7.2) is injected into the chick-egg chorioallantoic cavity. After 4 days' incubation at 35.4°C, eggs are chilled at 4°C for 6 hours. Proliferated HVJ is collected from the chorioallantoic fluid (5–10 ml fluid per egg) and the virus fluid stored preferably at 4°C. HVJ is stable for at least 3 months at 4°C. HVJ is then purified from chorioallantoic fluid by centrifugation and its turbidity measured by absorbance photometry. The virus titer and hemagglutinating unit, (HAU, OD_{540} = 1.0 corresponds to 15 000 HAU) which reflects index for fusion ability, are then calculated.

HMG-1 purification

HMG-1 is purified from about 1 kg of calf thymus tissue. The purification method for HMG-1 is described elsewhere (Goodwin, 1975). HMG-1 is normally prepared at a concentration of 1.3 mg/ml and stored at −70°C.

HVJ–liposome complex formation (Figure 4.6)

Plasmid DNAs (200 μg) are mixed with HMG-1 (65 μg) and the volume is adjusted to 200 μl with buffered salt solution buffer (137mM NaCl, 5.4mM KCl, 10mM Tris-HCl, pH 7.6). This mixture is incubated at 20°C for 1 hour to form a DNA–HMG-1 complex. This plasmid–HMG-1 complex is entrapped into liposomes by a vortex method and sonicated for 3–5 seconds. A 300-μl sample of BSS is then added to the mixture and shaken in

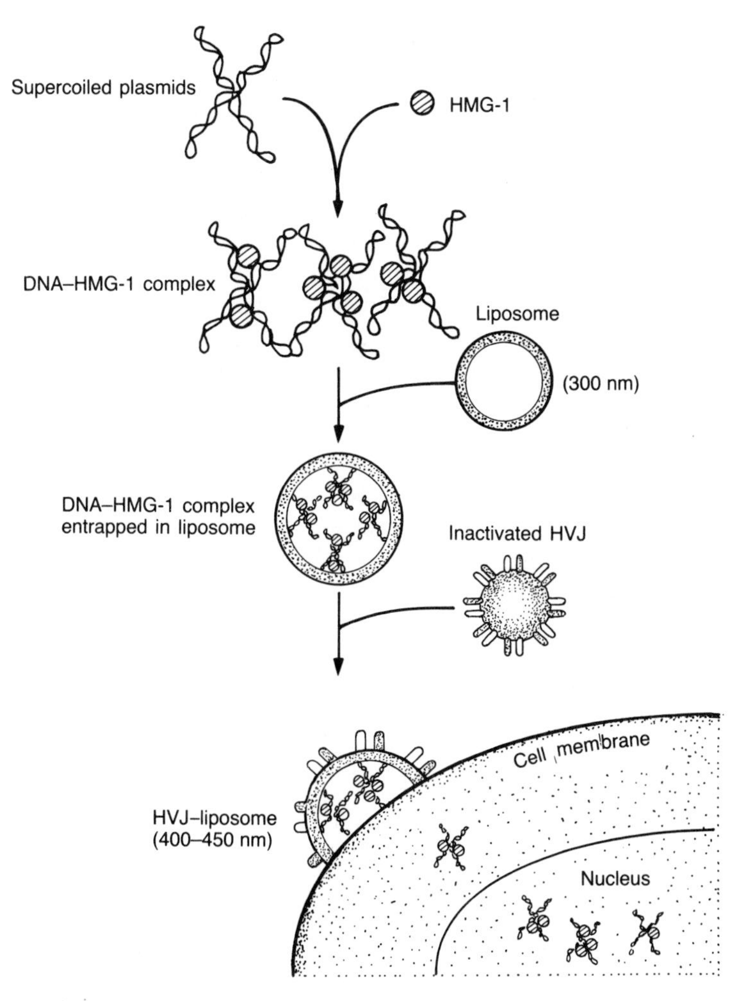

Figure 4.6 Schema of the HVJ–liposome gene transfer system.

the 37°C water bath for 30 minutes. Prepared HVJ (1.5 ml contains 45 000 HAU) is completely inactivated by ultra-violet irradiation (100 ergs/mm^2/s) for 3 minutes in a 10-cm diameter dish. Inactivated HVJ (30 000 HAU) is added to the lipid–DNA mixture on ice for 10 minutes and the solution is shaken at 37°C for 1 hour. BSS is added and the total complex volume is adjusted to 4 ml. Free HVJ is removed by density gradient ultracentrifugation (62 800g) for 3 hours. After centrifugation, about 2 ml of the HVJ–liposome sample is collected. Finally, about 10–30% of the prepared DNA (up to 20 kb) is included in the HVJ–liposome. Further details of the preparation are described by Kaneda (1994).

In vivo transfection by HVJ–liposome

Various *in vitro* and *in vivo* gene transfections using HVJ–liposomes have been reported (Table 4.1). In the early studies, the target organ was the liver. For example, the human insulin gene was co-introduced with HMG-1 or bovine serum albumin (BSA) into adult rat liver by HVJ using red blood cell ghosts instead of liposomes. The amount of transcript of the human insulin gene with HMG-1 protein was ten-fold greater than with BSA complexes. DNA and RNA of the human insulin and expressed insulin was detected for 10–14 days after injection but were rapidly decreased thereafter (Kaneda *et al.*,1989b). Hepatitis B virus surface antigen gene (HBsAg) injected into adult rat perisplanchnic membrane was expressed in the liver by the HVJ–liposome method. HBsAg was detected in the serum for 9 days, with maximum levels of HBsAg observed on day 2 (Kato *et al.*, 1991a). Moreover, antibody against HBsAg was produced in the rat serum, resulting in focal necrosis and degeneration of the hepatic cells with lymphocyte infiltration. This system provides an animal hepatitis model (Kato *et al.*, 1991b). Human renin gene was expressed in the adult rat liver by the HVJ method; on day 5, active human renin was detected in the plasma of rats, resulting in hypertension which could be abolished by a specific human renin inhibitor (Tomita *et al.*, 1993).

HVJ–liposome gene transfer seems also to have the ability to mediate trans-arterial delivery. Administration via the renal artery, resulted in SV40 large T antigen expression *in vivo* in the kidney of adult rats, with expression maximal on day 4 and thereafter gradually decreasing (Tomita *et al.*, 1992). In order to better understand glomerulosclerosis, transforming factor-β (TGF-β) gene and platelet-derived growth factor-B (PDGF-B) gene have been separately introduced and overexpressed in rat kidney via the renal artery. Overexpression of both genes resulted in glomerulosclerosis, but TGF-β affected extracellular matrix (ECM) proliferation preferentially whereas PDGF affected more the mesangial cell proliferation (Isaka *et al.*, 1993). Angiotensin-converting enzyme (ACE) gene was introduced into cultured carotid artery or vascular smooth muscle cells (VSMC). The increase of the vascular ACE was parallel to the increase in DNA synthesis in VSMC. In studies in COS cells, HVJ–liposomes were superior to Lipofectin; however, ACE activity in VSMC was virtually identical between the Lipofectin and HVJ–liposome methods. With Lipofectin a 24-hour incubation was required to achieve the same activity that HVJ–liposomes produced in only 35 minutes (Morishita *et al.*, 1993a). Furthermore, local overexpression of ACE *in vivo* in the rat carotid artery caused local generation of angiotensin II (paracrine/autocrine factor), resulting in direct vascular hypertrophy (Morishita *et al.*, in press).

Thus many genes have been transferred by the HVJ–liposome method to various animals *in vivo*, without organ damage being observed. Also, some

Table 4.1 *In vivo* expression by HVJ–liposome

Organ (animal)	*Gene*	*Duration*	*Effect of gene expression*
Liver (rat)	Insulin	10–14 days	Decrease of blood glucose level
	HB virus surface antigen	7 days	Liver cell degeneration and infiltration
	Renin	7 days	High blood pressure
Kidney (rat)	TGF-β, PDGF	10 days	Glomerulosclerosis
Heart (rat)	TGF-β	> 2 weeks	Cardiac hypertrophy
Heart (dog)	CAT	> 2 weeks	No change
Skeletal muscle (mouse)	Dystrophin	2 weeks	Dystrophin expression in *mdx* mouse muscle
Artery (rat)	Renin, ACE	> 2 weeks	Hyperplasia and hypertrophy of vascular smooth muscle cells
	Antisense oligonucleotide	> 2 weeks	Inhibition of neointima formation
Lung (rat)	TGF-β	> 2 weeks	Fibrosis
Joint space	SVT	n.d.	Gene expression at synovia and cartilage
Testis	CAT	> 8 months	No change

ACE, angiotensin converting enzyme; CAT, chloramphenicol acetyl transferase; PDGF, platlet-derived growth factor; TGF-β, transforming factor-β; SVT, SV40, large T antigen; HB, hepatitis B; n.d. not determined preliminary data show safety for repeated HVJ–liposome injection in animals. Consequently, although HVJ–liposome expression is transient, it is relatively safe because no integration occurs into chromosomes.

Antisense oligonucleotide delivery

Impressive studies on transfer of antisense oligonucleotides (ODNs) have been reported with HVJ–liposome. We found that using HVJ–liposomes, FITC-labelled ODNs were concentrated in the cell nucleus 5 minutes after transfer and could be detected for more than 3 days. Moreover, the inhibitory effect on DNA synthesis in VSMC by antisense ODNs inhibiting basic fibroblast growth factor was 50 times more efficient using the HVJ–liposome method than using Lipofectin.

More recently antisense ODNs for cdc3 kinase and proliferating-cell nuclear antigen (PCNA) were examined. *In vitro*, antisense cdc 2 kinase ODNs or antisense PCNA ODNs alone did not inhibit VSMC growth, but growth was inhibited when antisense cdc 2 kinase ODNs and antisense PCNA ODNs were co-transfected. *In vivo* co-transfection of antisense cdc 2 kinase ODNs and antisense PCNA ODNs into angioplasty injured arteries resulted in marked decrease in cdc2 and PCNA mRNA expression and complete inhibition of neointima formation for up to 8 weeks (Morishita *et al.*, 1993b).

In a similar fashion, the cell cycle regulatory enzyme, cyclin-dependent kinase 2 (cdk 2) is known to increase in the balloon angioplasty injury. With HVJ–liposome, antisense cdk 2 ODNs were introduced into the balloon injured artery. cdk 2 mRNA was significantly reduced with inhibition (60%) in neointima formation. Finally, antisense cdk 2 and antisense cdc 2 ODNs co-transfection showed almost complete inhibition of the neointima formation, while antisense cdc 2 ODNs alone inhibited up to 40% (Morishita *et al.*, 1994).

4.5 FUTURE PROSPECTS FOR LIPOSOME-MEDIATED GENE THERAPY

For *in vivo* gene therapy using liposome-mediated gene transfer, much effort is being exerted to develop methods for specific targeting of particular organs, increasing efficient uptake of entrapped genes into cytoplasm, targeting of genes into nucleus, and achieving long-term gene expression.

Gene therapy can be broadly classified into two categories: compensatory therapy to replace deficient enzymes or proteins; and preventive therapy such as suppressing hypertension or cancer. Compensatory therapies usually require high levels of systemic gene expression, whereas preventive therapies need more organ- or site-specific treatments to suppress the expression of unregulated genes. With the HVJ–liposome method, we can target organs by changing the route of administration, but more precise organ targeting may be accomplished through liposome lipid bilayers decorated with antibodies, receptor-binding proteins, or some type of cell adhesion recognition molecule. In addition, we can prepare immunosomes and virosomes by putting immunogen or viral protein into the lipid bilayer.

To enhance gene transfer into the cytoplasm, we have utilized the HVJ fusion activity. Although we have no evidence for circulating antibody against the HVJ in transfected animals, it may be important to reduce the antigenicity of viral proteins. It might be ideal that only the fusion protein of the virus be inserted in lipid bilayer for liposome–cell fusion. As a result, it will be important to purify, characterize and synthesize viral fusion protein systems by recombinant technologies.

The use of HMG-1 has successfully allowed translocation of DNA complexes into the nucleus from cytoplasm. Although the mechanism involved is not yet fully understood, the elucidation of the uptake mechanism and an understanding of the contribution that activated gene transcription plays will be important. If more efficient nuclear proteins and/or more specific transcriptional factors can be applied, these will be powerful tools for designing novel gene expression systems.

For long-term gene expression, it is important to stabilize, regulate, and replicate transgenes. To realize these goals, there are three probable routes. Firstly, because liposomes can entrap extremely large nucleic acids and complexed proteins, it may be possible to transfer (artificial) chromosome-like structures into the nucleus, which have promoters, enhancers and transcriptional factors, are protected from nucleases, and ideally contain elements allowing replication of the transgene along with the host cell proliferation cycle.

Secondly, targeting and replacing damaged gene fragments will be useful for natural regulation of corrected gene. For this therapy, site-specific integration or homologous recombination activities *in vivo* would have to be incorporated into the liposome gene transfer method.

Thirdly, some viral genomes are known to replicate in the cytoplasm. This replication has limitations in the host cell because overproliferation results in host cell death. If we can characterize and harness the cytoplasmic replication machinery, liposomes would become efficient vectors for cytoplasmic gene therapy.

In conclusion, the solution and adaptation of some of these complex molecular processes may lead in the future to the liposome becoming an ideal vector system for gene therapy.

REFERENCES

Alton, E.W.F.W., Middeleton, P.G., Caplan, N.J. *et al.* (1993) Non-invasive liposome-mediated gene delivery can correct the ion transport defect in cystic fibrosis mutant mice. *Nature Genetics*, **5**, 135–42.

Bangham, A.D., Standish, M.M. and Watkins, J.C. (1965) Diffusion of univalent ions across the lamellae of swollen phospholipids . *J. Mol. Biol.*, **13**, 238–52.

Bianchi, M.E., Beltrame, M. and Paonessa, G. (1989) Specific recognition of cruciform DNA by nuclear protein HMG1. *Science*, **243**, 1056–9.

Canonico, A.E., Conary, J.T., Meyrick B.O. and Brigham, K.L. (1994) Aerosol and intravenous transfection of human α1-antitrypsin gene to lungs of rabbits. *Am. J. Respir. Cell. Mol. Biol.*, **10**, 24–9.

Connor, J., Yatvin, M.B. and Huang, L. (1984) pH-sensitive liposomes: acid-induced liposome fusion. *Proc. Natl Acad. Sci. USA*, **81**, 1715–18.

Eggens, I., Fenderson, B., Toyokuni, T. *et al.* (1989) Specific interaction between Le^x and Le^x determinants. *J. Biol. Chem.*, **264**, 9476–84.

Felgner, P.L. and Ringold, G.M. (1989) Cationic liposome-mediated transfection. *Nature*, **337**, 387–8.

Felgner, P.L., Gadek, T.R., Holm, M. *et al.* (1987) Lipofection: a highly efficient, lipid-mediated DNA-transfection procedure. *Proc. Natl Acad. Sci. USA*, **84**, 7413–17.

Gaillard, C. and Strauss, F. (1994) Association of poly(CA).poly(TG) DNA fragments into four-stranded complexes bound by HMG1 and 2. *Science*, **264**, 433–6.

Goodwin, G.H., Nicolas, R.H. and Johns, E.W. (1975) An improved large scale fractionation of high mobility group non-histone chromatin proteins. *Biochim. Biophys. Acta*, **405**, 280–91.

Hyde, S.C., Gill, D.B., Higgins, C.F. and Trezise, A.E.O. (1993) Correction of the ion transport defect in cystic fibrosis transgenic mice by gene therapy. *Nature*, **362**, 250–5.

Isaka, Y., Fujiwara, Y., Ueda, N., Kaneda, Y., Kamada, T. and Imai, E. (1993) Glomerulosclerosis induced by *in vivo* transfection of transforming growth factor-β or platelet derived growth factor gene into the rat kidney. *J. Clin. Invest.*, **92**, 2597–601.

Kaneda, Y. (1994) Virus (sendai virus envelopes)-mediated gene transfer, in *Cell Biology: A Laboratory Handbook*, Academic Press, Inc., New York.

Kaneda, Y., Uchida, T., Kim, J., Ishiura, M. and Okada, Y. (1987) The improved efficient method for introducing macromolecules into cells using HVJ (Sendai virus) liposomes with gangliosides. *Exp. Cell Res.*, **173**, 56–69.

Kaneda, Y., Iwai, K. and Uchida, T. (1989a) Increased expression of DNA cointroduced with nuclear protein in adult rat liver. *Science*, **243**, 375–9.

Kaneda, Y., Iwai, K. and Uchida, T. (1989b) Introduction and expression of the human insulin gene in adult rat liver. *J. Biol. Chem.*, **264**, 12126–9.

Kato, K., Nakanishi, M., Kaneda, Y., Uchida, T. and Okada, Y. (1991a) Expression of hepatitis B virus surface antigen in adult rat liver. *J. Biol. Chem.*, **266**, 3361–4.

Kato, K., Kaneda, Y., Sakurai, M., Nakanishi, M. and Okada, Y. (1991b) Direct injection of hepatitis B virus DNA into liver induced hepatitis in adult rats. *J. Biol. Chem.*, **266**, 22071–4.

Koshizaka, T., Hayashi, Y. and Yagi, K. (1989) Novel liposomes for efficient transfection of β-galactosidase gene into COS-1 cells. *J. Clin. Biochem. Nutr.*, 7, 185–92.

Lamb, R.A. (1993) Paramyxovirus fusion: a hypothesis for changes. *Virology*, **197**, 1–11.

Lipsky, N.G. and Pagano, R.E. (1985) Intracellular translocation of fluorescent sphingolipids in cultured fibroblasts: endogenously synthesized sphingomyelin and glucocerebroside analogues pass through the Golgi apparatus en route to the plasma membrane. *J. Cell. Biol.*, **100**, 27–34.

Losordo, D.W., Pickering, J.G., Takeshita, S. *et al.* (1994) Use of the rabbit ear artery to serially assess foreign protein secretion after site-specific arterial gene transfer *in vivo*. *Circulation*, **89**, 785–92.

Mizuno, M., Yoshida, J., Sugita, K. *et al.* (1990) Growth inhibition of glioma cells transfected with the human β-interferon gene by liposomes coupled with monoclonal antibody. *Cancer Res.*, **50**, 7826–9.

Morishita, R., Gibbons, G.H., Kaneda, Y., Ogihara, T. and Dzau, V.J. (1993a) Novel and effective gene transfer technique for study of vascular renin angiotensin system. *J. Clin. Invest.*, **91**, 2580–5.

Morishita, R., Gibbons, G.H., Ellison, K.E. *et al.*, (1993b) Single intraluminal delivery of antisense cdc 2 kinase and proliferating-cell nuclear antigen oligonucleotides results in chronic inhibition of neointimal hyperplasia. *Proc. Natl Acad. Sci. USA*, **90**, 8474–8.

Morishita, R., Gibbons, G.H., Ellison, K.E. *et al.* (1994) Intimal hyperplasia after vascular injury is inhibited by antisense cdk 2 kinase oligonucleotides. *J. Clin. Invest.*, **93**, 1458–64.

Morishita, R., Gibbons G.H., Ellison K.E. *et al.* (in press) Evidence for direct local effect of angiotensin in vascular hypertrophy; *in vivo* gene transfer of angiotensin converting enzyme. *J. Clin. Invest.*,

Nabel, G.J., Nabel, E.G., Yang, Z-Y. *et al.* (1993) Direct gene transfer with DNA–liposome complexes in melanoma: expression, biologic activity, and lack of toxicity in humans. *Proc. Natl Acad. Sci. USA*, **90**, 11307–11.

San, H., Yang, Z-Y., Pompili V.J. *et al.* (1993) Safety and short-term toxicity of a novel cationic lipid formulation for human gene therapy. *Human Gene Ther.*, **4**, 781–8.

Schaefer-Ridder, M., Wang, Y. and Hofschneider, P.H. (1982) Liposomes as gene carriers: efficient transformation of mouse L cells by thymidine kinase gene. *Science*, **215**, 166–8.

Singer, S.J. and Nicolson, G.L. (1972) The fluid mosaic model of the structure of cell membranes. *Science*, **175**, 720–31.

Smith, J.G., Walzem R.M. and German J.B. (1993) Liposomes as agent of DNA transfer. *Biochim. Biophys. Acta*, **1154**, 327–40.

Strauss, W.M. and Jaenisch, R. (1992) Molecular complementation of a collagen mutation in mammalian cells using yeast artificial chromosomes. *EMBO J.*, **11**, 417–22.

Tomita, N., Higaki, J., Morishita, R. *et al.* (1992) Direct *in vivo* gene introduction into rat kidney. *Biochem. Biophys. Res. Commun.*, **186**, 129–34.

Tomita, N., Higaki, J., Kaneda, Y. *et al.* (1993) Hypertensive rats produced by *in vivo* introduction of the human renin gene. *Circ. Res.*, **73**, 898–905.

Tsuneoka, M., Imamoto, N.S. and Uchida, T. (1986) Monoclonal antibody against non-histone chromosomal protein high mobility group 1 co-migrates with high mobility 1 into the nucleus . *J. Biol. Chem.*, **261**, 1829–34.

Weiner, A.L. (1989) Liposomes as carriers for polypeptides. *Adv. Drug Deliv. Rev.*, 3, 307–41.

5

Intramuscular injection of plasmid DNA

DOMINIC J. WELLS

5.1 INTRODUCTION

Striated muscle has recently been shown to be an excellent target tissue for gene transfer *in vivo*. Both skeletal and cardiac muscle are post-mitotic highly stable structures; consequently gene transfer into these tissues produces long-lasting gene expression. A number of gene transfer systems work effectively with skeletal muscle, including retroviral gene vectors (Dunckley *et al*., 1993) and adenoviral vectors (Ragot *et al*., 1993) but the simplest method is the direct injection of naked plasmid DNA into the muscle. Gene uptake and expression following *in vivo* transfection with plasmid DNA is substantially higher in striated muscle compared with any other tissue (Acsadi *et al*., 1991a). This methodology, first reported by Wolff and co-workers (1990), has attracted considerable interest as a means of *in vivo* gene therapy for a wide range of diseases. Skeletal muscle can serve as a platform for systemic gene expression; for example the production of Factor VIII for the treatment of hemophilia A, as well as being the target for treatment, as in the case of Duchenne muscular dystrophy. Plasmid DNA can also be used as a means of delivering genetic vaccination, an approach that potentially can give rise to effective long-lasting humoral and cell-mediated immunity following a single treatment (see Chapter 18).

Application of this simple, cheap and safe plasmid DNA methodology to the treatment of human disease is currently constrained by the low efficiencies and poor reproducibility achieved to date. This chapter discusses the various factors affecting the efficiency of naked DNA gene transfer, the

Molecular and Cell Biology of Human Gene Therapeutics
Edited by George Dickson
Published in 1995 by Chapman & Hall. ISBN 0 412 62550 4

mechanisms involved in the uptake of plasmid DNA and the longevity of subsequent expression. Finally the potential applications of this methodology are briefly reviewed.

5.2 PLASMID PREPARATION AND TRANSFER

5.2.1 Construct design

In rodents the best expression in short-term studies has been obtained using one of three viral promoters: cytomegalovirus promoter/enhancer (CMV), Rous sarcoma virus long terminal repeat (RSV) and simian virus 40 (SV40). CMV is the most active in our hands and was chosen as the most effective in a recent study of a wide range of promoters (Manthorpe *et al.*, 1993). There are considerable differences when the promoter activity is examined in the context of long-term expression from injected plasmids. Surprisingly, expression from the RSV promoter actually appears to increase over time (Wolff *et al.*, 1992a). Davis *et al.* (1993a) have shown that RSV continues to drive expression at an increasing rate over long periods but expression from the SV40 promoter peaks at day 3 following injection and then falls rapidly to very low levels. A CMV promoter/intron A construct produced threefold higher luciferase expression than a similar construct driven by the RSV promoter at 7 days following intramuscular injection, but expression from the CMV promoter then declined to about 1% of the maximal value at 120 days after injection (Manthorpe *et al.*, 1993). Other viral promoters and a range of eukaryotic promoters have been used successfully for *in vivo* gene transfer via intramuscular plasmid injection (for examples see Table 5.1). In general, eukaryotic promoters produce substantially less expression than viral promoters.

A variety of genes have been expressed following direct injection including the reporter genes chloramphenicol acetyl transferase (CAT), bacterial β-galactosidase (β-*gal*) and luciferase (luc). Quantification of luciferase expression provides the best system for accurate assessment of expression levels over a very wide range. In contrast, β-*gal* histochemistry allows enumeration of the number of fibers that take up and express the injected plasmid (Figure 5.1). Other genes expressed include recombinant human dystrophin (Acsadi *et al.*, 1991b; Dunckley *et al.*, 1994) and a variety of immunogenic peptides (see reviews by Ulmer *et al.*, 1993 and Chapter 18).

5.2.2 Quality of DNA

The majority of studies have used plasmid DNA prepared by alkaline lysis and double cesium chloride gradient centrifugation. However, several authors have reported the successful use of plasmid prepared by the quicker resin-based purification procedures, such as that produced by Qiagen

Table 5.1 Promoters used in plasmid gene transfer into mammalian muscle *in vivo*

Promoter source	*Reference*
Viral	
Rous sarcoma virus	*Wolff *et al.* (1990, 1991)
	Acsadi *et al.* (1991a,b)
	Lin *et al.* (1990); Kitsis *et al.* (1991)
Cytomegalovirus	*Acsadi *et al.* (1991a,b)
	Davis *et al.* (1993b)
	Manthorpe *et al.* (1993)
Simian virus 40	Wells and Goldspink (1992); Wells (1993)
	Davis *et al.* (1993a)
Maloney murine leukemia virus	Dunckley *et al.* (1994)
	Gal *et al.* (1993)
Herpes simplex virus	Vitadello *et al.* (1994)
Murine sarcoma virus	Acsadi *et al.* (1991)
	von-Harsdorf *et al.* (1993)
SRa of HTLV-1	Manthorpe *et al.* (1993)
Adenovirus promoter	Manthorpe *et al.* (1993)
Eukaryotic	
Phosphoglycerol kinase	Acsadi *et al.* (1991a)
Albumin	Acsadi *et al.* (1991a)
α-Cardiac myosin heavy chain	Kitsis *et al.* (1991)
	Buttrick *et al.*(1992)
β-Cardiac myosin heavy chain	LeBlanc *et al.* (1992)
	Edwards *et al.* (1992)
	von-Harsdorf *et al.* (1993)
α_1 anti-trypsin	LeBlanc *et al.* (1992)
Cardiac troponin C	Parmacek *et al.*(1992)
Muscle creatine kinase	Vincent *et al.* (1993)
	Manthorpe *et al.* (1993)
DHF*R*	Manthorpe *et al.*(1993)
Skeletal actin	Manthorpe *et al.* (1993)
α-Globin	Manthorpe *et al.* (1993)
Muscle-specific dystrophin	D.J. Wells *et al.*(unpublished data)
Myoglobin	Bassel-Duby *et al.* (1993)

* This promoter has been used in many studies, only a few of which are cited in this table.

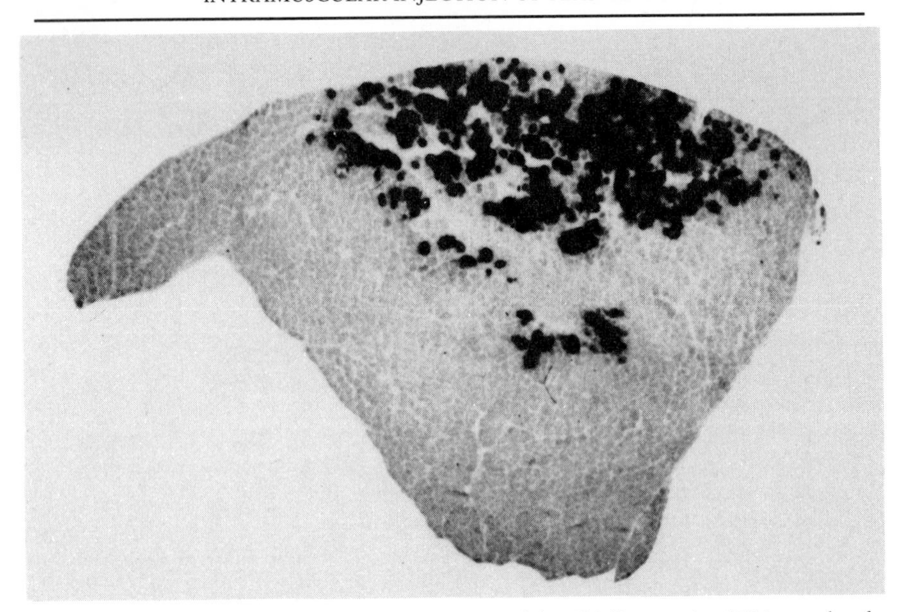

Figure 5.1 Photomicrograph of a cross-section of the tibialis anterior (TA) muscle of a 10-week-old C57Bl/10 male mouse showing high reporter gene expression in a large number of myofibers. The muscle was pre-treated with 50 μl of a 1.2% solution of barium chloride 5 days before the injection of 50 μg of pCMV-LacZ in 50 μl of phosphate-buffered saline. The animal was sacrificed 7 days later and the muscle stained by conventional histochemical techniques (e.g. Wells, 1993).

(Wells and Goldspink, 1992; Davis and Jasmin, 1993; Fazio *et al.*, 1994). The latter system has the advantage that there is no risk of preparations being contaminated with ethidium bromide, a known mutagen. Additionally, trace amounts of cesium chloride can induce myofiber necrosis. Plasmid DNA prepared by polyethylene glycol precipitation appears to generate substantial scarring in the heart when compared with hearts injected with plasmid DNA prepared by alkaline lysis, and so does not appear to be a useful method for preparing plasmid DNA for *in vivo* use (Gal *et al.*, 1993).

Injection of linear DNA produces less efficient gene expression than covalently closed-circular DNA (Buttrick *et al.*, 1992; Wolff *et al.*, 1992b). Manthorpe and colleagues (1993) have reported similar results with a 10–20-fold decrease in gene expression following digestion with a restriction enzyme. Our unpublished observations have noted that efficiency of gene expression appears to be inversely proportional to the ratio of nicked to covalently closed plasmid but Wolff and colleagues (1992b) noted no difference in expression following injection of open-circular compared with injection of covalently closed-circular plasmid DNA.

While considerable variation has been observed between individual

Table 5.2 Relationship between DNA injected and subsequent expression

Tissue (species)	*Dose (μg)*	*Result*	*Reference*
Skeletal (mouse)	10–100	Linear increase	Wolff *et al.* (1990)
Skeletal (monkey)	100–500	Linear increase	Jiao *et al.* (1992)
Skeletal (mouse)	0.1–50	Linear increase after sucrose pre-treatment	Davis *et al.* (1993a)
Skeletal (mouse)	1–500	No increase after 25 μg	Manthorpe *et al.* (1993)
Heart (rabbit)	10–100	Linear increase	Gal *et al.* (1993)
Heart (dog)	10–200	Linear increase	von-Harsdorf *et al.* (1993)

animals, much of which is likely to be due to differences in the precise delivery of each injection, many authors have reported considerable batch-to-batch variation for the same plasmid construct and method of preparation (e.g. Wolff *et al.*, 1991; Manthorpe *et al.*, 1993). Batch-to-batch variation can be substantial, an approximately 12-fold difference in average values is reported by Manthorpe and colleagues (1993). The same group undertook numerous studies of intrabatch variation, recording a maximum 15-fold difference in experimental averages on different days with the same batch but a less than twofold difference between experiments performed with the same batch on the same day. The origin of these differences is not clear but to avoid experimental artefacts comparison of different constructs should be made over several batches of plasmid with experiments being performed concurrently. This will be particularly important when using plasmid injection to analyze promoter activity *in vivo*.

5.2.3 Quantity of DNA

There are varied reports regarding the relationship between the quantity of DNA injected into striated muscle and the resultant gene expression (Table 5.2). The consensus of these studies demonstrates an approximately linear relationship between the quantity of DNA administered and foreign gene expression in both skeletal and cardiac muscle. The majority of other studies have used 50–100 μg of plasmid DNA per injection. It should be noted that in nearly all cases relatively small plasmids (4–7 kb) have been used and it is not yet clear what effect plasmid size will have on uptake by striated muscle. Access of high-molecular weight DNA to the muscle membrane is restricted by the external lamina (Wolff *et al.*, 1992b). However, plasmids containing the full-length dystrophin cDNA (16.7 kb) have been successfully transferred into *mdx* mouse muscle using intramuscular plasmid injection (Acsadi *et al.*, 1991b). We are currently examining the effect of plasmid size on the efficiency of DNA uptake by skeletal muscle.

5.2.4 Injection vehicle

A comprehensive study undertaken by Wolff and colleagues (1991) examined the effects of different solvents on subsequent expression of plasmid DNA injected intramuscularly. They demonstrated that DNA in isotonic saline solutions produced the highest expression. Plasmid injected in isotonic solutions lacking sodium chloride, hypertonic or hypotonic solutions produced lower levels of gene expression. Manthorpe *et al.* (1993) have reported very low gene expression when plasmid DNA dissolved in water is injected into muscle compared to similar injections with the DNA dissolved in normal saline. Consequently, the majority of recent studies have used normal saline or phosphate-buffered saline as the injection vehicle for the DNA. More complex solutions appear to offer no benefit.

Gal *et al.* (1993) have reported a significant increase in reporter gene expression when injecting a standard quantity of DNA in increasing volumes of solution into the myocardium of rabbits. Davis and colleagues (1993a) reported a similar response. In contrast both Wolff *et al.* (1991) and Manthorpe *et al.* (1993) found no statistically significant effect of volume on the level of gene expression. While large-volume injections may produce more reproducible, and in some cases higher, gene expression, successful gene transfer has been achieved by the implantation of pellets of plasmid DNA into striated muscle (Wolff *et al.*, 1991). In one study the multiple implantation of plasmid DNA pellets was more efficient than the injection of DNA in normal saline, as judged by luciferase expression (Jiao *et al.*, 1992).

The addition of a number of pharmacologically active drugs to the DNA solution failed to alter the level of plasmid gene expression (Manthorpe *et al.*, 1993). Agents tested included three muscle depolarizing drugs (acetylcholine, succinylcholine, potassium chloride), the metabolic stimulant ATP and bacterial endotoxin. Prevention of hematoma formation with epinephrine (adrenaline) decreased gene expression but the result was not significant. The presence of EDTA in the solution did not affect DNA expression in this study, in marked contrast to the strong inhibitory effect observed by Wolff and co-workers (1991).

5.2.5 Needle size and injection technique

Injection of plasmid DNA into skeletal muscle is a very simple affair. A narrow-gauge needle, 25–30-gauge, together with a small-volume syringe is used to deliver the concentrated plasmid DNA solution into the body of the muscle. Initially injections were made following a skin incision in order to accurately localize the site of injection (e.g. Wolff *et al.*, 1990, 1991) but subsequently many investigators have adopted a percutaneous approach that avoids any inflammatory responses attributable to the surgery and also ensures that connective tissue structures remain intact helping to retain the plasmid DNA within the muscle (e.g. Wells and Goldspink, 1992; Davis *et al.*, 1993a; Manthorpe *et al.*, 1993). In the majority of studies the animals have been anesthetized, mostly with long-acting injectable preparations that produce minimal muscle movement for up to several hours (e.g. Avertin, sodium pentobarbitol or a Hypnorm/Hypnovel mixture). In contrast, Manthorpe and colleagues (1993) injected restrained awake mice. They reported that relaxation of the muscle with metophane (a short-acting inhalation anesthetic) did not significantly affect gene expression but they did not examine the effect of a longer-acting anesthetic as used in many other studies. Stimulation of muscle contraction in anesthetized mice produced reductions in gene expression in studies carried out by Wolff and co-workers (1991). Massage of muscles following injection also markedly reduced reporter gene expression (Davis *et al.*, 1993a). In contrast, injections into the myocardium must by necessity take place into an active muscle mass. Although it appears that long-acting anesthesia may not be essential, intramuscular injections of large volumes are inherently painful and consequently wherever possible these procedures should be carried out with the use of appropriate anesthetics or analgesics.

In most cases injection into cardiac muscle has been undertaken via a thoracotomy (e.g. Lin *et al.*, 1990; Kitsis *et al.*, 1991, Acsadi *et al.*, 1991a; Vincent *et al.*, 1993) but plasmid gene transfer has been successfully achieved in Yucatan micropigs using a percutaneous approach (Gal *et al.*, 1993). In the latter case accurate localization of the injection was achieved using fluoroscopy. Treatment of the diaphragm with plasmid DNA is also an invasive procedure requiring a laparotomy (Davis and Jasmin, 1993).

Needle size does not appear to be a critical factor (Wolff *et al.*, 1991) and no significant effects were noted with needles ranging from 18G to 30G (Manthorpe *et al.*, 1993). There is some dispute over the effect of intentional muscle trauma; Wolff and co-workers (1991) reported a reduced gene expression in damaged muscles whereas Manthorpe *et al.* (1993) recorded no difference. This disparity may be the result of different protocols, muscles being damaged 2 hours prior to injection (Manthorpe *et al.*, 1993) or at the time of injection (Wolff *et al.*, 1991). None of the studies has reported any effect from varying the rate of injection of plasmid solutions into skeletal or cardiac muscle. Accurate localization of the plasmid delivery appears much more important than the rate of delivery.

From studies of plasmid injections into the mouse rectus femoris muscle, using Hoescht 33258 to visualize the DNA, Wolff and co-workers (1992b) concluded that plasmid distribution was relatively uniform and not a limitation to gene transfer. In contrast, Davis *et al.* (1993a), working with injections of Indian ink into the mouse tibialis anterior muscle, concluded that poor distribution was a major factor determining the success of gene transfer for individual injections. Gene transfer into striated muscle appears to be an inherently variable process. Substantial variation in expression has been recorded by a number of authors [standard error of 25–60% of the mean (Wolff *et al.*, 1991); average standard error of 30% (Manthorpe *et al.*, 1993)]. Pre-treatment of muscles with a sucrose solution appears to reduce the extent of this variation, probably by aiding a more uniform distribution of the injected DNA (Davis *et al.*, 1993a).

5.3 PHYSIOLOGICAL STATUS OF RECIPIENT MUSCLE

5.3.1 Tissues and species

Uptake and expression of plasmid DNA has been documented for both skeletal and cardiac muscle of rodents, skeletal muscle and cardiac muscle of various other mammals, and the skeletal muscle of fish (Table 5.3). Intramuscular injection of plasmid DNA has also been carried out in a number of other species (ferrets, cattle and chickens) during studies of the potential for genetic vaccination (for reviews, see Ulmer *et al.*, 1993 and Chapter 18). The physiological status of the recipient can markedly affect the uptake and expression of plasmid DNA. The age of the treated animal has a significant effect on the success of plasmid gene transfer in mice, with younger animals exhibiting significantly higher expression than older animals (Wells and Goldspink, 1992) (Figure 5.2a). This higher expression in young animals is matched by a greater number of fibers exhibiting gene expression (Figure 5.2b). Hence the increased expression appears to be due to increased uptake and retention of plasmid DNA in younger myofibers. Additionally, young male mice exhibited higher gene expression than females of the same age (Wells and Goldspink, 1992). As it seems likely that age, and perhaps sex, might have a similar influence in other species, interspecies comparisons should only be made with individuals of similar physiological condition. For example, the experiments of Jiao *et al.* (1992) suggested that primate muscle shows substantially reduced uptake and expression of plasmid DNA compared with rodent muscle. However, the comparison was made between adult full-sized primates and young rapidly growing mice and so differences may have been due, at least in part, to age-related rather than species-related differences in expression. More unfortunately, many reports provide insufficient information to allow any assessment of the contribution of the animal age or sex to the final experimental results. The only other study to examine the effect of age did so comparing 5 and 15 month old animals (Manthorpe *et al.*, 1993) and not surprisingly observed no difference between the two groups of adults. Immature muscles are also capable of taking up and expressing plasmid DNA *in vivo*, as demonstrated by reporter gene activity after plasmid injection into the muscle of neonatal rats (Wolff *et al.*, 1991).

Some differences in gene expression have been linked to the strain of mouse used. The *mdx* dystrophic mouse mutant exhibits greater gene expression than the original C57Bl/10 (B10) strain (Wells, 1993) (Figure 5.3a). Danko and colleagues (1994) also recorded differences in gene expression following induced regeneration in a variety of strains (see below). In contrast, Manthorpe *et al.* (1993) observed no difference in expression in normal Balb/c compared with C57Bl/6 muscles.

5.3.2 Regeneration

The presence of dystrophic or chemically induced (bupivacaine) muscle regeneration has marked effects on successful gene transfer, with increased expression being observed in regenerating muscle (Wells, 1993) (Figure 5.3a). Other workers have also reported improved gene transfer into regenerating muscle (Davis *et al.*, 1993b; Danko *et al.*, 1994; Vitadello *et al.*, 1994). Foreign gene expression can be increased approximately 80 times in regenerating rat soleus compared with control (Vitadello *et al.*, 1994) and two to six times control values in regenerating rat quadriceps (Danko *et al.*, 1994). Differences between the two studies might relate to the age of the animals tested. In regenerating mouse tibialis anterior muscle expression is increased five to ten times the control values (Davis *et al.*, 1993b; Wells, 1993) and 10–40 times control values in regenerating mouse quadriceps (Danko *et al.*, 1994). In all studies the best time for plasmid injection was in the range 3–7 days following induction of regeneration, with 5 days post-treatment as the optimal in most cases. Very high transfection efficiencies can be achieved following regeneration with up to 21–25% of muscle fibers in the mouse tibialis anterior muscle showing β-galactosidase activity after injection of the CMV-LacZ plasmid (Davis *et al.*, 1993b; D.J. Wells, unpublished data) (Figure 5.1).

The choice of myotoxic agent significantly affects foreign gene expression, increases being proportional to the amount of the muscle undergoing regeneration. This increased expression is linked to an increased number of fibers taking up and expressing the plasmid, similar to the relationship observed between mouse age and the degree of gene expression (Figure 5.3b). Good results have been obtained using bupivacaine (Wells, 1993; Danko *et al.*, 1994; Vitadello *et al.*, 1994), cardiotoxin (Davis *et al.*, 1993b) and barium chloride (D.J. Wells, unpublished data) (Figure 5.3c). Comparative data dealing with some of these agents can be found in Danko *et al.* (1994). In our experience the use of bupivacaine carries a higher risk of toxicity-related problems in mice than does the use of barium chloride (D.J. Wells, unpublished data) or cardiotoxin (H. Davis, personal communication). The increase in gene expression associated with regeneration appears to decline over time, possibly as a consequence of declining transcription/translation activity (Danko *et al.*, 1994) or as a result of destruction of transfected fibers by elements of the immune system

Table 5.3 Tissue, species, age and sex of experimental animals injected with plasmid DNA

Reference	*Tissue*	*Species (strain)*	*Age*	*Sex*
Lin *et al.* (1990)	Cardiac muscle	Rat (SD)	6–11 weeks (250 g)	?
Wolff *et al.* (1990)	Skeletal muscle	Mouse (Balb/c)	5–6 weeks	Mixed
Acsadi *et al.* (1991a)	Skeletal muscle	Mouse (mdx)	4–6 weeks	?
Acsadi *et al.* (1991b)	Skeletal muscle	Mouse (Balb/c)	?	?
	Skeletal/Cardiac muscle	Rat (SD)	Adult	?
	Cardiac muscle	Rat (nude SD)	?	?
Hansen *et al.* (1991)	Skeletal muscle	Carp	Young (10 cm)	?
	Skeletal muscle	Carp	Old (20 cm)	?
Kitsis *et al.* (1991)	Cardiac muscle	Rat (Wistar)	Adult	Female
Wolff *et al.* (1991)	Skeletal muscle	Mouse (Balb/c)	6 weeks	Mixed
	Skeletal muscle	Rat	5–7 weeks	?
	Skeletal muscle	Rat	1 day	?
Buttrick *et al.* (1992)	Cardiac muscle	Rat	Adult	?
Jiao *et al.* (1992)	Skeletal muscle	Rhesus monkey	5–11 years	?
	Skeletal muscle	Cat	?	?
	Skeletal muscle	Rat	?	?
Parmacek *et al.* (1992)	Cardiac muscle	Rat (SD)	6 weeks (250 g)	?
Wells and Goldspink (1992)	Skeletal muscle	Mouse (C57Bl/10)	2–24 weeks	Male
	Skeletal muscle	Mouse (C57Bl/10)	2–24 weeks	Female
Wolff *et al.* (1992a)	Skeletal muscle	Mouse	6–10 weeks	?
Wolff *et al.* (1992b)	Skeletal muscle	Mouse (Balb/c)	?	?
	Skeletal muscle	Rhesus monkey	?	?

Table 5.3 *Continued*

Reference	*Tissue*	*Species (strain)*	*Age*	*Sex*
Edwards *et al.* (1992)	Cardiac muscle	Rat	Adult	?
Davis *et al.* (1993a)	Skeletal muscle	C57Bl/6	6–7 weeks	Male
Davis *et al.* (1993b)	Skeletal muscle	C57Bl/6	5–6 weeks	Male
Davis and Jasmin (1993)	Skeletal muscle	C57Bl/6	6–7 weeks	Male
von-Harsdorf *et al.* (1993)	Skeletal/Cardiac muscle	Dog	Adult	Mixed
Gal *et al.* (1993)	Cardiac muscle	Rabbit (NZ White)	?	?
	Cardiac muscle	Pig (Yucatan)	?	?
Manthorpe *et al.* (1993)	Skeletal muscle	Mouse (Balb/c)	4–12 weeks 5 and 15 months	Female
	Skeletal muscle	Mouse (C57Bl/6)	4–12 weeks	Female
Vincent *et al.* (1993)	Skeletal/Cardiac muscle	Rat (SD)	8 weeks (350 g)	Male
Wells (1993)	Skeletal muscle	Mouse (C57Bl/10)	8, 11, 14 weeks	Male
	Skeletal muscle	Mouse (mdx)	8 weeks	Male
Bassel-Duby *et al.* (1993)	Cardiac muscle	Rat (SD)	8 weeks	Male
Danko *et al.* (1994)	Skeletal muscle	Mouse (C57Bl/6)	?	?
	Skeletal muscle	Mouse (Balb/c)	?	?
	Skeletal muscle	Mouse (ICR)	?	?
	Skeletal muscle	Mouse (mdx)	?	?
	Skeletal muscle	Rat (SD)	?	?
Vitadello *et al.* (1994)	Skeletal muscle	Rat (Wistar)	Adult (220–250 g)	Male
Fazio *et al.* (1994)	Skeletal muscle	Rat (Wistar)	8 weeks	?
	Skeletal muscle	Rat (Yoshida)	8 weeks	?

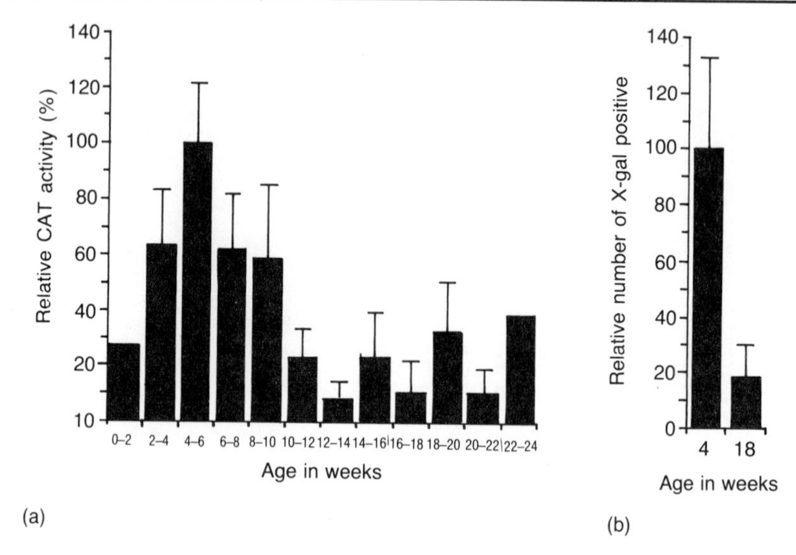

Figure 5.2 (a) The quadriceps muscles of male and female C57Bl/10 mice of various ages were injected with 100 μg of pSV40-CAT in 100 μl of 20% sucrose. CAT activity was assessed from muscle extracts prepared 72 hours after injection and is plotted normalized to the level of expression at 4–6 weeks of age. T-bars represent the standard error of the mean. Expression is significantly lower in the 10–22-week-old groups compared with the 4–6-week-old group ($P<0.05$). Modified from Wells and Goldspink (1992). (b) 4-week-old and 18-week-old C57Bl/10 male mice were injected with 50 μg of pRSV-LacZ in 50 μl of PBS into the tibialis anterior muscle. There was significantly higher uptake and expression of the RSV-LacZ plasmid in the younger animals when the muscles were examined one week later ($P<0.05$).

(Vitadello *et al.*, 1994). In the latter case, foreign gene expression was still 10–20 times higher in the regenerated muscle compared with the control after 60 days. Immune system-mediated destruction of transfected cells was also blamed for the rapid loss of foreign gene expression in rat hearts, as nude rats and cyclosporin-treated rats maintained the gene expression in cardiac muscle (Acsadi *et al.*, 1991a). However, other studies have not observed a similar rapid loss of gene expression in hearts injected with plasmid DNA (e.g. Leinwand and Leiden, 1991; Leiden and Barr, 1994).

The increased gene expression in regenerating muscle may be due to increased uptake of the plasmid, increased expression of the DNA or a combination of both. Histochemical staining for β-galactosidase suggests that uptake is enhanced in regenerating muscle (Davis *et al.*, 1993a; Wells, 1993). Alternatively, the increased synthetic activity associated with the regenerating fibers might induce a greater degree of transcription and/or translation of the plasmid DNA along with endogenous genes, and this

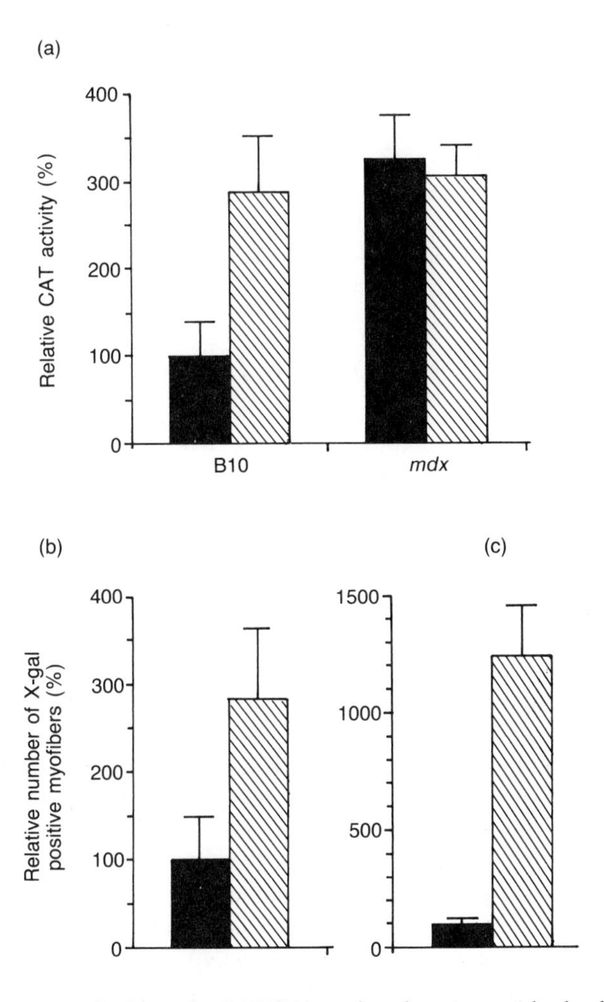

Figure 5.3 (a) 8-week-old male C57Bl/10 and *mdx* mice, with the left tibialis anterior muscle pre-treated with 50 μl of bupivacaine 5 days earlier, were injected with 50 μg of pSV40-CAT in 50 μl of phosphate-buffered saline. A significant difference in CAT activity was observed between the untreated muscles (solid bars) of the C57Bl/10 and *mdx* mice ($P<0.05$). CAT expression increased with bupivacaine pre-treatment (hatched bars) in the C57Bl/10 mice but not in the *mdx* mice. Modified from Wells (1993). (b and c) 11-week-old C57Bl/10 male mice, with the left tibialis anterior muscle pre-treated 5 days earlier with (b) 50 μl of bupivacaine or (c) 50 μl of a 1.2 solution of barium chloride in normal saline, were injected with 50 μg of pRSV-LacZ in 50 μl of phosphate-buffered saline. A significant increase in the number of fibers expressing β-galactosidase were observed in the pre-treated muscle (hatched bar) compared with the contralateral control (solid bar; $P<0.05$). Pre-treatment of the muscle with barium chloride produced a greater increase in gene expression than with the use of bupivacaine.

could show up fibers that take up DNA but fail to express it under normal conditions. Even those fibers that are not regenerating may have increases in synthesis associated with the higher concentration of growth factors in the local environment. Such explanations might well account for the greater fiber transfection and foreign gene expression detected in rapidly growing young mice (Wells and Goldspink, 1992). Plasmid distribution within the muscle and access to the muscle membrane is likely to be improved in regenerating muscle associated with some disruption of the extracellular matrix. Connective tissue barriers to plasmid distribution may also be less impenetrable in younger animals.

5.4 THE MECHANISM OF PLASMID UPTAKE BY MYOFIBERS

Skeletal and cardiac muscle have a greater ability to take up and express plasmid DNA compared with other tissues (Acsadi *et al.*, 1991a). The poor results with other tissues suggest *in vivo* uptake of DNA may be linked to unique features associated with striated muscle. Similarly, differentiated skeletal muscle cells (myotubes) take up and express naked plasmid DNA more efficiently than other cell types *in vitro* (Wolff *et al.*, 1992b; Dowty and Wolff, 1994). This is probably associated with an energy-dependent active transport process as uptake is inhibited at 4°C both in myotubes and muscle explants (Wolff *et al.*, 1992b). Electron microscopy examination of the distribution of colloidal gold complexed plasmid DNA showed no evidence of receptor-mediated endocytosis or pinocytosis (Wolff *et al.*, 1992b). The addition of hydroxychloroquine, an agent known to release substances internalized by cells through the endocytic pathway, did not increase gene expression but rather actually inhibited expression (Manthorpe *et al.*, 1993). The electron microscopic studies demonstrate that plasmid DNA is associated with T tubules and caveolae, suggesting that these structures may play a role in the uptake of foreign DNA. Furthermore, it was not possible to remove the injected DNA by washing unfixed tissues, which may indicate specific interactions between DNA and these structures (Wolff *et al.*, 1992b). Caveolae are associated with potocytosis, a mechanism for cellular uptake of low-molecular-weight molecules. Plasmid DNA may be able to exploit this mechanism by passing through the small transporter pore. If so, the size restriction might explain why the colloidal gold complexed plasmid DNA was not observed inside the muscle fibers (Wolff *et al.*, 1992b). Finally, once inside the myofiber the DNA must translocate to the nucleus before the foreign genes can be expressed. Preliminary *in vitro* studies reported by Dowty and Wolff (1994) suggest that translocation of cytoplasmically injected DNA may be enhanced in myotubes compared with all other cell types.

An alternative to the active transport hypothesis is that plasmid DNA may enter myofibers by gross membrane disruptions. However, protocols that might be expected to increase myofiber damage, for example the injection of depolarizing agents or deliberate muscle damage at the time of injection, do not increase uptake and expression of plasmid DNA (sections 5.2.4 and 5.2.5). Additionally, transfected fibers are often detected some distance away from the site of injection. Osmotic effects do not appear to be involved as injection of DNA in hypertonic solutions gives rise to lower expression than the use of normal saline (Wolff *et al.*, 1991).

In either case, before the plasmid DNA can reach the muscle membrane it must traverse the external lamina. This does not appear to be the rate-limiting step for plasmid DNA gene transfer in mice but does appear to present a barrier for positively charged macromolecules as well as other negatively charged molecules such as polyglutamic acid (Wolff *et al.*, 1992b). In other species the injected plasmid DNA may show a more restricted distribution (Jiao *et al.*, 1992; Wolff *et al.*, 1992b) although it is not clear if this effect is due to the species or the age of the animal. The thickness of connective tissue structures and differences in the volume of the perimyseal space in different animals may limit the amount of DNA that is available for contact with the myofibers (Dowty and Wolff, 1994). Interestingly, foreign gene expression in rats is 10–100 times higher in the heart compared with skeletal muscle (Lin *et al.*, 1990; Acsadi *et al.*, 1991a; Kitsis *et al.*, 1991) and may be a consequence of differences in the access of the plasmid DNA to the myofibers.

5.5 LONGEVITY OF PLASMID EXPRESSION

Foreign gene expression in mouse skeletal muscle can be detected as early as 2 minutes after plasmid DNA injection (Manthorpe *et al.*, 1993) and can persist for at least 19 months (Wolff *et al.*, 1992a). Foreign DNA longevity may be reduced in cardiac compared with skeletal muscle (Acsadi *et al.*, 1991a), although other studies have reported expression in cardiac tissues up to 6 months (Leinwand and Leiden, 1991; Gal *et al.*, 1993) and 8 months after injection (Leiden and Barr, 1994). As discussed earlier, the choice of promoter will be important in ensuring longevity of expression. The use of reporter genes such as bacterial β-galactosidase may attract the attention of the immune system and hence be responsible for the gradual decline in detectable gene expression over time. Further studies are required to fully evaluate the absolute longevity of plasmid expression in muscle, but it appears likely that gene expression can be life-long, albeit at reduced levels. This is a particularly attractive and initially unexpected consequence of gene transfer into striated muscle.

Plasmid DNA maintains its bacterial pattern of methylation which demonstrates that there is no replication of this foreign DNA in the muscle fibers (Wolff *et al.*, 1992a). Southern blot analysis has shown that the majority of the plasmid is present in the circular form with no evidence of integration

(Wolff *et al.*, 1990). In a more sensitive assay for the presence of integrated plasmid Wolff and colleagues cloned, digested and ligated cellular DNA extracted from injected muscle but could not detect any chromosomal/ plasmid junctions in over 1800 colonies examined (Wolff *et al.*, 1992a). Therefore it appears likely that plasmid DNA exists only as an episome in striated muscle although there remains a very small probability of occasional integration events. Hence plasmid gene transfer into muscle carries very little risk of insertional mutagenesis following treatment.

5.6 GENE THERAPY APPLICATIONS FOR INTRAMUSCULAR PLASMID GENE TRANSFER

Gene transfer into muscle via the injection of plasmid DNA offers an attractive method of gene therapy, not only for the treatment of muscle disorders such as Duchenne muscular dystrophy, but also as a site for the expression of systemically active products such as Factor VIII for the treatment of hemophilia A. The potential of this latter approach has recently been demonstrated by Fazio *et al.* (1994) who report the secretion of an apolipoprotein into the plasma of rats following intramuscular injection of naked DNA encoding the human apolipoprotein-E. The episomal nature of the plasmid minimizes the genetic risks associated with gene transfer yet long-term persistence and expression is possible in the stable environment provided by the muscle. With careful preparation the DNA should be free of potentially immunogenic protein contaminants. Indeed, no immune reactions or adverse side effects were noted in a study using non-human primates, even after repeated treatments (Jiao *et al.*, 1992). The simple preparation and administration of the plasmid DNA should ensure that this approach remains relatively cheap. While the efficiency of gene transfer into normal muscle is low, significant improvements can be made by injecting the plasmid into regenerating muscle. This improvement in efficiency is so marked that it has been claimed that plasmid DNA is superior to viral vectors for direct gene transfer into muscle (Davis *et al.*, 1993b). A further advantage of using muscle as a platform for systemic expression lies in the accessible nature and good regenerative capacity of the tissue as, in a worst-case scenario, it would be possible to resect the muscle containing the foreign genes if problems arose with the treatment.

The use of strong viral promoters does not appear to carry any significant risks and the poor uptake and expression of plasmid DNA by other tissues will limit possible problems associated with gene transfer into inappropriate tissues. Application is currently limited by the low efficiencies and high variability. Experiments with mice and rats have demonstrated some of the factors that influence the efficiency of gene uptake and expression, but it is not yet clear what problems may be encountered when scaling up this procedure for gene transfer into human muscle.

Recently, direct DNA injection has been used successfully as a means of vaccinating mice against influenza and a number of groups are now using this methodology for genetic vaccination in a wide range of species (see reviews by Ulmer *et al.*, 1993 and Chapter 18). Direct intramuscular plasmid injection can also be used to modulate humoral and cellular immune responses *in vivo* through the expression of cytokines (Raz *et al.*, 1993).

5.7 OTHER APPLICATIONS FOR INTRAMUSCULAR PLASMID GENE TRANSFER

The ability to easily examine gene expression and regulation *in vivo* without the need to construct new vectors provides an important extension to cell culture studies. While *in vitro* studies are an important first step in the examination of gene function, a monolayer of cells on a plastic dish cannot replicate the complex environment that cells encounter in the whole animal. Plasmid DNA can avoid some of the problems associated with other *in vivo* gene transfer systems, for example the absence of foreign proteins should avoid stimulation of the immune system. Gene transfer into muscle using plasmid DNA is analogous to the transient transfection assay *in vitro* as in both cases there is no integration of the transferred gene. However, in muscle the DNA is subject to a complex hormonal and neural environment and protein products can be examined in the context of an intact extracellular matrix. For example, thyroid hormone modulation of foreign gene transcription has been reported following injection of plasmid DNA carrying the α-cardiac myosin heavy chain promoter into rat myocardium (Kitsis *et al.*, 1991). Gene transfer by direct plasmid injection can in some cases substitute for the expensive and time-consuming production of transgenic animals, while still providing the more complex *in vivo* environment in which to study regulation of gene activity. Possibly this process of substitution could be extended to fetal studies by transuterine intramuscular injections of plasmid DNA.

Direct injection of plasmid DNA has been used to examine skeletal- and cardiac-specific control elements of striated muscle genes. Parmacek and coworkers (1992) have used plasmid gene transfer *in vivo* to define a cardiac-specific transcriptional unit in the slow/cardiac troponin C gene. Vincent and colleagues (1993) have used a similar approach to define important promoter and enhancer sequences that direct cardiac-specific expression in adult striated muscles. Edwards *et al.* (1992) and von-Harsdorf *et al.* (1993) have both examined the regulation of the β-cardiac myosin heavy chain promoter by plasmid gene transfer *in vivo*, and Bassel-Duby and colleagues (1993) have performed similar experiments with the myoglobin promoter. The intramuscular plasmid injection is a particularly useful technique for the study of cardiac gene regulation as no good permanent cell lines are available for *in vitro* work. An essential control in all such studies is the co-injection of a separate reporter gene construct to compensate for variations inherent in the injection process. It has been suggested that gene transfer into regenerating muscle can be used for promoter analysis (Vitadello *et al.*, 1993) but this approach is likely to give confusing results due to the rapid recapitulation of embryonic and neonatal gene expression.

Finally, this gene transfer technique can be used to examine the localization of products of recombinant human dystrophin cDNAs in *mdx* muscles *in vivo* (Acsadi *et al.*, 1991b) and we have recently examined the distribution of the products of additional recombinant dystrophin genes and have demonstrated independent localization of the C- and N-termini of dystrophin to the sarcolemma (Dunckley *et al.*, 1994). We also use direct intramuscular injections of plasmid DNA to test the integrity of constructs intended for expression in the muscles of transgenic mice, as an adjunct to tests performed *in vitro*.

5.8 CONCLUSIONS

Although there are considerable differences in the detail of the results of individual studies, a broad consensus can be drawn from this review of the factors affecting gene transfer via intramuscular injection of plasmid DNA. Maximal short-term expression can be obtained with the use of a strong viral promoter such as CMV, injecting moderate quantities of covalently closed circular plasmid DNA in a large enough volume of normal saline to cause marked muscle swelling. This expression may be increased if young, rapidly growing muscles are treated. Additionally, multiple injections may increase gene expression (Wolff *et al.*, 1991; Manthorpe *et al.*, 1993). For longer term strong expression the RSV LTR appears to be the promoter of choice.

Gene transfer by intramuscular injection of plasmid DNA is suitable for a wide variety of applications from *in vivo* analysis of the regulation of gene expression through to treatment of genetic diseases. The mechanism involved in the uptake of plasmid DNA by the myofibers remains poorly understood, although an active uptake process associated with caveolae currently appears to be the most likely (Wolff *et al.*, 1992b). Further work is required to examine barriers to the distribution of the DNA, the route of entry into the myofiber, and limits to the successful passage through the cytoplasm to the nucleus. An increased knowledge of limitations to gene expression imposed by different physiological states will be important when applying this gene transfer technology to mixed populations. Understanding the mechanisms involved may allow development of strategies for increasing uptake and expression of naked plasmid DNA, decreasing the variation associated with individual treatments and hence improving the prospects for clinical application of this attractively simple method of gene transfer into striated muscle.

REFERENCES

Acsadi, G., Jiao, S., Jani, A. *et al.* (1991a) Direct gene transfer and expression into rat heart *in vivo*. *New Biologist*, **3**, 71–81.

Acsadi, G., Dickson, G., Love, D.R. *et al.* (1991b) Human dystrophin expression in mdx mice after intra-muscular injection of DNA constructs. *Nature*, **352**, 815–18.

Bassel-Duby, R., Grohe, C.M., Jesson, M.E. *et al.* (1993) Sequence elements required for transcriptional activity of the human myoglobin promoter in intact myocardium. *Circ. Res.*, 73, 360–6.

Buttrick, P.M., Kass, A., Kitsis, R.N., Kaplan, M.L. and Leinwand, L.A. (1992) Behavior of genes directly injected into the rat heart *in vivo*. *Circ. Res.*, **70**, 193–8.

Danko, I., Fritz, J.D., Jiao, S., Hogan, K., Latendresse, J.S. and Wolff, J.A. (1994) Pharmacological enhancement of *in vivo* foreign gene expression in muscle. *Gene Therapy*, **1**, 114–21.

Davis, H.L. and Jasmin, B.J. (1993) Direct gene transfer into mouse diaphragm. *FEBS Lett.*, **333**, 146–50.

Davis, H.L., Whalen, R.G. and Demeneix, B.A. (1993a) Direct gene transfer into skeletal muscle *in vivo*: factors affecting efficiency of transfer and stability of expression. *Human Gene Ther.*, **4**, 151– 9.

Davis, H.L., Demeneix, B.A., Quantin, B., Coulombe, J. and Whalen, R.G. (1993b) Plasmid DNA is superior to viral vectors for direct gene transfer into adult mouse skeletal muscle. *Human Gene Ther.*, **4**, 733–40.

Dowty, M.E. and Wolff, J.A. (1994) Possible mechanisms of DNA uptake in skeletal muscle, in *Gene Therapeutics: Methods and Applications of Direct Gene Transfer*, (ed. J.A. Wolff), Birkhauser, Boston, pp. 82–98.

Dunckley, M.G., Wells, D.J., Walsh, F.S. and G. Dickson (1993) Direct retroviral-mediated transfer of a dystrophin minigene into mdx mouse muscle *in vivo*. *Hum. Mol. Genet.*, **2**, 717–23.

Dunckley, M., Piper, T., Wells, K.E., Wells, D.J. and Dickson, G. (1994) Independent localisation of dystrophin N- and C-terminal regions to the sarcolemma of mdx mouse myofibres *in vivo*. *J. Cell Sci.*, **107**, 1469–75.

Edwards, J.G., Bahl, J.J., Flink, I., Milavetz, J., Goldman, S. and Morkin, E. (1992) A repressor region in the human beta-myosin heavy chain gene that has a partial position dependency. *Biochem. Biophys. Res. Commun.*, **189**, 504–10.

Fazio, V.M., Fazio, S., Rinaldi, M. *et al.* (1994) Accumulation of human Apolipoprotein-E in rat plasma after *in vivo* intramuscular injection of naked DNA. *Biochem. Biophys. Res. Commun.*, **200**, 298–305.

Gal, D., Weir, L., Leclerc, G., Pickering, J.G., Hogan, J. and Isner, J.M. (1993) Direct myocardial transfection in two animal models: evaluation of parameters affecting gene expression and percutaneous gene delivery. *Lab. Invest.*, **68**, 18–25.

Hansen, E., Fernandes, K., Goldspink, G., Butterworth, P., Umeda, P. and Chang, K-C. (1991) Strong expression of foreign genes following direct injection into fish muscle. *FEBS Lett.*, **290**, 73– 6.

Jiao, S.S., Williams, P., Berg, R.K. *et al.* (1992) Direct gene transfer into nonhuman primate myofibres *in vivo*. *Human Gene Ther.*, **3**, 21–33.

Kitsis, R.N., Buttrick, P., McNally, E., Kaplan, M. and Leinwand, L.A. (1991) Hormonal modulation of a gene injected into rat heart *in vivo*. *Proc. Natl Acad. Sci. USA*, **88**, 4138–42.

LeBlanc, J.M., Kitsis, R.N., Buttrick, P.M. and Leinwand, L.A. (1992) Molecular genetic manipulation of cardiac myosin, in *Neuromuscular Development and Disease*, (eds A.M. Kelly and H.M. Blau), Raven Press, New York, pp. 223–37.

Leiden, J.M. and Barr, E. (1994) *In vivo* gene transfer into the heart, in *Gene Therapeutics: Methods and Applications of Direct Gene Transfer*, (ed. J.A. Wolff), Birkhauser, Boston, pp. 363–81.

Leinwand, L.A. and Leiden, J.M. (1991) Gene transfer into cardiac myocytes *in vivo*. *Trends Cardiovasc. Med.*, **1**, 271–6.

Lin, H., Parmacek, M.S., Morle, G., Bolling, S. and Leiden, J.M. (1990) Expression of recombinant genes in myocardium *in vivo* after direct injection of DNA. *Circulation*, **82**, 2217–21.

Manthorpe, M., Cornefert-Jensen, F., Hartikka, J. *et al.* (1993) Gene therapy by intramuscular injection of plasmid DNA: studies on firefly luciferase gene expression in mice. *Human Gene Ther.*, **4**, 419–31.

Parmacek, M.S., Vora, A.J., Shen, T., Barr, E., Jung, F. and Leiden, J.M. (1992) Identification and characterization of a cardiac-specific transcriptional regulatory element in the slow/cardiac troponin C gene. *Mol. Cell. Biol.*, **12**, 1967–76.

Ragot, T., Vincent, N., Chafey, P. *et al.* (1993) Efficient adenovirus-mediated transfer of a human minidystrophin gene to skeletal muscle of mdx mice. *Nature*, **361**, 647–50.

Raz, E., Watanabe, A., Baird, S.M. *et al.* (1993) Systemic immunological effects of cytokine genes injected into skeletal muscle. *Proc. Natl Acad. Sci. USA*, **90**, 4523–7.

Ulmer, J.B., Donnelly, J.J. and Liu, M.A. (1993) Polynucleotide vaccines. *Curr. Opin. Invest. Drugs*, **2**(9), 983–9.

Vincent, C.K., Gualberto, A., Patel, C.V. and Walsh, K. (1993) Different regulatory sequences control creatine kinase-M gene expression in directly injected skeletal and cardiac muscle. *Mol. Cell. Biol.*, **13**, 1264–72.

Vitadello, M., Schiaffino, M.V., Picard, A., Scarpa, M. and Schiaffino, S. (1994) Gene transfer in regenerating muscle. *Human Gene Ther.*, **5**, 11–18.

von-Harsdorf, R., Schott, R.J., Shen, Y.T., Vatner, S.F., Mahdavi, V. and Nadal-Ginard, B. (1993) Gene injection into canine myocardium as a useful model for studying gene expression in the heart of large mammals. *Circ. Res.*, **72**, 688–95.

Wells, D.J. (1993) Improved gene transfer by direct plasmid injection associated with regeneration in mouse skeletal muscle. *FEBS Lett.*, **332**, 179–82.

Wells, D.J. and Goldspink, G. (1992) Age and sex influence expression of plasmid DNA directly injected into mouse skeletal muscle. *FEBS Lett.*, **306**, 203–5.

Wolff, J.A., Malone, R.W., Williams, P. *et al.* (1990) Direct gene transfer into mouse muscle *in vivo*. *Science*, **247**, 1465–8.

Wolff, J.A., Williams, P., Acsadi, G., Jiao, S., Jani, A. and Wang, C. (1991) Conditions affecting direct gene transfer into rodent muscle *in vivo*. *Biotechniques*, **11**, 474–85.

Wolff, J.A., Ludtke, J.J., Acsadi, G., Williams, P. and Jani, A. (1992a) Long-term persistence of plasmid DNA and foreign gene expression in mouse muscle. *Hum. Mol. Genet.*, **1**, 363–9.

Wolff, J.A., Dowty, M.E., Jiao, S. *et al.* (1992b) Expression of naked plasmids by cultured myotubes and entry of plasmids into T tubules and caveolae of mammalian skeletal muscle. *J. Cell Sci.*, **103**, 1249–59.

6

Yeast artificial chromosome vectors

STEWART A. FABB and JIANNIS RAGOUSSIS

6.1 INTRODUCTION

The ability to clone DNA fragments is vital to molecular biology. Until fairly recently the most commonly used vectors were plasmids, λ bacteriophage and cosmids (propagated in the bacterial host *Escherichia coli*), the latter of which could accommodate up to 50 kilobase pairs (kb) of DNA (Burke *et al.*, 1987). The P1 based vectors doubled this maximum size to about 100 kb (Sternberg, 1990). However, many problems of eukaryote molecular biology, such as isolation of very large genes (e.g. the human dystrophin and cystic fibrosis genes), analysis of large transcription units, and physical mapping of ordered fragments from large genomes, require analysis of hundreds or even thousands of kb of contiguous DNA. The introduction of the yeast artificial chromosome (YAC) (Burke *et al.*, 1987) as a cloning vehicle for fragments of DNA, that are at least one order of magnitude larger than was previously achieved using other systems, has opened these areas for investigation.

This chapter discusses some of the ways in which YACs (vectors and insert DNA) can be manipulated and subsequently transferred for analysis into mammalian cells. It also describes some of the developments leading towards the creation of a mammalian artificial chromosome.

6.2 YAC VECTORS

For a vector to be able to propagate itself both in bacteria as a plasmid, and in yeast as an artificial chromosome a number of sequences are required. For propagation and selection in bacteria all YAC vectors must contain first, a

Molecular and Cell Biology of Human Gene Therapeutics
Edited by George Dickson
Published in 1995 by Chapman & Hall. ISBN 0 412 62550 4

bacterial origin of replication (*ori*) and second, a resistance gene (usually ampicillin). For propagation as an artificial chromosome the vector requires: (1) a centromere derived from yeast (*CEN4*); (2) a yeast autonomously replicating sequence (*ARS1*); (3) two *Tetrahymena* telomere (*TEL*) sequences (*Tetrahymena TEL* sequences were cloned before yeast sequences, and although not identical, function perfectly in yeast); and (4) yeast genes for the selection of YAC transformants (usually *URA3* and *TRP1*).

The original series of YAC vectors (Burke *et al.*, 1987) differ only in the cloning sites present in the *SUP4* intron. pYAC2 has a *Sma*I site (also present in subsequent pYAC vectors); pYAC3 a *Sna*BI site (for blunt-end cloning); pYAC4 (the most commonly used vector) (Figure 6.1) an *Eco*RI site; and pYAC55 a *Not*I site. (pYAC55 was not the original nomenclature of Burke *et al.* (1987); this vector was originally designated 'pYAC5', however some distributed versions of pYAC5 were found to be structural variants of the original and have subsequently been replaced with pYAC55, which is identical to the original pYAC5 (Hieter *et al.*, 1990)). Other vectors include pYAC-RC (Marchuk and Collins, 1988) which has a polylinker containing multiple rare-cutter restriction sites; pYAC-4 neo which contains a *pSV2neo* cassette (Cooke and Cross, 1988); pCGS996 which permits copy number amplification of YACs (Smith *et al.*, 1990) (see section 6.3.3, Amplification vectors); and pJS97 and pJS98 (which represent the centric and acentric arms respectively) allow riboprobes from the rescued ends of exogenous DNA to be generated from the T7 bacteriophage promoter adjacent to the multiple cloning site (Shero *et al.*, 1991).

6.3 YAC MODIFICATION

6.3.1 Homologous recombination

A great deal of progress in molecular biology has resulted from examining the effects, on gene expression, of well-defined alterations to the genome. As an experimental organism, yeast is particularly suited to this kind of manipulation since site-directed integration of exogenous DNA occurs almost exclusively by the highly efficient process of homologous recombination (Scherer and Davis, 1979). Since exogenous DNA segments cloned in YACs, as well as the YAC vectors themselves, can also be subjected to this kind of manipulation (via vectors that target homologous recombination) the ability to introduce changes is greatly enhanced.

6.3.2 Resistance marker vectors

To examine the expression and fate of YACs in detail it is often necessary to establish them in recipient cells. For this, a selectable marker is usually

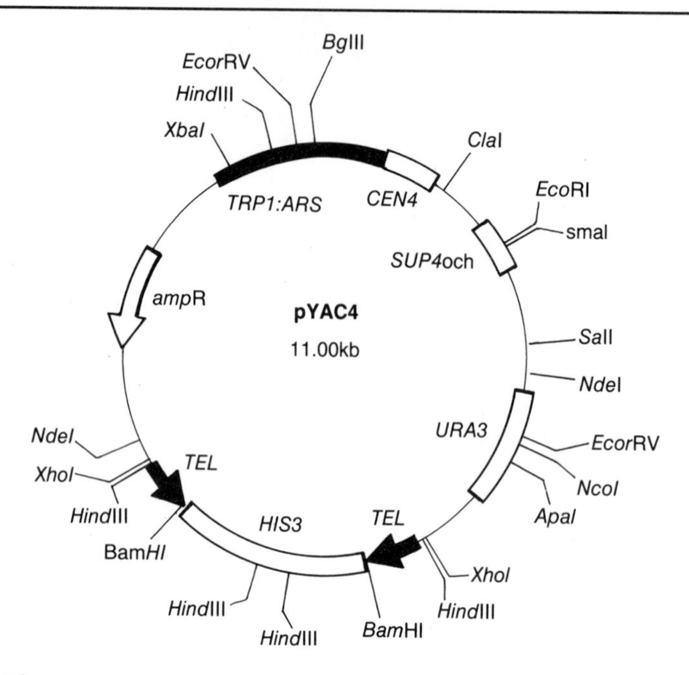

Figure 6.1 The pYAC4 cloning vector (Burke *et al.*, 1987) comprises *cis*-acting yeast chromosomal sequences joined to segments of the bacterial plasmid pBR322 which allow propagation in bacteria as well as yeast (see text). The vector is first digested with *Bam*HI to expose active telomere (*TEL*) sequences. It is then digested at the unique *Eco*RI or *Sma*I cloning sites and source DNA cloned in.

necessary. This could be accomplished either by constructing new YAC libraries with vectors carrying a selectable marker (Traver *et al.*, 1989), or more simply by modifying the YAC *in vivo* by homologous recombination. Because many of the libraries currently in use are constructed in the original, and still widely used pYAC4 vector (Burke *et al.*, 1987), which contains no selectable marker, altering the vector by homologous recombination is the quickest and easiest way to fulfil this requirement.

The most commonly used selectable marker in mammalian cells is aminoglycoside phosphotransferase (more commonly referred to as neo, G418 or APH) which blocks protein synthesis by interfering with ribosomal function (Southern and Berg, 1982). A number of different vectors have been designed using this marker. These vectors, which fall into three main categories: left arm, right arm, and integrating vectors, have advantages as well as disadvantages. Left and right arm vectors suffer from the fact that when YACs are introduced into mammalian cells approximately 90% of those clones carrying YACs suffer loss of terminal sequences (Pavan *et al.*, 1990a; Gnirke *et al.*, 1991; Huxley *et al.*, 1991) and as a result lose their selectable marker. Integrating vectors on the other hand, while generally overcoming this problem may cause an interruption of gene expression since they are able to integrate at a number of different sites. These problems aside, a number of mammalian cell lines have been successfully transformed (see section 6.4) using a variety of methods with these vectors as the basis for selection.

Left arm vectors

pLNA (Davies *et al.*, 1992) and pLNA-1 (Riley *et al.*, 1992) both contain a *LYS2* gene, a *neoR* gene and an *Alu* repetitive element. This vector is targeted to the left arm by linearizing within the *AmpR* gene.

Right arm vectors

pRV1 (*LYS2*, *neoR*, *URA3*) (Srivastava and Schlessinger, 1991), pLUNA (*LYS2*, *neoR*, *URA3*, *Alu*) (Davies *et al.*, 1992), pLNT (*LYS2*, *neoR*, *tetR*) (Riley *et al.*, 1992) and pRAN4 (*ADE2*, *neoR*, *URA3*) (Markie *et al.*, 1993) are all able to target the right arm. In the case of pRV1 and pRAN4 the *neoR* and marker (*LYS2*, *ADE2*) genes are bracketed by fragments of the *URA3* gene whereas in pLUNA the *URA3* gene is intact. In these three vectors homologous recombination is achieved by releasing the recombinogenic *URA3* ends using a single restriction enzyme. (The main advantage of pRAN4 over the other vectors is that it can also be used as a color marker to follow the segregation of YACs within yeast colonies.) Targeting of pLNT on the other hand is achieved by linearization of the vector within an incomplete *tetR* gene.

Integrating vectors

The vectors pBP47 (Pavan *et al.*, 1990a), pLNA (Davies *et al.*, 1992) and pLNB-1 (Riley *et al.*, 1992) all contain a *neoR* gene and an *Alu* repetitive element. (pBP47 contains a *HIS3* gene and pLNA and pLNB-1 a *LYS2* gene.) Targeting into human DNA is facilitated by linearizing each vector within the *Alu* repeat sequence.

pRV2 (Srivastava and Schlessinger, 1991) is designed to be used in conjunction with the vector TKLU2 (Eliceiri *et al.*, 1991) to introduce a positive selectable marker into the insert of a YAC and a negative selectable marker into the vector arm. The right arm of the YAC is first modified with TKLU2 which contains the thymidine kinase (*TK*) gene and *LYS2* gene inserted into a *URA3* gene, and can be used on its own in TK-deficient mammalian cells. Further modification can be achieved with pRV2, which contains a *neoR* gene, a *URA3* gene and *LINE* (*L1*) repetitive elements. This vector is targeted to *L1* repetitive elements in human DNA by removing the fragment between these sequences.

6.3.3 Other modifying vectors

Amplification vectors

Methods that require a rich source of YAC chromosomal DNA would be greatly enhanced by elevating the number of YAC copies present within the yeast cell. Not only would this be useful in preparing sufficient quantities of gel-purified DNA for introduction into cell lines (see Schedl *et al.*, 1992, 1993a, b, section 6.4.3), but also may be beneficial in the direct sequencing of insert ends, subcloning of insert fragments or cytogenetic localization by fluorescent *in situ* hybridization (Moir *et al.*, 1993).

Smith *et al.* (1990) described a YAC vector that permits selection for elevated numbers of YAC copies. This vector (pCGS966) contains a conditional centromere that is turned on or off by changing the carbon source from glucose to galactose, and a heterologous *TK* gene which is activated by adding exogenous thymidine in the presence of methotrexate and sulfanilamide. By growing YACs under these conditions a 10–20-fold amplification can be achieved. (This vector also contains a *neoR* gene, thereby eliminating the need for further manipulation before introducing YAC clones into mammalian cells.) At present only one library (human) has been constructed in this vector (Moir *et al.*, 1993); however the average insert size is relatively small (200 kb) and the library covers just one genome equivalent. Smith *et al.* (1993) described two modification vectors (pCGS990 and pCGS1002) derived from pCGS966 which allow YACs constructed in pYAC4 or related vectors to be modified. Because this vector replaces the normal centromere with a conditional centromere targeting is to the left arm. pCGS990 and pCGS1002 differ from the original vector in that *LYS2* replaces *TRP1*, and the *HIS3*, *URA3*, *neoR* and one of the two *ARS1* genes have all been deleted (in the case of pCGS1002 both *ARS1* genes have been deleted).

Fragmentation and integration vectors

Chromosome fragmentation was originally designed for physical mapping of yeast chromosomes (Volrath *et al.*, 1988; Gerring *et al.*, 1990) but was readily adapted to the analysis of YACs and has recently been used to map genes based on complementation of gene function (Cook *et al.*, 1994).

Chromosome fragmentation vectors contain a targeting sequence that can

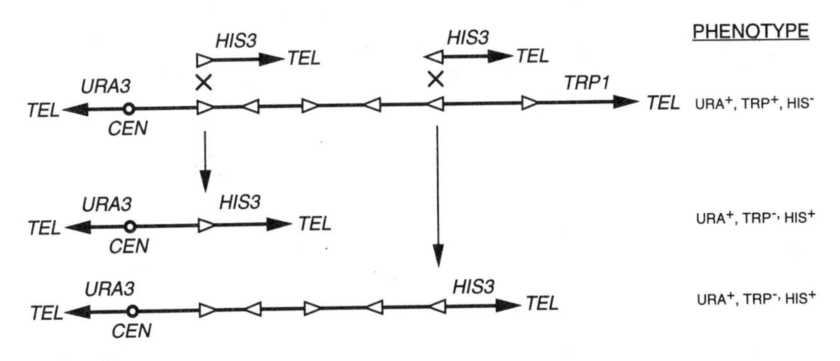

(a) Acentric YAC fragmentation

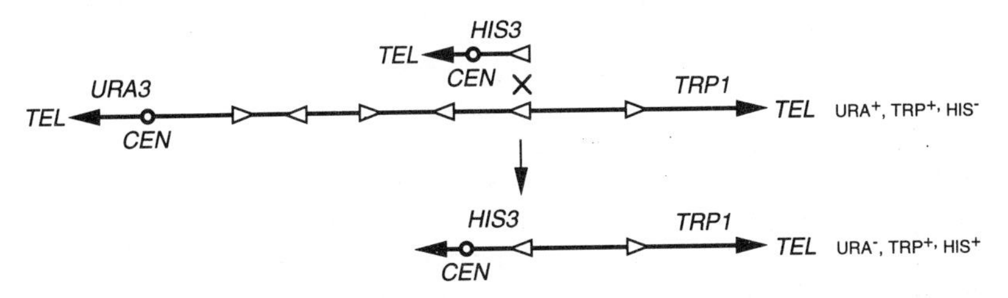

(b) Centric YAC fragmentaton

Figure 6.2 YAC fragmentation model. Homologous recombination between repetitive elements in the YAC with an acentric fragmentation vector (containing the corresponding sequences in opposite directions) (a), and (b) a centric fragmentation vector. This results in the substitution of *TRP1* with *HIS3* and subsequent loss of all sequences distal to the site of insertion in situation (a), and in (b), the substitution of *URA3* with *HIS3* and subsequent loss of sequences proximal to the site of insertion. Open arrows represent repetitive elements; closed arrows and circles represent telomere (*TEL*) sequences and centromeric (*CEN*) sequences respectively and *TRP1*, *URA3* and *HIS3* are yeast selectable markers. (Adapted from Pavan *et al.*, 1990, 1991).

be any DNA with homology to the YAC. The most commonly used sequences are those from repetitive elements (Figure 6.2). *Alu* elements are the most frequently occurring repetitive sequence in the human genome (approximately every 4 kb) and for this reason form the basis of most fragmentation vectors. *L1* elements are also used but are less frequent targets.

More precise fragmentation and manipulation may be accomplished by using cDNA sequences as the targeting segment. This procedure has several potential applications which include exon mapping of large genes, determination of gene orientation, production of 'minigenes' that contain any desired number of introns and the introduction of reporter genes or mutations into the desired YAC sequence (Hieter *et al.* 1990; Das Gupta *et al.* 1993) (Figure 6.3).

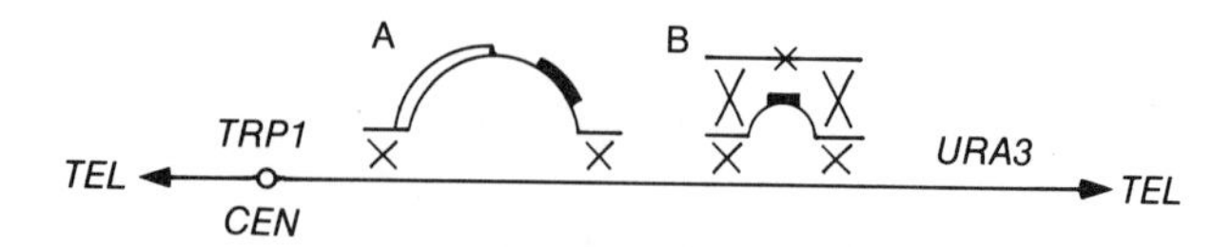

Figure 6.3 Diagram showing two ways in which DNA cloned into YACs can be manipulated. A, Replacement of the endogenous gene with a reporter gene (□) and a yeast-selectable marker (■). B, Introduction of a subtle mutation: a selectable marker (■) is inserted and then replaced by the desired mutation (×). (*TRP1* and *URA3* are yeast-selectable markers; *TEL* and *CEN* represent telomeric and centromeric sequences respectively). (Adapted from Huxley and Gnirke, 1991).

Pavan *et al.* (1990b) described two vectors (pBP63A and pBP63B) that contain *Alu* elements in both orientations and target insertion of a yeast-selectable marker (*HIS3*) plus a sequence capable of forming a telomere (Y′) into repetitive elements distributed along human-derived YAC inserts. Homologous recombination with the YAC deletes all sequences distal to the site of insertion. As a result the fragmented derivative loses the *TRP1* marker and gains the *HIS3* marker. Unfortunately only 4.6% of the yeast transformed with this vector were of the correct phenotype (Ura$^+$, His$^+$, Trp$^-$) with the majority of transformants maintaining the vector as a small replicating linear fragment. The reasoning given was that the vector probably acquired an *ARS* by gene conversion between the Y′ element and an endogenous Y′ sequence, several of which are known to display *ARS* activity (Zakian, 1989).

To increase the efficiency of successful targets Pavan *et al.* (1991) created a new series of fragmentation vectors which contained a functional telomere that is incapable of acquiring *ARS* activity in yeast. As a result between 45% and 78% of *Alu*-targeted transformants contained a fragmented YAC. pBP108 and pBP109 target *Alu* repetitive sequences in both orientations (3′–5′ and 5′–3′ respectively) and pBP110 and pBP111 target *L1* repetitive elements in both orientations (5′–3′ and 3′–5′ respectively) with all four resulting in loss of sequences distal to the site of insertion (acentric YAC fragmentation) (Figure 6.2a). [pBP109 has been further modified in this laboratory (S. Fabb, unpublished results) by ligating from the vector pREP4 (Invitrogen) the resistance marker hygromycin B (*HygB*), *TK* promoter and poly A tail. Thus with one transformation a YAC is fragmented, with each product containing a selectable marker.] pBP81 is a centromere-containing vector which contains two head-to-tail *Alu* sequences. This vector results in the loss of all sequences proximal to the site of insertion (centric YAC fragmentation) (Figure 6.2b).

Another pair of vectors (pF9L1a and pF9L1b) that use *L1* repetitive elements were designed by Campbell *et al.* (1991). These vectors (which differ only in the orientation of their *L1* element) were used to generate a nested series of interstitial deletions in a 650 kb human YAC. To accomplish this they contain a sequence that is uniquely represented in the YAC (in this instance the complete gene for clotting factor IX [*F9*]) coupled with a *L1* repetitive element. Therefore, if one arm of the vector pairs with the unique sequence and the other arm pairs with any of its homologous elements, recombination involving a double crossover will result in the loss of sequences between the unique sequence and the repetitive element. Although this vector was very effective in generating deletions, it was specific to the YAC clone used and would therefore need to be modified to examine other YAC clones.

One of the problems with all the above vectors is that they are based on auxotrophies not present in the background of most recombinant libraries. The majority of yeast strains commonly used for YAC cloning are auxotrophic for *HIS3* which is the selection employed in fragmentation vectors. Unfortunately most libraries are constructed in the yeast strain AB1380 which is auxotrophic for *HIS5*. As a result it is usually necessary to change the auxotrophic background of the YAC either through DNA-mediated transformation of yeast (Connelly *et al.*, 1991), standard genetic cross (Green and Olson, 1990), or by a method just developed that takes advantage of the properties of *kar1* mutants which are unable to undergo normal karyogamy (nuclear fusion) during mating (Hugerat *et al.*, 1994; Spencer *et al.*, 1994). Lewis *et al.* (1992) have circumvented the need to change yeast strains by designing a YAC deletion vector (pBCL) that is based on the *LYS2* auxotrophy and is therefore fully compatible with AB1380. Based on transformation of three unrelated YACs with pBCL containing an *Alu* repetitive element, derivative clones with the expected phenotype were observed at rates of 27–49%.

Although fragmentation vectors have been used primarily to generate physical maps of YACs based on the distribution of repetitive elements (for example Ragoussis *et al.*, 1992 who determined the genomic organization of the human folate receptor genes on chromosome 11q13), Das Gupta *et al.* (1993) developed a method for generating expressed-sequence maps of human chromosomes using specific cDNA sequences cloned into the fragmentation vector pBP103 (the basis of the pBP series of vectors developed by Pavan *et al.*, 1991). This method initially involved construction of libraries of short cDNAs using random oligomers as primers. Using flanking vector primers the cDNAs were amplified by the polymerase chain reaction and then hybridized to immobilized DNA of a specific defined YAC. Those cDNAs that hybridized to the YAC DNA after two rounds of selection were amplified again and subsequently cloned into pBP103. Sequences in the cDNAs then undergo homologous recombination with corresponding sequences present in the YAC, yielding a series of truncated products, with each truncated YAC specifying the location of an exon in the YAC. Because all the truncation events end with the same vector sequence it was possible to sequence these ends to generate expressed sequence tags. This method was used successfully to target individual exons within the *F9* gene on a 650 kb YAC and to rescue and sequence an exon–intron junction from a novel gene within a 330 kb YAC from the human major histocompatibility complex (MHC).

6.4 INTRODUCTION OF YACS INTO MAMMALIAN CELLS

There are a number of methods for transferring DNA into mammalian cells; however, with the exception of microinjection all are very inefficient. Because of this, a selectable marker is essential (except in those instances where the YAC contains a selectable gene such as phosphoribosylglycinamide formyltransferase (*GART*) or hypoxanthine phosphoribosyltransferase (*HPRT*) which can be selected for in mutant mammalian cell lines, or when transient expression is all that is required). There are five main methods of non-viral gene transfer available. These are fusion of yeast spheroplasts with mammalian cells, lipofection, microinjection, calcium phosphate co-precipitation and electroporation.

6.4.1 Fusion of yeast spheroplasts with mammalian cells

Of all the methods of DNA transfer, fusion is the least invasive (and most widely used) since the DNA is never extracted and thus is not sheared. For this reason the size of the molecule that can be transferred is potentially unlimited. Apart from the low transformation frequency, there are however two main drawbacks. First, not all cell types are amenable to cell fusion (human cell lines and cells of lymphoblastoid origin have proved very difficult to transform) and second, a large amount of the yeast genome is also transferred and integrated into the host genome along with the YAC (which may or may not be problematic).

A number of studies have shown that cell fusion can be used to transfer YAC DNA to mammalian cells. Traver *et al.* (1989) demonstrated the transfer of a small YAC (75 kb) containing mouse DNA into mouse 3T3 cells by spheroplast fusion and subsequent G418 selection. From 10^6 cells fused, three independent G418-resistant colonies were isolated and shown to have integrated into the mouse genome. Pachnis *et al.* (1990) transferred a 450 kb human DNA-containing YAC clone into mouse L cells and achieved a frequency of approximately 50 G418 resistant colonies/10^6 L cells used. Interestingly, they also found a similar frequency of G418 resistant colonies when L cells were fused with yeast that contained a multicopy plasmid carrying the *neoR* gene, suggesting that the transfer of YAC sequences is no less efficient than the transfer of plasmid sequences. Similarly, Pavan *et al.* (1990a) introduced a 360 kb human-derived YAC into a mouse embryonal carcinoma (EC) cell line. As with the previous study, they demonstrated the stable uptake and maintenance of a large segment of human DNA in cultured cells. This was inferred not only from the 40+ doublings without selection, but also from the observations that the YAC was present in low copy number and that the sequences near the end of the YAC vectors were linked to flanking sequences. They also showed that the yeast × rodent EC cell hybrids retained the ability to differentiate when grown in the presence of retinoic acid.

In contrast to the above studies that all used anonymous DNA, several groups have transferred YACs containing the human *GART* and *HPRT* genes into mammalian cell lines deficient for these genes and assayed for successful transfer on the basis of complementation. These studies show that for human genetic diseases that display easily scored cellular phenotypes, complementation tests may prove to be very useful in the identification of YACs containing the mutated gene. It may also be possible using these YACs to generate novel mouse strains that serve as models for human disease.

Gnirke *et al.* (1991) used two methods to introduce YAC DNA into their mutant Chinese hamster ovary (CHO) cell lines; fusion and lipofection. With both methods, five to ten positive colonies were isolated from 2×10^6 cells plated for selection. In both cell lines complementation of the purine defect was observed, suggesting that both YACs containing the *GART* gene (360 kb and 600 kb) were functioning properly in their new environment. Similarly, Huxley *et al.* (1991) transferred a 680 kb YAC containing the human *HPRT* gene into a mutant mouse L cell line and observed complementation of the defect. mRNA expression was also observed, suggesting that the $HPRT^+$ phenotype of the fusion lines is as a result of expression of the transferred human *HPRT* gene and not by an indirect mechanism such as marker rescue of the mutant mouse gene by gene conversion from the YAC.

To assess whether YACs could be maintained and amplified extrachromosomally (and thus ultimately eliminate the need for integration) Featherstone and Huxley (1993) introduced linear and circular forms of a 660 kb *HPRT* containing YAC into mutant mouse L cells. After selection, two characteristic forms of amplified DNA were observed, integrated and extrachromosomal. Although these latter elements, which resembled double minutes, were maintained extrachromosomally they segregated poorly. (Copy number differences between independent clones was also noticed by

Markie *et al.*, 1993, after fusing yeast containing a 550-kb YAC from the human MHC class II region with CHO cells. Whether this was due to multiple YAC insertions or to amplification following YAC integration is unknown.) To further delineate the region(s) that allow extrachromosomal replication and maintenance of foreign DNA it may be possible to dissect the YAC by directed homologous recombination. Similarly, if DNA that confers segregation is added to the YAC, then the segregation of these extrachromosomal elements would be expected to stabilize in the absence of selection.

The above studies show conclusively that YACs (along with large amounts of endogenous yeast DNA) can be introduced into different cell lines using fusion with no observable detrimental effects. However, Pavan *et al.* (1990a) raised the doubt whether this method of transfer could be used to modify embryonic stem (ES) cells and subsequently re-colonize the mouse germ line. To answer this, Jakobovits *et al.* (1993) transferred a human *HPRT*-containing YAC into mouse ES cells and subsequently microinjected these into mouse blastocysts. The chimeric mice generated showed that germ-line transmission and expression of the *HPRT*-containing YAC was not impaired by cointegration of a large portion of the yeast genome along with the YAC.

Other studies of interest using fusion were conducted by Demmer and Chaplin (1993) and Cook *et al.* (1994). Demmer and Chaplin (1993) transferred a YAC spanning 330 kb containing several linked genes from the human MHC class II region into CHO and mouse L cells. This study provided an opportunity to examine the expression of these genes in an environment approximating their normal chromosomal context. They noticed that transfer of the YAC in an apparently intact form appeared to occur with high frequency, even when the mammalian selectable marker was placed at the end of the YAC in the right vector arm. After fusion, the genes were shown to be transcriptionally active in the L cell lines and in the case of the *HLA-DQ* genes, able to express serologically detectable cell surface HLA-DQ protein. Also noticed was that genes near the middle of the YAC insert (*TAP1* and *TAP2*) were expressed in a higher fraction of transfected clones than genes encoded near the end of the YAC (*HLA-DQA* and *HLA-DQB*). This suggests that genes close to the site of integration may be influenced more by the surrounding chromosomal sequences than genes further away. Thus, the closer a gene is to its natural environment the more likely it will be regulated at its normal level.

Cook *et al.* (1994) used a chromosomal fragmentation procedure to localize the human interferon-γ receptor accessory gene (*AF-1*) known to reside on a 540 kb YAC which also encodes the *GART* gene. A set of deletions from the acentric end were produced and introduced into CHO cells. Clones were selected for in G418 (rather than by complementation). When the deletions exceeded 390 kb accessory factor activity ceased (as judged by MHC class I antigen inducibility). Thus the *AF-1* gene was localized to a 150 kb region at the centric end of the parental YAC.

Several studies using spheroplast fusion that are potentially important in the development of mammalian artificial chromosomes (see section 6.5) have recently been described (Nonet and Wahl, 1993; Brown *et al.*, 1994; Larin *et al.*, 1994; Taylor *et al.*, 1994). Nonet and Wahl (1993) transferred linear and circularized YACs either containing or lacking a biochemically defined origin of replication (70 kb and 120 kb respectively) into a dihydrofolate reductase-deficient CHO cell line. They noticed that all eight clones that lacked the origin of replication and seven of nine clones with the origin region contained single copies of the transfected YAC integrated into single but different chromosomal sites. By contrast, two transformants derived from circularized YACs containing the putative replication of origin showed very heterogeneous YAC copy number and numerous integration sites following many generations of *in vitro* propagation. Also noticed, soon after fusion, was that these cell lines contained variously sized extrachromosomal elements found in some cells harboring amplified genes. These observations are consistent with the interpretation that YACs containing a putative origin of replication can initially replicate autonomously, followed by integration into multiple chromosomal locations, as has been reported to occur in a number of examples of gene amplification in mammalian cells.

Similarly, to investigate the DNA sequences required for mammalian centromere function, Larin *et al.* (1994) re-introduced human YAC clones (120 kb and 420 kb) containing either centromeric (alphoid DNA) or non-centromeric sequences from the Y chromosome into hamster and human cells using both spheroplast fusion and lipofection. Spheroplast fusion was appropriate for the hamster CHO cells only, since no G418 resistant colonies were seen when human HT1080 cells were used. They were, however, successful when YAC DNA was purified by pulsed field gel electrophoresis (PFGE) and subsequently used to transform the human cell line by lipofection. In all cell lines examined, a constriction was seen at the site of alphoid DNA integration (although this was difficult to distinguish in the hamster cell lines due to the large amounts of yeast DNA present) but not when non-centromeric DNA was used. This study is important, not only as an another step forward in understanding what sequences are important for centromere function, and subsequently may be crucial in the development of mammalian artificial chromosomes, but also to illustrate the differences in effectiveness of transfection techniques between different cell lines.

6.4.2 Lipofection

Unfortunately, fusion between mammalian cells and yeast spheroplasts is not appropriate to all cell types (see above, Larin *et al.*, 1994). This may be due to a number of factors, including the sensitivity of some cells to polyethylene glycol, the presence of yeast in the culture media, or perhaps the product of introducing large amounts of yeast DNA into the genome along with YAC DNA. Lipofection is a method of transfection that is gaining in popularity due to its relative efficiency and ease of use.

With the exception of the study by Gnirke *et al.* (1991) mentioned above, all subsequent studies using lipofection have purified the YAC DNA away from yeast chromosomes via PFGE, followed by digestion of the agarose using β-agarase. Although this manipulation can potentially damage the DNA it does have the advantage of removing all extraneous yeast sequences (unless the YAC co-migrates with a yeast chromosome). Strauss and Jaenisch (1992) successfully introduced intact 150 kb of homogeneous gel-purified YAC DNA spanning the murine *Col1aI* collagen locus back into murine fibroblasts carrying a mutation at this locus. In 10% of the transfected clones the DNA was integrated as a stable intact unit and RNA was expressed. Strauss *et al.* (1993) transfected murine ES cells using the same construct as above but modified the methodology to improve efficiency by isolating the DNA in 100 μm spermine, adding poly-L-lysine before the addition of lipid and exposing the cells to the DNA–lipid complex in suspension instead of growing as a monolayer. Chimeric founder mice were derived from two transfected ES cell clones. These chimeras transmitted the full length transgene through the germ line, generating two transgenic mouse strains. Expression levels of the transgene, as determined by visualization of nascent transcripts in interphase nuclei and ribonuclease protection analysis, was found to be comparable to the endogenous collagen gene.

Two other studies that produced transgenic mice via lipid-mediated DNA transfer are Choi *et al.* (1993) and Lamb *et al.* (1993). Choi *et al.* (1993) co-lipofected a small (85 kb) YAC carrying a fragment of the human heavy chain immunoglobulin gene, along with an unlinked selectable marker, into a mouse ES cell line. Chimeric founder animals were produced by blastocyst injection, and transgenic offspring expressing the YAC-borne immunoglobulin gene fragment were obtained. This study also showed that it is not necessary to retrofit the YAC before transfer. This offers several advantages: first, the ability to introduce different selectable markers into cell lines without prior retrofitting; second, it enables the use of YACs carried in recombination-deficient hosts (as is sometimes necessary with many mammalian genomic fragments that contain tandem repeat sequences which are unstable in yeast); and third, it eliminates the need to transfer the entire YAC in order to transfer the selectable marker.

One of the potential problems perceived with lipid-mediated transfer was the size of the YAC that could be transferred. Lamb *et al.* (1993), however, showed that very large DNA fragments can be transferred using this method by introducing a 650 kb YAC that contains an entire unrearranged 400 kb human β-amyloid precursor protein gene (*APP*) and 250 kb of flanking sequence into mouse ES cells. In addition to this, germ line transmission of the *APP* YAC in transgenic mice and expression of human APP mRNA and protein at levels were found to be comparable to endogenous APP.

6.4.3 Microinjection

The main advantage of microinjection over other methods is the high frequency of transformation, which allows functional experiments on a small number of cells with limited amounts of DNA. There are, however two main disadvantages. As with lipofection, the DNA used in microinjection requires extensive handling (not the least of which involves transferring the DNA through a very fine needle) which increases the chances of mechanical shearing. Also, the method requires specialized equipment and a great deal of skill to successfully microinject. Despite these drawbacks, microinjection is gaining in popularity as a method for transferring YAC DNA into mammalian cells.

Microinjection of a YAC into mammalian cells was first successfully accomplished by Gnirke and Huxley (1991). In this study, they gel purified, phenol extracted and ethanol precipitated their YACs which sheared the DNA from the original sizes of 660 kb (*HPRT*) and 590 kb (*GART*) down to about 100 kb. Since the genes encoded by these YACs were only 45 kb and 40 kb respectively it was assumed that some of the fragments would contain entire copies of the genes. They achieved a frequency of one transformed cell per 250 injected (which is approximately fivefold lower than that obtained with plasmids), and DNA fragments up to 110 kb were maintained intact. This indicates that the microinjection procedure was not limiting the size of the DNA fragments transferred.

Schedl *et al.* (1992, 1993a, b) showed that transgenic mice expressing the tyrosinase gene could be generated by pronuclear injection of YAC DNA into mutant mouse oocytes. In these studies, to increase the concentration of DNA, the YACs were retrofitted with a modification of the amplification vector pCGS966, described by Smith *et al.* (1990), grown under conditions selective for amplification, and subsequently separated from yeast DNA using PFGE. YAC DNA was concentrated further either by ethanol precipitation (Schedl *et al.*,1992), or by electrophoresing the YAC from low percentage agarose into high percentage agarose (Schedl *et al.*, 1993a, b). (The different methods of final concentration are a result of the size of the YAC used; 35 kb in the first instance and 250 kb in the second.) Mice expressing the transgene were identified at birth by eye and skin pigmentation. Of those injected, 3% containing the small YAC were pigmented, whereas only 1% using the larger YAC showed pigmentation. The reduced frequency of pigmentation observed by Schedl *et al.* (1993a, b) may be a reflection of the sevenfold increase over the first study in the size of the DNA injected. Schedl *et al.* (1993b) also showed that the transgene was expressed at levels comparable with that of the endogenous gene, was copy number dependent and position independent.

Peterson *et al.* (1993) also generated transgenic mice by direct microinjection into fertilized eggs of a purified 248 kb YAC bearing an intact 82 kb human β-globin locus and 148 kb of its flanking DNA. Seventeen of the 148 newborn (11.5%) were transgenic. Of these, 13 contained both γ- and β-globin sequences, and 10 were full length. Throughout development the functional genes of the human β-globin locus displayed the proper controls. There was ϵ- and γ-globin, but not β-globin mRNA expression from egg-yolk samples; only minor γ and mostly β expression in the 14-day-old liver; and only β-mRNA in the blood of adult animals. Gaensler *et al.* (1993) essentially duplicated these results by successfully integrating a 150 kb YAC into the germ line of transgenic mice as a single unrearranged fragment that contained the entire β-globin locus and 30 kb of 3′ flanking sequence. As found in the study by Peterson *et al.* (1993), expression of the transgenic human β-globin locus was both tissue- and developmental stage-specific and closely followed the pattern of expression of the endogenous mouse β-globin locus. These two studies indicate that it is now feasible to perform mutagenesis, via homologous recombination in yeast, to dissect regulatory elements within the context of an entire locus *in vivo*.

To assess the size limit of DNA fragments that can be transferred into mammalian cells by microinjection Gnirke *et al.* (1993) injected a 590 kb YAC containing the *GART* gene into CHO cells. Approximately 150 injections were necessary to generate one expressing colony. Nine of 12 lines analyzed contained an intact stretch of at least 110 kb, one cell line contained at least 480 kb but in no instance was the entire 590 kb YAC transferred intact. Similarly, mouse L cells were injected with a 230 kb YAC containing the human β-globin gene cluster. Of the 10 resulting lines, three contained the entire YAC with the remaining seven containing at least 140 kb of YAC DNA. They also noticed that a sodium chloride concentration of 100 mM was necessary to protect the DNA from shearing during these procedures. Whether these sizes represent the upper limit for microinjection remains to be determined; however it should be possible to transfer intact YACs of 200–300 kb routinely or, with very careful handling, YACs of up to 480 kb.

6.4.4 Calcium phosphate co-precipitation and electroporation

Calcium phosphate co-precipitation has been used primarily for introducing plasmids into mammalian cells. Eliceiri *et al.* (1991), however, reported the successful transfer of unpurified YACs of 40–50 kb into a TK-deficient mouse cell line. The YACs (retrofitted with a *TK*-containing plasmid) contained the genes for human glucose-6-phosphate dehydrogenase (*G6PD*) and ribosomal RNA (*rDNA*). Analysis of transformants showed that YACs containing *rDNA* and *G6PD* were incorporated essentially intact into the mammalian cell DNA. In the case of *G6PD*, fully expressing, single copies of the transfected YAC were found in each of two transformants. Although the YACs in this study were successfully transferred, they were very small. In addition to this, the paucity of studies using this relatively straightforward method suggests that it may be suitable only for the transfer of small YACs.

Similarly, electroporation which is gaining in popularity as a method of transfecting plasmids into mammalian cells (primarily because it yields a high frequency of both stable and transient transfectants, and also because it requires fewer steps than alternative techniques), has not been used to any extent for YAC transfer. This may be due to the specialized equipment required, or perhaps, as with calcium phosphate coprecipitation, to the size of the YAC that can be successfully transferred. Fernandez-Luna *et al.* (1991) transfected a 310 kb YAC clone containing the human leukocyte common antigen gene (*CD45*) into a mouse B-cell line. Although no permanent cell line was established transient expression of the low-molecular-weight isoform of the protein was observed on the cell surface. From this, they inferred the size of the human *CD45* gene to be approximately 120 ± 10 kb. Therefore, at least 110–130 kb of DNA must have been transferred intact (although not necessarily integrated) using this method.

6.5 MAMMALIAN ARTIFICIAL CHROMOSOMES

One of the problems in introducing YACs back into mammalian cell lines is that in order for the gene(s) to be stably expressed over long periods of time the construct must integrate back into the genome. Not only is integration a very rare event but when it does happen it is impossible to predict where it will occur. One way of potentially overcoming both of these problems is to use a mammalian artificial chromosome (MAC). Like YACs, MACs would require an origin of replication, and functional centromeric and telomeric sequences. To date, the only one of these elements that has been isolated from mammals and shown to unequivocally function *in vitro* is the telomere.

Origins of DNA replication have been identified in a number of mammalian DNA viruses. Perhaps the best characterized of these is that from the Epstein–Barr virus (*oriP*) which allows the efficient maintenance and replication of extrachromosomal DNA in human cells expressing the EBNA1 protein. Although the corresponding mammalian origins of replication remain elusive, potential regions have been located within mammalian chromosomes (Hamlin, 1992) and even used to generate apparently autonomous extrachromosomal structures in some transformants containing these sequences (Nonet and Wahl, 1993; section 6.4.1). This latter study may prove to be very valuable not only in the further analysis of mammalian

igins, but also for structure–function analyses if replication
juires *cis*-acting elements dispersed across large molecular

an centromeres are larger and more complicated structures than
st. In contrast to the single microtubule in yeast, mammalian
bind between 10 and 100 microtubules (Rattner, 1991). Although all the elements required for a functional mammalian centromere have not yet been determined, several centromeric proteins (CENP-A, -B, -C and -E) as well as centromeric DNA has been isolated. Of these, CENP-B and alphoid DNA are strongly implicated in centromere function. Larin *et al.* (1994) showed that several hundred kb of human Y alphoid DNA transfected back into human and hamster cell lines was sufficient to direct the formation of at least three properties of a human centromere: a cytogenetic constriction, disruption of anaphase movement in a dicentric chromosome, and calcinosis/Raynaud's phenomenon/esophageal dysmotility/sclerodactyly/telangiectasia (CREST) antiserum binding. The absence of frequent chromosome breakage suggests that the new centromere may either not have acquired full activity (perhaps because other sequences are also required) or may have been partially suppressed. Similarly, Brown *et al.* (1994) using telomeric DNA to break the human Y chromosome found that the two derivative chromosomes which contain 140 kb and 480 kb of alphoid DNA were able to segregate accurately at mitosis, although the long acrocentric derivative lagged with respect to the other chromosomes in about 10% of anaphase cells, suggesting that other sequences may be required for orderly segregation. In contrast to this, the short arm acrocentric chromosome which contained 140 kb of alphoid DNA and was about 10–12 mb in length segregated normally, and may with further deletions be small enough to be used as a prototype MAC vector.

Mammalian telomere sequences on the other hand have been isolated and used successfully to seed telomere formation in linear molecules introduced into mammalian cell lines (Farr *et al.*, 1991). (Human telomere DNA consists of about 5 kb of the repeat (TTAGGG)*n*, although only small stretches of this sequence are required to seed telomere formation.) Taylor *et al.* (1994) have taken this process one step further by constructing retrofitting vectors to replace YAC telomeres with cloned human telomeric DNA, thereby opening the possibility of developing MACs directly from YACs. They found that after transfer to CHO cells the telomere adjacent to the selectable marker gene was functional in 11% of clones while the telomere 200 kb away at the other end of the YAC was functional in 2% of clones. Thus, despite being modified with yeast telomere sequences, the human telomeric DNA was still functional, suggesting that the construction of MACs is feasible by this route. The seeding of new telomeres by cloned telomeric DNA may, however, be cell-line dependent since they were either undetectable or highly inefficient in non-tumorigenic human or mouse cell lines, indicating that other factors may be required to protect against integration (Barnett *et al.*, 1993).

6.6 CONCLUSIONS

The ability to clone very large fragments of DNA in YACs, and manipulate these, via homologous recombination, to make virtually any change is an extremely powerful method for studying the function of cloned DNA. A number of studies have now demonstrated the successful transfer of large DNA fragments in the form of YACs into mammalian cells. Potential uses for the techniques mentioned above include the isolation of genes within these DNA segments through functional assays, and the characterization of long-range control elements and *cis*-acting factors. This allows the establishment of cell lines, correction of defects, alteration of phenotypes and generation of transgenic animals.

The ultimate combination of the above is the MAC which can be stably propagated in mammalian cells. Although our understanding of the requirements for replication and segregation of chromosomal DNA in mammalian cells is far from complete, the ability to build a MAC would not only provide much information about how mammalian chromosomes work, but may also be a powerful new tool in gene therapy.

REFERENCES

Barnett, M.A., Buckle, V.J., Evans, E.P. *et al.* (1993) Telomere directed fragmentation of mammalian chromosomes. *Nucleic Acids Res.*, **21**, 27–36.

Brown, K.E., Barnett, M.A., Burgtorf, C. *et al.* (1994) Dissecting the centromere of the human Y chromosome with cloned telomeric DNA. *Hum. Mol. Genet.*, **3**, 1227–37.

Burke, D.T., Carle, G.F. and Olson, M.V. (1987) Cloning of large segments of DNA into yeast by means of artificial chromosome vectors. *Science*, **236**, 806–12.

Campbell, C., Gulati, R., Nandi, A.K. *et al.* (1991) Generation of a nested series of interstitial deletions in yeast artificial chromosomes carrying human DNA. *Proc. Natl Acad. Sci. USA*, **88**, 5744–8.

Choi, T.K., Hollenbach, P.W., Pearson, B.E. *et al.* (1993) Transgenic mice containing a human heavy chain immunoglobulin gene fragment cloned in a yeast artificial chromosome. *Nature Genet.*, **4**, 117–23.

Connelly, C., McCormick, M.K., Shero, J. and Hieter, P. (1991) Polyamines eliminate an extreme size bias against transformation of large yeast artificial chromosome DNA. *Genomics*, **10**, 10–16.

Cook, J.R., Emanuel, S.L., Donnelly, R.J. *et al.* (1994) Sublocalization of the human interferon-γ receptor accessory factor gene and characterization of accessory factor activity by yeast chromosomal fragmentation. *J. Biol. Chem.*, **269**, 7013–18.

Cooke, H. and Cross, S. (1988) pYAC-4 Neo, a yeast artificial chromosome vector which codes for G418 resistance in mammalian cells. *Nucleic Acids Res.*, **16**, 11817.

Das Gupta, R., Morrow, B., Marondel, I. *et al.* (1993) An integrated approach for identifying and mapping human genes. *Proc. Natl Acad. Sci. USA*, **90**, 4364–8.

Davies, N.P., Rosewell, I.R. and Brüggemann, M. (1992) Targeted alterations in yeast artificial chromosomes for inter-species gene transfer. *Nucleic Acids Res.*, **20**, 2693–8.

Demmer, L.A. and Chaplin, D.D. (1993) Simultaneous transfer of four functional genes from the HLA class II region into mammalian cells by fusion with yeast spheroplasts carrying an artificial chromosome. *J. Immunol.*, **150**, 5371–8.

Eliceiri, B., Labella, T., Hagino, Y. *et al.* (1991) Stable integration and expression in mouse cells of yeast artificial chromosomes harboring human genes. *Proc. Natl Acad. Sci. USA*, **88**, 2179–83.

Farr, C., Fantes, J., Goodfellow, P. and Cooke, H. (1991) Functional reintroduction of human telomeres into mammalian cells. *Proc. Natl Acad. Sci. USA*, **88**, 7006–10.

Featherstone, T. and Huxley, C. (1993) Extrachromosomal maintenance and amplification of yeast artificial chromosome DNA in mouse cells. *Genomics*, **17**, 267–8.

Fernandez-Luna, J.L., Matthews, J., Brownstein, B.H. *et al.* (1991) Characterization and expression of the human leukocyte-common antigen (CD45) gene contained in yeast artificial chromosomes. *Genomics*, **10**, 756–64.

Gaensler, K.M.L., Kitamura, M. and Kan, Y.W. (1993) Germ-line transmission and developmental regulation of a 150-kb yeast artificial chromosome containing the human β-globin locus in transgenic mice. *Proc. Natl Acad. Sci. USA*, **90**, 11381–5.

Gerring, S., Connelly, C. and Hieter, P. (1990) Positional mapping of cloned genes by chromosome blotting and chromosome fragmentation. *Methods Enzymol.*, **194**, 57–77.

Gnirke, A. and Huxley, C. (1991) Transfer of the human HPRT and GART genes from yeast to mammalian cells by microinjection of YAC DNA. *Somat. Cell Mol. Genet.*, **17**, 573–80.

Gnirke, A., Barnes, T.S., Patterson, D. *et al.* (1991) Cloning and *in vivo* expression of the human GART gene using yeast artificial chromosomes. *EMBO J.*, **10**, 1629–34.

Gnirke, A., Huxley, C., Peterson, K. and Olson, M.V. (1993) Microinjection of intact 200- to 500-kb fragments of YAC DNA into mammalian cells. *Genomics*, **15**, 659–67.

Green, E.D. and Olson, M.V. (1990) Chromosomal origin of the cystic fibrosis gene in yeast artificial chromosomes: A model for human gene mapping. *Science*, **250**, 94–8.

Hamlin, J.L. (1992) Mammalian origins of replication. *BioEssays*, **14**, 651–9.

Hieter, P., Connelly, C., Shero, J. *et al.* (1990) Yeast artificial chromosomes: promises kept and pending, in *Genome Analysis Vol 1: Genetic and Physical Mapping*, Cold Spring Harbor Laboratory Press, Cold Spring Harbor, New York, pp. 83–120.

Hugerat, Y., Spencer, F., Zenvirth, D. and Simchen, G. (1994) A versatile method for efficient YAC transfer between any two strains. *Genomics*, **22**, 108–17.

Huxley, C., and Gnirke, A. (1991) Transfer of yeast artificial chromosomes from yeast to mammalian cells. *BioEssays*, **13**, 545–50.

Huxley, C., Hagino, Y., Schlessinger, D. and Olson, M.V. (1991) The human HPRT gene on a yeast artificial chromosome is functional when transferred to mouse cells by cell fusion. *Genomics*, **9**, 742–50.

Jacobovits, A., Moore, A.L., Green, L.L. *et al.* (1993) Germ-line transmission and expression of human-derived yeast artificial chromosome. *Nature*, **362**, 255–8.

Lamb, B.T., Sisodia, S.S., Lawler, A.M. *et al.* (1993) Introduction and expression of the 400 kilobase *precursor amyloid protein* gene in transgenic mice. *Nature Genet.*, **5**, 22–9.

Larin, Z., Fricker, M.D. and Tyler-Smith, C. (1994) *De novo* formation of several features of a centromere following introduction of a Y alphoid YAC into mammalian cells. *Hum. Mol. Genet.*, **3**, 689–95.

Lewis, B.C., Shah, N.P., Braun, B.S. and Denny, C.T. (1992) Creation of a yeast artificial chromosome fragmentation vector based on lysine-2. *Genet. Anal. Tech. Applic.*, **9**, 86–90.

Marchuk, D. and Collins, F.C. (1988) pYAC-RC, a yeast artificial chromosome vector for cloning DNA cut with infrequently cutting restriction endonucleases. *Nucleic Acids Res.*, **16**, 7743.

Markie, D., Ragoussis, J., Senger, G. *et al.* (1993) New vector for transfer of yeast artificial chromosomes to mammalian cells. *Somat. Cell Mol. Genet.*, **19**, 161–9.

Moir, D.T., Dorman, T.E., Smyth, A.P. and Smith, D.R. (1993) A human genome YAC library in a selectable high-copy-number vector. *Gene*, **125**, 229–32.

Nonet, G.H. and Wahl, G.M. (1993) Introduction of YACs containing a putative mammalian replication of origin into mammalian cells can generate structures that replicate autonomously. *Somat. Cell Mol. Genet.*, **19**, 171–92.

Pachnis, V., Pevny, L., Rothstein, R. and Costantini, F. (1990) Transfer of a yeast artificial chromosome carrying human DNA from *Saccharomyces cerevisiae* into mammalian cells. *Proc. Natl Acad. Sci. USA*, **87**, 5109–13.

Pavan, W.J., Hieter, P. and Reeves, R.H. (1990a) Modification and transfer into an embryonal carcinoma cell line of a 360-kilobase human-derived yeast artificial chromosome. *Mol. Cell. Biol.*, **10**, 4163–9.

Pavan, W.J., Hieter, P. and Reeves, R.H. (1990b) Generation of deletion derivatives by targeted transformation of human-derived yeast artificial chromosomes. *Proc. Natl Acad. Sci. USA*, **87**, 1300–4.

Pavan, W.J., Hieter, P., Sears, D. *et al.* (1991) High-efficiency yeast artificial chromosome fragmentation vectors. *Gene*, **106**, 125–7.

Peterson, K.R., Clegg, C.H., Huxley, C. *et al.* (1993) Transgenic mice containing a 248-kb yeast artificial chromosome carrying the human β-globin locus display proper developmental control of human globin genes. *Proc. Natl Acad. Sci. USA*, **90**, 7593–7.

Ragoussis, J., Senger, G., Trowsdale, J. and Campbell, I.G. (1992) Genomic organization of the human folate receptor genes on chromosome 11q13. *Genomics*, **14**, 423–30.

Rattner, J.B. (1991). The structure of the mammalian centromere. *BioEssays*, **13**, 51–6.

Riley, J.H., Morten, J.E.N. and Anand, R. (1992) Targeted integration of neomycin into yeast artificial chromosomes (YACs) for transfection into mammalian cells. *Nucleic Acids Res.*, **20**, 2971–6.

Schedl, A., Beermann, F., Thies, E. *et al.* (1992) Transgenic mice generated by pronuclear injection of a yeast artificial chromosome. *Nucleic Acids Res.*, **20**, 3073–7.

Schedl, A., Larin, Z., Montoliu, L. *et al.* (1993a) A method for the generation of YAC transgenic mice by pronuclear microinjection. *Nucleic Acids Res.*, **21**, 4783–7.

Schedl, A., Montoliu, L., Kelsey, G. and Schütz, G. (1993b) A yeast artificial chromosome covering the tyrosinase gene confers copy number-dependent expression in transgenic mice. *Nature*, **362**, 258–61.

Scherer, S. and Davis, R. (1979) Replacement of chromosome segments with altered DNA sequences constructed *in vitro*. *Proc. Natl Acad. Sci. USA*, **76**, 4951–5.

Shero, L., McCormick, M., Antonarakis, S. and Hieter, P. (1991) Yeast artificial chromosome vectors for efficient clone manipulation and mapping. *Genomics*, **10**, 505–8.

Smith, D.R., Smyth, A.P. and Moir, D.T. (1990) Amplification of large artificial chromosomes. *Proc. Natl Acad. Sci. USA*, **87**, 8242–6.

Smith, D.R., Smyth, A.P., Strauss, W.M. and Moir, D.T. (1993) Incorporation of copy-number control elements into yeast artificial chromosomes by targeted homologous recombination. *Mamm. Genome*, **4**, 141–7.

Southern, P.J. and Berg, P. (1982) Transformation of mammalian cells to antibiotic resistance with a bacterial gene under control of the SV40 early region promotor. *J. Mol. Appl. Gen.*, **1**, 327–41.

Spencer, F., Hugerat, Y., Simchen, G. *et al.* (1994) Yeast *kar1* mutants provide an effective method for YAC transfer to new hosts. *Genomics*, **22**, 118–26.

Srivastava, A.K. and Schlessinger, D. (1991) Vectors for inserting selectable markers in vector arms and human DNA inserts of yeast artificial chromosomes (YACs). *Gene*, **103**, 53–9.

Sternberg, N. (1990) Bacteriophage P1 cloning system for the isolation, amplification and recovery of DNA fragments as large as 100kb. *Proc. Natl Acad. Sci. USA*, **87**, 103–7.

Strauss, W.M. and Jaenisch, R. (1992) Molecular complementation of a collagen mutation in mammalian cells using yeast artificial chromosomes. *EMBO J.*, **11**, 417–22.

Strauss, W.M., Dausman, C.B., Johnson, C. *et al.* (1993) Germ line transmission of a yeast artificial chromosome spanning the murine α_1(I) collagen locus. *Science*, **259**, 1904–7.

Taylor, S.S., Larin, Z. and Smith, C.T. (1994) Addition of functional human telomeres to YACs. *Hum. Mol. Genet.*, **8**, 1383–6.

Traver, C.N., Klapholz, S., Hyman, R.W. and Davis, R.W. (1989) Rapid screening of a human genomic library in yeast artificial chromosomes for single-copy sequences. *Proc. Natl Acad. Sci. USA*, **86**, 5898–902.

Volrath, D., Davis, R., Connelly, C. and Hieter, P. (1988) Physical mapping of large DNA by chromosome fragmentation. *Proc. Natl Acad. Sci. USA*, **85**, 6027–31.

Zakian, V.A. (1989) Structure and function of telomeres. *Annu. Rev. Genet.*, **23**, 579–604.

7

Gene transfer and cancer chemotherapy

DONALD W. MOORMAN and KENNETH W. CULVER

7.1 INTRODUCTION

Gene transfer as applied to human cancers is a rapidly evolving and encouraging discipline. As the experience with these new therapeutic techniques grows, so does the awareness of the tolerable toxicities and the potential therapeutic opportunities. Currently many laboratories are aggressively pursuing new vector delivery systems and new therapeutic genes thereby increasing the spectrum of tumor types in clinical trials. With this explosive investigative effort, two early major themes have emerged. First, increasing evidence indicates that the gene products can have anti-tumor efficacy. Second, the transfer of genes to target tumor cells for *in vivo* expression has so far been found safe with acceptable toxicity. Certainly this emerging technology is creating great enthusiasm and apparent promise for cancer therapeutics.

The application of gene transfer to oncology is currently directed in five basic areas (Table 7.1). These include: (i) marker gene transduction of tumor infiltrating lymphocytes (TIL), hematopoietic stem cells (HSC) or tumor cells. The *NeoR* gene, which encodes the bacterial enzyme neophosphotransferase II and has no therapeutic effect, allows the opportunity to track cell movement, growth and survival *in vivo*; (ii) gene transfer is being utilized to produce a vigorous immune response against autologous tumor cells. These early experiments have focused on the insertion of cytokine genes (e.g. *IL-2, TNF, GM-CSF*), foreign MHC antigens (HLA-B7) and expression of the B7 co-stimulatory protein into tumor cells that are then injected subcutaneously into patients; (iii) a gene can be introduced that is not an immune response enhancer, but rather a direct cytotoxic agent. One

Molecular and Cell Biology of Human Gene Therapeutics
Edited by George Dickson
Published in 1995 by Chapman & Hall. ISBN 0 412 62550 4

Table 7.1 Approved types of gene transfer for cancer clinical trials

- *NeoR* gene 'marking' of T-lymphocytes, hematopoietic stem cells or tumor cells
- Up-regulation of an immune response to autologous tumor cells
- Transfer of directly tumoricidal genes into tumor cell deposits
- Insertion of a tumor suppressor gene or down-regulation of oncogene expression
- Protection of hematopoietic stem cells from chemotherapy toxicity by insertion of the *MDR1* gene

example is the herpes simplex–thymidine kinase (*HS-tk*) gene which confers a selective sensitivity to minimally toxic anti-herpes drugs; (iv) gene delivery can potentially be used to correct a loss of a tumor suppressor gene such as *p53*, or to block oncogene function such as occurs with *K-ras* overexpression; and (v) the insertion of a resistance gene (i.e. multiple drug resistance, *MDR1*) into hematopoietic stem cells to protect them from the toxic effects of chemotherapy.

While the investigations in all these areas are progressing through phase I and/or II clinical trials, it remains critical that the student of gene therapy realizes that the current vector systems, target genes and delivery specificity remain primitive to the eventual vision. It is with great labor and scientific methodology amid burdensome financial demands, that will bring novel, new cancer therapies to patients.

7.2 MARKING CELLS TO STUDY SAFETY, SURVIVAL AND TRAFFICKING

The first published application of *ex vivo* gene transfer with subsequent reintroduction of the 'foreign' gene was reported in 1990 (Rosenberg *et al.*, 1990). This study was an effort to learn about the survival, cell trafficking and principal cellular phenotype mediating the anti-tumor effect seen in about 30% of patients treated with TIL. By utilizing Moloney murine leukemia virus (MoMLV) vectors, the *NeoR* gene was inserted into TIL grown from tumor biopsies. These retroviral vectors have been disabled by excision or truncation of the *gag*, *pol* and *env* viral genes to minimize the risk of recombination and the development of replication-competent retroviruses. Infusions of the *NeoR*-containing TIL demonstrated survival of the marked TIL in the blood and tumor deposits for several weeks. Importantly, this experiment demonstrated that *ex vivo* retroviral-mediated gene transfer into T-lymphocytes was safe in humans.

Additional marker protocols are currently in progress that are an attempt to track the survival of transplanted autologous HSC or tumor cells harbored in autologous hematopoietic stem cell transplants. Early results were published in 1993 (Brenner *et al.*, 1993). They used retroviral vectors for *ex vivo* transfer of the *NeoR* gene into autologous HSC from patients with acute lymphocytic leukemia (ALL) in remission. They determined that the vectors successfully marked residual marrow tumor cells. Even though post-purging analysis could not identify residual tumor cells in the transplanted marrow, the ALL relapse contained *NeoR*-marked tumor cells. These results confirm that the relapse was derived from residual tumor cells in the transplanted marrow, not extra-marrow locations. This study produced one of the first fruits of human gene transfer and now physician–scientists can use marker genes to compare different marrow purging methods. This is a substantial advance toward the development of optimal, curative autologous bone marrow transplantation techniques for leukemia.

7.3 T-LYMPHOCYTE GENE MANIPULATION AND CANCER IMMUNOTHERAPY

The discovery of a variety of cytokine genes has allowed the production of large quantities of potent, new recombinant immunomodulating agents. Initial studies have focused on the injection of these agents directly into tumors, by systemic intravenous infusion or in tissue culture as a means to activate immune cells. For example, human recombinant tumor necrosis factor (TNF) has demonstrated potent anti-tumor activity with intravenous infusion but with significant concomitant systemic toxicities (Asher *et al.*, 1991). The possibility of using TIL that have been genetically altered to contain the TNF gene, as cellular vehicles to deliver TNF directly to the tumor are under investigation. A preliminary report has demonstrated that the human T-lymphocytes are difficult to genetically alter at high efficiency with murine retroviral vectors and that the TIL do not stably express the exogenous *TNF* gene (Hwu *et al.*, 1993). The same difficulties have been identified in murine TIL cells. While the delivery of cytokines by TIL may be helpful to certain groups of patients, there are many cancers from which TIL cannot be grown in sufficient numbers for therapy. Therefore, scientists have looked at alternative methods for the stimulation of TIL and other effector populations *in vivo* through the genetic manipulation of tumor cells.

7.4 TUMOR CELL GENE MANIPULATION AND CANCER IMMUNOTHERAPY

In early 1989, Leder and colleagues opened a new chapter in cancer immunotherapy (Tepper *et al.*, 1989). Their studies demonstrated that the direct injection of tumor cells genetically-modified to express interleukin-4

Table 7.2 Approved human cytokine gene therapy trials for cancer in the USA

Gene	*Transduced cell type*	*Type of cancer*
IL-2	Tumor cells	Colorectal Melanoma Neuroblastoma Renal cell carcinoma Small cell lung cancer
IL-4	Transduced fibroblasts + tumor cells Tumor cells Transduced fibroblasts + tumor cells	Colorectal Melanoma Breast Colorectal Melanoma Renal cell carcinoma
TNF	T-cells Tumor cells	Melanoma Colorectal Melanoma Renal cell carcinoma
Interferon γ	Tumor cells	Melanoma Neuroblastoma

IL, interleukin
TNF, tumor necrosis factor

(IL-4) resulted in rejection of the tumor and resulting systemic immunity against reinjection of the wild-type tumor cells. Since their publication, there have been a series of papers published with numerous tumors modified with a variety of cytokine genes (for a review, refer to Colombo and Forni, 1994). Based upon the growing body of information that the expression of cytokine genes in non-immunogenic or weakly immunogenic tumors can induce rejection and systemic immunity, a series of human clinical trials have been approved in the USA (Table 7.2).

The most common tumors involved in these trials have been metastatic melanoma, renal cell carcinoma and colorectal cancer. These tumor types are the most extensively investigated because they are thought to be more amenable to immunomodulation compared with other tumors. This is based upon reactions to biological response modifiers and the rare observation that melanomas can disappear spontaneously.

The clinical trials have taken seven forms. The most prevalent is the surgical removal of the tumor, growth in tissue culture, transfer of the cytokine gene (e.g. *IL-2*, *TNF* or *GM-CSF*) with retroviral vectors *in vitro* and the reimplantation of the tumor subcutaneously in the same patient (Brenner *et al.*, 1992; Rosenberg *et al.*, 1992). Most investigators irradiate the tumor before reinjection. In one approved trial for metastatic renal cell carcinoma, scientists are comparing lethally radiated inoculums of autologous tumor cells with and without *GM-CSF* gene transfer. The trial will also focus on escalating doses of both transduced and non-transduced cells to determine if a critical level is necessary for a clinical response and if greater doses will magnify the anti-tumor response.

The second method involves the use of one allogeneic tumor cell line for all patients instead of growing tumor cells from each patient (Gansbacher *et al.*, 1992a, b). These studies have used a melanoma tumor cell line that is matched at the class I HLA A2 locus to facilitate antigen recognition and the development of an anti-tumor response. One variation on this strategy is the mixture of an allogeneic melanoma cell line transduced with the *IL-2* gene that is mixed with progressively higher quantities of autologous non-transduced melanoma cells. The tumor cell populations are irradiated before subcutaneous re-implantation.

The third approach involves a mixture of genetically-altered autologous fibroblasts and autologous tumor cells (Lotze *et al.*, 1994). First, skin fibroblasts are removed from the patient. These early approaches utilize the insertion of either an *IL-2* or *IL-4* gene into the fibroblasts that are then admixed with autologous tumor cells. The cell mixture is irradiated and injected subcutaneously into the patient. Since fibroblasts can be grown in culture more reliably than tumor cells, this offers greater possibilities for patient treatment. The intent of the cytokine-transduced fibroblasts is to stimulate an anti-tumor immune response to the surrounding tumor cells. IL-4 is thought to recruit and activate macrophages, while IL-2 is a potent T-cell growth factor. Animal experiments suggest that this form of vaccination with *IL-2* may be less effective than tumor cells that are actively secreting the cytokine (Tsai *et al.*, 1993). No human trial data are available at this time.

The fourth idea is the insertion of the B7 co-stimulatory gene into tumor cells *ex vivo* (Townsend and Allison, 1993). The development of a maximal cellular immune response to tumor antigens requires two signals (Figure 7.1). One through MHC class I : T-cell receptor interaction and a second through a B7 surface molecule that binds to the CD28 surface receptor on T-lymphocytes. Transfection of the B7 surface molecule into melanoma cells resulted in greatly enhanced antigen presentation and tumor eradication *in vivo*. The anti-tumor response was mediated by CD8+ T cells independent of CD4+ T cells. This mechanism for the up-regulation of antigen presentation ability by tumor cells may prove an efficient and effective way to enhance tumor immunogenicity across various histologic tumor cell lines.

Another approach is to increase tumorigenicity by blocking mechanisms through which tumor cells hide from the immune system. One example of this strategy utilizes an anti-sense insulin growth factor type-1 (*IGF-1*) gene to block IGF-1 expression (Figure 7.2). The introduction of an anti-sense

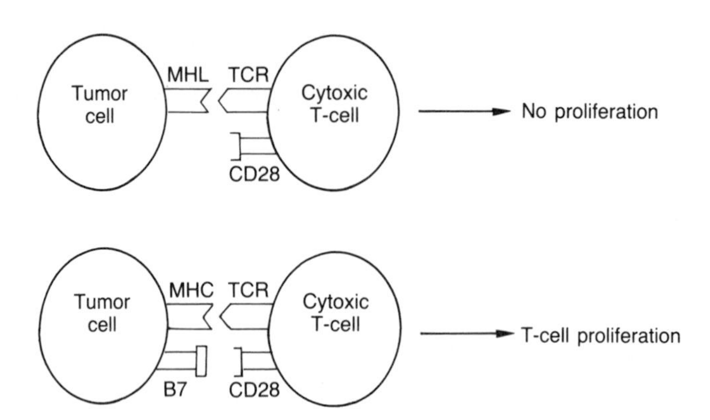

Figure 7.1 Tumor cells that lack the expression of the B7 surface molecule do not effectively induce a potent tumor-specific T-cell response. Interaction between the tumor-associated MHC and the T-cell receptor (TCR) alone is insufficient. The transfer of the *B7* gene into B7-deficient tumor cells can result in dual signaling and the development of a potent anti-tumor immune response.

cDNA of human IGF-1 into high IGF-1-expressing tumors appears to allow the tumor-bearing host to mount an effective CD8 anti-tumor immune response that mediates complete tumor destruction (Trojan *et al.*, 1993). The precise manner in which IGF-1 affords the tumor cells protection from immunologic destruction remains unclear. This approach proved successful in destroying rat brain tumors after the injection of anti-sense IGF-1 transfected cells into the leg, suggesting that the anti-tumor immune response can cross the blood–brain barrier. Therefore, this technique can also result in the development of systemic immunity.

The final two approaches utilize *in vivo* gene transfer. The first is the direct injection of murine retroviral vectors into tumor deposits instead of removing and growing tissues from patients. The initial clinical trials involve the *in vivo* gene transfer of the human γ-interferon gene into melanoma or neuroblastoma deposits. γ-Interferon is a potent stimulant of MHC class II molecule expression and effector cells such as natural killer cells. The second is the transfer of a foreign class I HLA B7 gene into tumor cells *in vivo* (Nabel *et al.*, 1992). In this protocol, liposomes (lipid bodies containing DNA) are utilized to deliver the gene into melanoma deposits *in vivo*. Tumor cells phagocytize the liposomes, resulting in transient expression of the exogenous gene. Upon expression of the foreign MHC in animal tumor models, cytotoxic T lymphocyte recruitment was achieved leading to tumor regression. The clinical trial began in 1992 and results have been published on five patients (Nabel *et al.*, 1993). None of the five demonstrated any toxicity related to the procedure. One patient had clear evidence of regression of the injected lesion and distant lesions. These

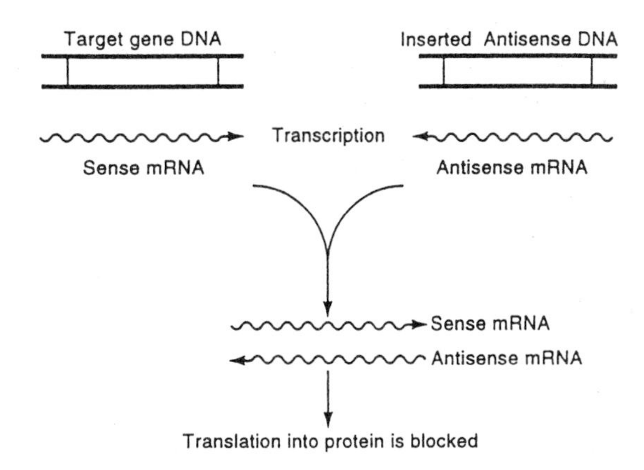

Figure 7.2 The inserted anti-sense gene produces a mirror image copy of the target mRNA (sense RNA). These mirror image copies bind together within the cell preventing translation of the mRNA into protein.

findings were accompanied by increased anti-tumor T-lymphocyte immunity in the patient. These studies have now been amended to include co-transfection of the HLA B7 and β2-microglobulin genes in an effort to further enhance the anti-tumor immune response.

Certainly the advent of clinical gene therapy applications brings an entirely new set of possibilities for the development of successful immunotherapy of cancer. These seven forms of investigation each have the primary goal of enhancing anti-tumor immunity through expression of exogenous genes. Combinations of these approaches may allow synergistic anti-tumor effects. As information is gained about *in vivo* gene expression, induction of anti-tumor immune responsiveness with different gene combinations, *in vivo* gene delivery, etc., we will have moved significantly closer to our goal of the vaccination of patients against neoplastic transformation and in the prevention of recurrence of their malignancy.

7.5 CYTOTOXIC GENE DELIVERY SYSTEMS

Although the use of gene transfer to alter tumorigenicity is promising, this approach may have limited utility in many patients as a result of their immunosuppressed condition from a disseminated tumor burden and prior immunosuppressive therapy. Therefore, methods that directly target tumor cell destruction may be advantageous. The introduction of 'sensitivity' or suicide genes into the tumor cell can be directly cytotoxic. This strategy utilizes genes that selectively alter the susceptibility of a tumor to non-toxic or minimally toxic pharmacologic agents. Two systems are currently under intensive study: (i) the herpes simplex–thymidine kinase gene (*HS–tk*) which confers a sensitivity to the anti-herpes drug ganciclovir (GCV); and (ii) the bacterial cytosine deaminase gene which confers a sensitivity to the antifungal drug 5-fluorocytosine (5-FC). In both systems, subsequent administration of the appropriate pharmacologic agent (GCV or 5-FC) results in selective tumor cell death. Available evidence suggests that the destruction of the tumor cells with these methods can result in the development of systemic immunity. The following sections describe the pertinent pre-clinical and clinical studies that support the use of this general approach to cancer therapy.

7.5.1 Herpes simplex–thymidine kinase gene therapy

The *HS–tk* gene is the most extensively studied of the sensitivity genes available. The HS–tk enzyme phosphorylates the anti-herpes drugs, acyclovir (ACV) and ganciclovir (GCV), to their monophosphate (MP) forms. Once converted, the MP forms accumulate within the cell. Other endogenous intracellular phosphorylases then convert the MP form to the triphosphate (TP) form. ACV-TP and GCV-TP inhibit DNA polymerase and incorporate into the DNA of dividing cells, causing chain termination and death of the HS–tk-positive cell (Elion, 1980). The endogenous *tk* gene has a much lower affinity for phosphorylation of ACV and GCV resulting in a differential effect in *HS–tk* gene-positive cells, limiting toxicity to transduced cells.

Insertion of the *HS-tk* gene into tumor cells has been shown to result in tumor cell death (Moolten, 1986). He demonstrated that the insertion of the *HS-tk* gene into tumor cells resulted in tumor cell death both *in vitro* and *in vivo* with the administration of GCV. In a series of experiments published in 1992, scientists demonstrated that the GCV treatment of 1 : 1 mixtures of *HS-tk* transduced and non-transduced tumor cells growing in mice resulted in complete tumor killing in 14 of 15 animals (Culver *et al.*, 1992). Interestingly, if only 10% of the tumor cells were *HS–tk*-positive, 9 out of 15 animals were tumor-free 5 weeks later. This phenomenon of destruction of adjacent HS–tk-negative tumor cells has been termed the 'bystander effect'.

The etiology of this bystander tumor cell-killing phenomenon has not been proven. The leading hypotheses are the intercellular transfer of the GCV-TP through gap junctions and the transfer of apoptotic vesicles from dying cells to the surrounding tumor cells through phagocytosis. Bi *et al.* (1993) have demonstrated that radiolabeled GCV will pass from cell to cell only when the cells are in contact. They concluded that either GCV or one of its phosphorylated derivatives is probably being transferred through gap junctions to the neighboring cells, conferring the toxic effects just as in the *HS–tk* transduced cells. Gap junctions will only permit molecules 1000 Da or less to pass, virtually eliminating the possibility that the HS–tk enzyme, which is >100 kDA, is passing between cells. Freeman *et al.* (1993) used flow cytometry to observe that cell fragments of *HS–tk*-positive cells are being internalized by *HS–tk*-negative cells. Their investigations led to the conclusion that neighboring tumor cells were phagocytizing fragments of the HS–tk-destroyed cells, suggesting that the material from the dying *HS–tk*-positive cells are mediating the bystander effect.

While the precise mechanism(s) of the bystander effect remains unclear, the presence of such a process means that a gene delivery method can be less than 100% efficient and still destroy an entire tumor mass. Since there are no gene delivery systems that are 100% efficient, the possibility that the bystander effect will allow complete tumor destruction with gene transfer efficiencies of ⩾10% or more of the cells makes the possibility of human application reasonable. As the investigations into the bystander effect continue in several laboratories, clinical trials are now in progress for the treatment of ovarian cancer and brain tumors with *HS–tk* and GCV.

The ovarian cancer trial is based upon the use of *HS–tk* transduced allogeneic human ovarian cancer cells that are irradiated and delivered into the peritoneal cavity. Scientists hope that the transduced cells will engraft adjacent to the patient's tumor cells and, with the administration of GCV, will produce a bystander effect that will destroy the patient's tumor and induce anti-tumor immunity. This therapeutic approach was evaluated in an intraperitoneal murine model in which mice bearing intraperitoneal (IP) wild-type KBALB sarcoma tumor cells were treated with an IP injection of *HS–tk* transduced KBALB tumor cells. Some of animals had a significant reduction in tumor burden and a significant increase in survival, but all mice developed KBALB non-transduced tumors at the isolated site of needle injection, indicating there is a need for intimate cell-to-cell contact to transfer the bystander factor. Although patients are currently accruing into the trial, there are no preliminary results to discuss at this time.

HS–tk has now been approved for use in five brain tumor protocols involving seven clinical centers (Culver *et al.*, 1993; Oldfield *et al.*, 1993). Each of these trial sites is using MoMLV-based vectors for selective gene transfer into tumor cells because these vectors require a proliferating target cell for integration and vector gene expression. These vectors are manufactured by vector producer cells (VPC) that have been engineered to contain the genes required for production of an infectious, replication-deficient virus particle. The vector particles bud from the genetically engineered cells and can be harvested from the overlying tissue culture medium (Miller, 1990). Initial studies focused on the direct injection of vector-containing supernate, but the gene transfer efficiency was very low (1–3%). However, the injection of the cells producing the retroviral vectors resulted in a gene transfer efficiency of 10–55% (Ram *et al.*, 1993a). Since most cells are not actively dividing at the same time, the direct injection of VPC that actively produces vectors for 7–14 days within the tumor offers a greater opportunity for gene transfer into more slowly dividing tumor cells. Each of the early human gene therapy trials using the injection of *HS–tk* VPC has used a modified NIH 3T3 cell (PA317) to minimize the risk of recombination to produce replication-competent viruses.

Initial animal studies using the subcutaneous implantation of *HS–tk* VPC mixed with tumor cells demonstrated substantial efficacy without local or systemic toxicity. Mixtures of *HS–tk* VPC and a weakly immunogenic fibrosarcoma (MCA205) resulted in complete elimination of the tumor with GCV administration. Next, the use of this *HS–tk in vivo* gene delivery system was evaluated in a Fischer 344 rat brain tumor model. The syngeneic 9L gliosarcoma was stereotactically implanted in the right frontal lobe. Some 5 days later, the *HS–tk* VPC were injected into the tumor using the same stereotactic coordinates. Treatment with GCV resulted in complete tumor destruction with 50–60% of the animals able to survive for more than 4 months, a finding consistent with a cure. Unlike the murine experiments, this treatment utilized xenogeneic VPC, suggesting that the brain was sufficiently immunologically privileged to allow sufficient survival and anti-tumor efficacy in a xenogeneic model. Equally important in these investigations was the high degree of safety afforded to the normal surrounding tissues. Further studies in rodents and rhesus monkeys demonstrated *in vivo* gene transduction, stable gene expression, selective tumor cell destruction and lack of local or systemic toxicity to normal tissues (Ram *et al.*, 1993b).

These pre-clinical data supported the submission of an application to initiate a human clinical trial for the treatment of recurrent brain tumors at the National Institutes of Health. The protocol, which would use magnetic resonance imaging (MRI)-guided stereotactic injection of *HS–tk* VPC into areas of gadolinium enhancement (a marker of blood–brain barrier abnormalities) was initiated on December 7, 1992 (Figure 7.3). Current accrual of patients in this trial has reached 14. Preliminary data are available on the first eight patients, seven with recurrent glioblastoma multiforme (GBM) and one with recurrent metastatic melanoma. As reported by Ram *et al.* at the International Conference of Brain Tumor Therapy in Norway, October, 1993, none of the patients has demonstrated any evidence of an adverse reaction to the injection of the cells or the infusions of GCV. Five with GBM had evidence of an anti-tumor effect related to the therapy as measured by MRI and positron emission tomography (PET) imaging. These anti-tumor effects were manifested as a decrease in size of the gadolinium-enhancing area and cystic formation in the area of *HS–tk* VPC injection. These early experiments also suggest that the larger the VPC inoculum that is delivered per gram of tissue, the greater the anti-tumor effect. Therefore, as with the development of many cancer therapies, the efficiency of delivery (e.g. gene, antibody, etc.) remains the greatest limiting aspect of the potential usefulness of this approach for the treatment of brain tumors.

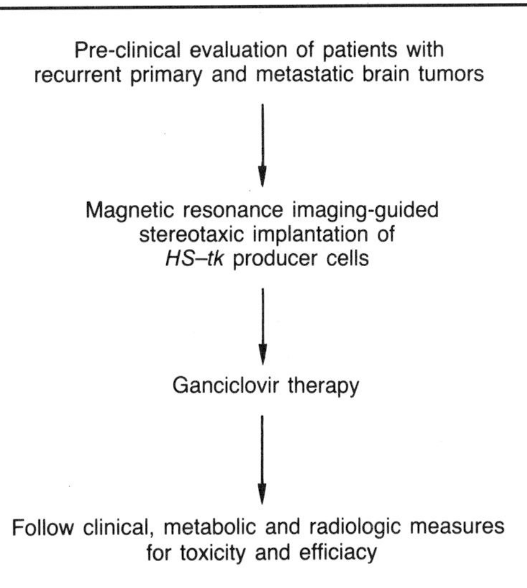

Figure 7.3 The National Institutes of Health (NIH) brain tumor gene therapy trial. Following a thorough assessment of eligibility of patients with recurrent glioblastoma or metastatic tumors (i.e. renal cell, melanoma, lung and breast cancers), the patients undergo the stereotactic implantation of *HS–tk* VPC in multiple locations within the gadolinium-enhancing portion of the tumor. Gene transfer is allowed to occur for 7 days after the injection of cells. GCV is then administered intravenously twice daily at 5 mg/kg/day for 14 days.

Based upon these encouraging results, Culver and colleagues have developed an approach to hopefully improve upon the delivery of the *HS–tk* gene into tumor cells. This trial is focused on the delivery of *HS–tk* into areas of tumor cell growth that cannot be surgically resected and on the repeated injection of VPC into the tumor cavity. (Figure 7.4) In their study, patients with recurrent glioblastoma multiforme tumors will be maximally debulked at surgery and the residual margin of the tumor will be injected with *HS–tk* VPC at 1-cm increments. An access reservoir (Ommaya) will be left in the tumor cavity to allow for sequential injections of VPC into the tumor bed. The fifth approved trial will inject *HS–tk* VPC into the cerebrospinal fluid ventricular system in an effort to treat meningeal carcinomatosis.

One other approach to the treatment of brain tumors with *HS–tk* is in pre-clinical development. In a recent report, recombinant adenovirus vectors were used to deliver the *HS–tk* gene into more than 50% of the C6 glioma cells in a rat brain tumor model (Chen *et al.*, 1994). Treatment with GCV had no discernible toxicities in normal surrounding tissues despite the fact that the *HS–tk* gene was delivered into normal tissues. This suggests that cell proliferation is required for *HS–tk*/GCV cell destruction. This feature appears to further limit the potential toxic effects of the *HS–tk*/GCV system to tumor cells. Efficacy studies have demonstrated that the tumors could be substantially reduced in size in eight of ten mice, with two of ten

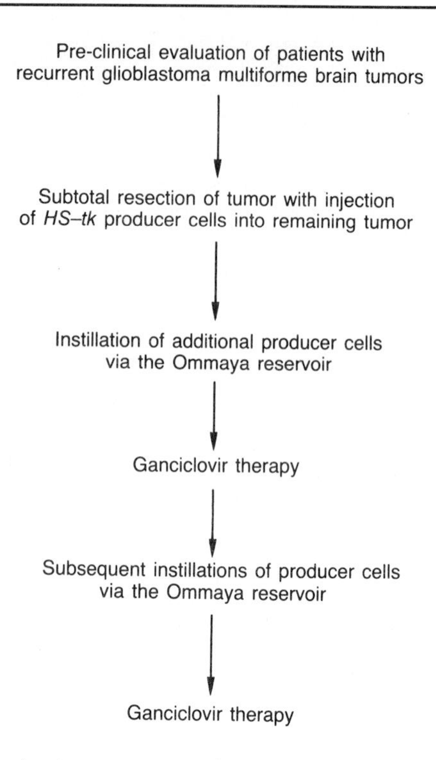

Figure 7.4 The Iowa brain tumor gene therapy trial. In an effort to improve the efficiency of gene delivery into recurrent glioblastoma tumors, the patients will undergo an attempted total tumor resection. Due to the infiltrative nature of these recurrent tumors, a total resection is considered impossible. Since recurrences are generally within 2 cm of the margin of the resection, *HS–tk* will then be injected 1 cm deep and 1 cm apart throughout the walls of the tumor resection. An Ommaya reservoir is left in place. At 7 days after the resection, additional *HS–tk* VPC are injected through the reservoir into the tumor bed; 14 days later, GCV is administered intravenously for 2 weeks. Repeated injections of *HS–tk* and GCV will be given via the reservoir in responding patients.

remaining tumor-free. These studies further document the fact that the *HS–tk*/GCV system can be used safely in the brain. As with retroviral vectors, further optimization of the efficiency and distribution of adenoviral vector delivery of the *HS–tk* gene into solid tumors is needed to effect complete tumor eradication.

Extending this technology to tumors outside of the brain may be possible. Moorman *et al.* (1994) have demonstrated that the intrahepatic injection of *HS–tk* VPC followed by GCV treatment is non-toxic to surrounding normal hepatic tissues. Using *LacZ* VPC, no local hepatocyte transduction was observed and VPC survival was documented to be 7–14 days in this xenogeneic model. Caruso *et al.* (1993) have reported that the direct injection of *HS–tk* VPC into established colon tumors in the rat liver resulted in significant regression of tumors with GCV treatment. Our group is currently investigating the safety and efficacy of the *HS–tk*/GCV *in vivo* gene transfer system in a variety of non-CNS tumor models.

7.5.2 Cytosine deaminase gene therapy

Cytosine deaminase is a bacterial enzyme that converts the antifungal drug 5-FC to the toxic drug 5-FU within the cell. This effect has been demonstrated both *in vitro* and *in vivo* with promising results in treatment of colorectal carcinoma cell lines (Huber *et al.*, 1993). The *CD* gene/5-FC approach does not result in the same degree of bystander effect as *HS–tk* and will require a much greater efficiency of gene transfer for complete tumor destruction. Ongoing pre-clinical studies are in progress at several centers.

7.6 TARGETING THE GENETIC BASIS OF MALIGNANT TRANSFORMATION

Our growing understanding of the genetic basis of cancer allows new opportunities for cancer therapy. These include the insertion of normal tumor suppressor genes (e.g. *p53*) and the introduction of anti-sense oncogene constructs (e.g. *K-ras*) (Baker *et al.*, 1990; Zhang *et al.*, 1993). Pre-clinical data have suggested that these manipulations reduce tumorigenicity leading to the approval of this strategy for the treatment of non-small cell lung cancers. Two protocols are in development. The first uses the repeated injection of retroviral vector supernate and the second, recombinant adenoviral vectors. As the efficiency of gene delivery systems improves, we expect that there will be a number of additional attempts to directly interrupt the genetic basis of malignant transformation.

7.7 CONCLUSIONS

There are new possibilities for the selective immunologic enhancement of an anti-tumor effect, the selective molecular destruction of tumor cells and the interruption of the genetic basis of cancer. As we enter this new era in medicine, we are now poised to begin the acquisition of new clinical tools for treatment of cancer. With further advances in the fields of gene discovery, gene delivery and the genetic basis of cancer, significant advances can be expected in the treatment and prevention of cancer that will have profound implications for the practice of medicine.

REFERENCES

Asher, A.L., Mule, J.J., Kasid, A. *et al.* (1991) Murine tumor cells transduced with the gene for tumor necrosis factor-α. *J. Immunol.*, **146**, 3227–34.

Baker, S.J., Markowitz, S., Fearon, E.R. *et al.* (1990) Suppression of human colorectal carcinoma cell growth by wild-type p53. *Science*, **249**, 912–15.

Bi, W.L., Parysek, L.M., Warnick, R. and Stambrook, P.J. (1993) *In vitro* evidence that metabolic cooperation is responsible for the bystander effect observed with HSV tk retroviral gene therapy. *Human Gene Ther.*, **4**, 725–32.

Brenner, M.K., Furman, W.L., Santana, V.M. *et al.* (1992) Phase I study of cytokine gene modified autologous neuroblastoma cells for treatment of relapsed/refractory neuroblastoma. *Human Gene Ther.*, **3**, 665–76.

Brenner, M.K., Rill, D.R., Moen, R.C. *et al.* (1993) Gene-marking to trace origin of relapse after autologous bone-marrow transplantation. *Lancet*, **341**, 85–6.

Caruso, M., Panis, Y., Gagandeep, S. *et al.* (1993) Regression of established macroscopic liver metastases after *in situ* transduction of a suicide gene. *Proc. Natl Acad. Sci. USA*, **90**, 7024–8.

Chen, S-H., Shine, H.D., Goodman, J.C., Grossman, R.G. and Woo, S.L.C. (1994) Gene therapy for brain tumors: regression of experimental gliomas by adenovirus-mediated gene transfer *in vivo*. *Proc. Natl Acad. Sci. USA*, **91**, 3054–7.

Colombo, M.P. and Forni, G. (1994) Cytokine gene transfer in tumor inhibition and tumor therapy: where are we now? *Immunology Today*, **15**, 48–51.

Culver, K., Ram, Z., Walbridge, S. *et al.* (1992) *In vivo* gene transfer with retroviral vector–producer cells for treatment of experimental brain tumors. *Science*, **259**, 1550–2.

Culver, K.W., Van Gilder, J., Link, C.J. *et al.* (1993) Gene therapy for the treatment of malignant brain tumors with *in vivo* tumor transduction with the herpes simplex thymidine kinase gene/ganciclovir system. *Human Gene Ther.*, **5**, 343–77.

Elion, G.B. (1980) The chemotherapeutic exploitation of virus-specified enzymes. *Adv. Enz. Regul.*, **18**, 53–66.

Freeman, S.M., Abboud, C.N., Whartenby, K.A. *et al.* (1993) The 'bystander effect': tumor regression when a fraction of the tumor mass is genetically modified. *Cancer Res.*, **53** 5274–83.

Gansbacher, B., Houghton, A., Livingston, P.*et al.*(1992a) A pilot study of immunization with HLA-A2 matched allogeneic melanoma cells that secrete interleukin-2 in patients with metastatic melanoma. *Human Gene Ther.*, **3**, 677–90.

Gansbacher, B., Motzer, R., Houghton, A. *et al.* (1992b) A pilot study of immunization with interleukin-2 secreting allogeneic HLA-A2 matched renal cell carcinoma cells in patients with advanced renal cell carcinoma. *Human Gene Ther.*, **3**, 691–703.

Huber, B.E., Richards, C.A. and Krenitsky, T.T. (1993) *In vivo* antitumor activity of 5-fluorocytosine on human colorectal cells genetically modified to express cytosine deaminase. *Cancer Res.*, **53**, 4619–26.

Hwu, P., Yannelli, J., Kriegler, M. *et al.* (1993) Functional and molecular characterization of tumor-infiltrating lymphocytes transduced with tumor necrosis factor-α cDNA for the gene therapy of cancer in humans. *J. Immunol.*, **150**, 4104–15.

Lotze, M.T., Rubin, J.T, Carty, S. *et al.*(1994) Gene therapy of cancer: a pilot study of IL-4-gene-modified fibroblasts admixed with autologous tumor to elicit an immune response. *Human Gene Ther.*, **5**, 41–55.

Miller, A.D. (1990) Retrovirus packaging cells. *Human Gene Ther.*, **1**, 5–14.

Moolten, F.L. (1986) Tumor chemosensitivity conferred by inserted herpes thymidine kinase genes: paradigm for a prospective cancer control strategy. *Cancer Res.*, **46**, 5276–81.

Moorman, D.W., Butler, D.A., Stanley, J.D. *et al.* (1994) Survival and toxicity of xenogeneic murine retroviral vector producer cells in liver. *J. Surg. Oncol.*, **57** (3), 152–6.

Nabel, G.J., Chang, A., Nabel, E.G. *et al.* (1992) Immunotherapy of malignancy by *in vivo* gene transfer into tumors. *Human Gene Ther.*, **3**, 399–410.

Nabel, G.J., Nabel, E.G., Yang, Z. *et al.* (1993) Direct gene transfer with DNA–liposome complexes in melanoma: expression, biologic activity, and lack of toxicity in humans. *Proc. Natl Acad. Sci. USA*, **90**, 11307–11

Oldfield, E.H., Ram, Z., Culver, K.W. and Blaese, R.M. (1993) A clinical protocol: gene therapy for the treatment of brain tumors using intra-tumoral transduction with the thymidine kinase gene and intravenous ganciclovir. *Human Gene Ther.*, **4**, 39–69.

Ram, Z., Culver, K.W., Walbridge, S. *et al.* (1993a) *In situ* retroviral-mediated gene transfer for the treatment of brain tumors in rats. *Cancer Res.*, **53**, 83–8.

Ram, Z., Culver, K.W., Walbridge, S., *et al.* (1993b) Toxicity studies of retroviral-mediated gene transfer for the treatment of brain tumors. *J. Neurosurg.*, **79**, 400–7.

Rosenberg, S.A., Aebersold, P., Cornetta, K. *et al.* (1990) Gene transfer into humans – immunotherapy of patients with advanced melanoma, using tumor-infiltrating lymphocytes modified by retroviral gene transduction. *N. Engl. J. Med.*, **323**, 570–8.

Rosenberg, S.A., Kasid, A., Anderson, W.F. *et al.* (1992) Immunization of cancer patients using autologous cancer cells modified by insertion of the gene for interleukin-2. *Human Gene Ther.*, **3**, 75–90.

Tepper, R.I., Pattengale, P.K. and Leder, P. (1989) Murine interleukin-4 displays potent anti-tumor activity *in vivo*. *Cell*, **57**, 503–12.

Townsend, S.E. and Allison, J.P. (1993) Tumor rejection after direct costimulation of CD8+ T cells by B7-transfected melanoma cells. *Science*, **259**, 368–70.

Trojan, J., Johnson, T.R., Rudin, S.D. *et al.* (1993) Treatment and prevention of rat glioblastoma by immunogenic C6 cells expressing antisense insulin-like growth factor 1 RNA. *Science*, **259**, 94–7.

Tsai, S-C.J., Gansbacher, B., Tait, L., Miller, F.R. and Heppner, G.H. (1993) Induction of antitumor immunity by interleukin-2 gene-transduced mouse mammary tumor cells versus transduced mammary stromal fibroblasts. *J. Natl Cancer Inst.*, **85**, 546–53.

Zhang, Y., Mukhopadhyay, T., Donehower, L.A. *et al.* (1993) Retroviral vector-mediated transduction of K-ras antisense RNA into human lung cancer cells inhibits expression of the malignant phenotype. *Human Gene Ther.*, **4**, 451–60.

8

Vaccine immunotherapy for cancer

JOHN BRIDGEWATER and MARY COLLINS

8.1 INTRODUCTION

Is it possible to be immunized against cancer? Although the epidemiology and natural history of infectious diseases has changed markedly through vaccination, vaccines for cancer have not been used apart from in specific trials. This is mainly because of the intrinsic lack of immunogenicity of most spontaneous human cancers. Recent advances in the understanding of tumor immunology and technical advances in the genetic modification of tumor has stimulated both experimental and clinical interest in tumor vaccines. However, several problems remain:

1. In contrast to infectious diseases, most tumor vaccines, initially, will be used in the context of advanced disease where the tumor has already escaped host immunity. The task of reversing anergy differs from that of preventing it: increased tumor load intuitively implies increased variability and immunological escape. Although many animal studies demonstrate protection from tumor by tumor vaccine few animal studies have addressed the efficacy of vaccine against established metastasis and hence the validity of using this strategy in advanced human disease.
2. Experimental animal tumors are mostly chemically, virally or radiation induced. Such tumors are inherently more immunogenic than spontaneously occurring tumors and may yield misleadingly optimistic results when used as vaccines. Experience with spontaneous human tumors has been less encouraging than that with original murine experiments.
3. Immunological stimulation is a paracrine process, in particular the interaction between cell and cytokine. Tumor heterogeneity, particularly in

Molecular and Cell Biology of Human Gene Therapeutics
Edited by George Dickson
Published in 1995 by Chapman & Hall. ISBN 0 412 62550 4

advanced metastatic disease of solid tumors, dictates that each tumor mass, indeed each individual metastasis, will have a unique antigen and cytokine environment and hence unique response to immune modulation. A differential response to immunotherapy is thus inevitable.

4. Many of the human trials described are not randomized and report survival or response against historical controls. Most tumors described are immunogenic human tumors: renal cell cancer and melanoma are most common. In order to avoid the well-described inter-patient variation and rare incidence of spontaneous response among patient samples as well as selection bias and changes in practice over time, randomized trials are required.
5. Immunological treatment is unlike conventional chemotherapy in its endpoint. Most chemotherapeutic regimes require a complete response or a good partial response for cure or good palliation. There are now many cases where immunotherapy has provided long-term palliation without massive tumor reduction. Immunity may be stimulated to a degree which holds tumorigenicity in check and most importantly, provides good palliation for the patient in a manner that differs essentially from chemotherapy.

8.2 BACKGROUND

The initial demonstrations of cell-mediated tumor immunity were performed by the Hellstroms and others (Boon *et al.*, 1980; Hellstrom and Hellstrom, 1993) using chemically induced tumors in mice. Tumors were grown subcutaneously following inoculation and resected before the development of metastases. Immunocompetent mice were protected when subsequently challenged with that tumor. This rejection was tumor-specific, and the ability to reject was transferred adoptively to other mice through lymphocytes but not serum. A specific antigen was able to regulate rejection: polyoma virus-induced tumors were rejected in mice previously immunized with polyoma virus. However, this model did not hold true for spontaneously arising murine tumors. Recent advances in the understanding of the cellular tumor response (Figure 8.1) might now be exploited to overcome human tumor anergy to immunotherapy.

Some tumor antigens are taken up by antigen-presenting cells and expressed in conjunction with the major histocompatibility complex class II antigens (MHC II) and a co-stimulatory molecule, B7 (Townsend and Allison, 1993). This combined stimulus selects clones of CD4+ T helper cells (Th_1 cells). Absence of B7 or extremes of antigen burden may induce anergy rather than immunity. In the presence of CD4+ clones and an appropriate cytokine milieu, clones of CD8+ T cells become stimulated following CD8+ recognition of tumor antigen in conjunction with class I major histocompatibility antigen (MHC I) (van der Bruggen *et al.*, 1991)

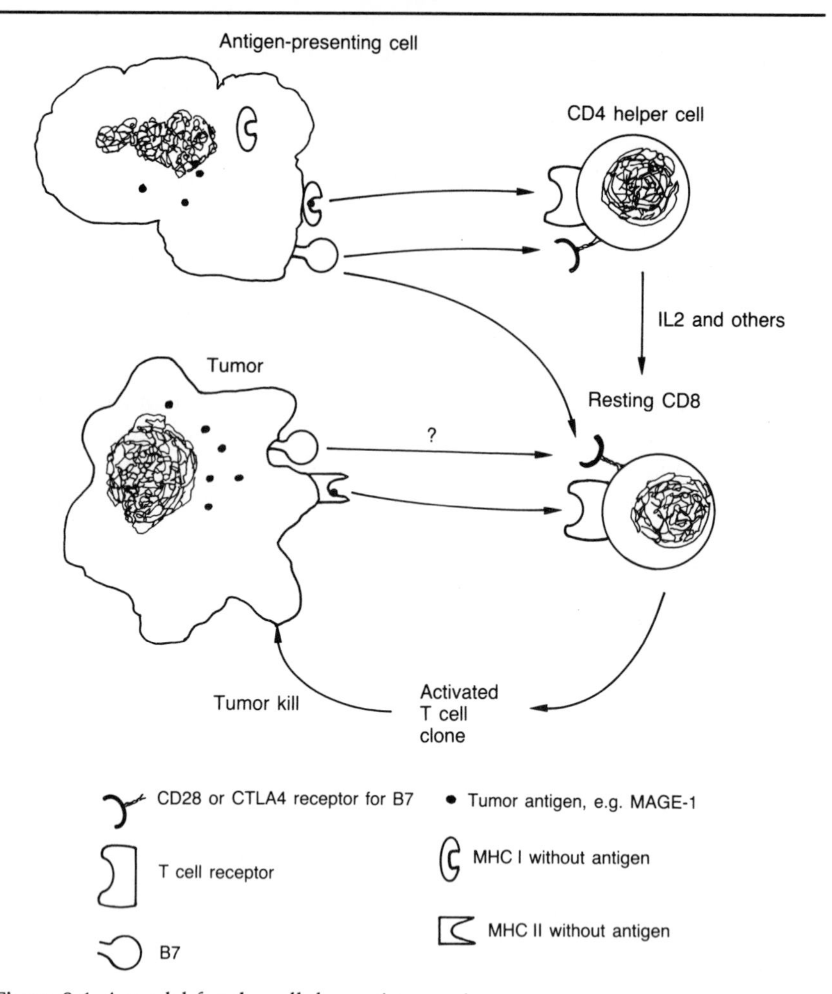

Figure 8.1 A model for the cellular anti-tumor immune response.

and kill tumor (Mukherji *et al.*, 1989). Conversely, in the presence of immunosuppressive cytokines such as interleukin 10 (IL-10) and transforming growth factor-β (TGF-β), CD8+ cells are not activated (de Waal-Malefyt *et al.*, 1993). Many tumors have been found to secrete immunosuppressive cytokines (Gastl *et al.*, 1993; Weiskirch *et al.*, 1994), emphasizing the role of tumor immune escape through local cytokine variation.

8.3 VIRUS AND HAPTEN MODIFIED VACCINE

In 1990, Schirrmacher's group reported results of tumor vaccine modified with a virus not infectious in man, the Newcastle disease virus (NDV)

(Cassel *et al.*, 1983; Bohle *et al.*, 1990). Phase II studies have been performed in patients with colorectal tumors of stage II or III disease which had been fully resected. Irradiated whole tumor cells (10^7) which had been incubated with virus were used as vaccine. A total of 23 patients were treated and have shown a survival advantage over historical matched controls after a short follow-up. Clinical trials using a lytic form of NDV incubated with autologous tumor oncosylates as vaccine have shown improved survival of colon cancer patients against historical controls (Hanna *et al.*, 1993). In Australia, a similar non-randomized trial has used allogeneic melanoma tumor lysate modified with vaccinia (VMCL) in patients with high-risk stage I and II melanoma. Results show a significant trend towards improved survival after 7 years of follow-up both with and without low-dose cyclophosphamide as adjuvant. A randomized trial using VMCL and low-dose cyclophosphamide is in progress (Hersey *et al.*, 1987).

Berd *et al.* (1990) have used irradiated autologous melanoma with Bacille Calmette-Guerin (BCG) and dinitrophenyl benzene (DNP) in patients with metastatic melanoma. A hapten response when using DNP was induced 2 weeks before vaccination. Low-dose cyclophosphamide (CYA) was also used to modify T suppressor cell response. Responses among patients with metastatic disease are approximately 20% and those with a positive delayed hypersensitivity reaction (DTH) show a much improved long-term survival among those treated. Histological examination of tumors revealed a dense CD3+ T cell infiltrate which exhibited clonal restriction for autologous tumor in killing assays. The same vaccine has been used in patients with resected stage I and II high-risk melanoma. Results after 3 years of follow-up suggest an improvement in disease-free survival when compared with historical controls. A randomized study of BCG modified autologous colon cancer has been used by Hanna *et al.* (1993) based on animal work using a hepatocellular cell line. A total of 98 patients were surgically cleared of Dukes' B or C disease and then randomized to receive vaccine or no treatment. A significant survival advantage was present for those who received vaccine and improved survival was correlated with HLA and intercellular adhesion molecule (ICAM) expression.

Mitchell *et al.* (1990) have used a novel tumor vaccine composed of DETOX, a chemical adjuvant, and HLA-matched melanoma cell lysates. Following the demonstration of a specific HLA restricted anti-autologous cytotoxic T lymphocyte (CTL) response in melanoma (Mukherji *et al.*, 1989), this vaccine was refined to act more specifically through this autologous response. Two melanoma cell lines were chosen for their extensive expression of HLA types. Mitchell *et al.* (1993) have used this vaccine in patients with advanced melanoma in phase I and II trials. The overall response rate was 20% and some remissions were durable. The cellular response has been carefully characterized. Tumor-specific T cell clones have been isolated, both CD4+ and CD8+. Vβ chain restriction and specific amplification has been shown both in peripheral blood lymphocytes and in tumor infiltrating lymphocytes (Kan-Mitchell *et al.*, 1993). HLA-A2/28, B 12/44/45 were associated with an improved response, and Mitchell proposes that these HLA types predispose to CTL recognition in this model. If vaccination was followed by treatment with IFN-α in those who failed to respond, the consequent response rate was 44%, with some long-lasting remission (Mitchell *et al.*, 1993). Overall survival was not prolonged.

Several groups have been working on melanoma-specific antigens expressed on allogeneic cultured cell lines. Morton and colleagues (Hoon *et al.*, 1990; Morton *et al.*, 1993) have isolated six antigens from three different melanoma cell lines. Three of these are gangliosides and elicit primarily an IgM response in sera following immunization in patients. The remainder are two glycoproteins and a lipoprotein which elicit both IgG and IgM responses. These cell lines were chosen for their ability to elicit the most powerful antibody response specific to these antigens. Vaccination was given at 2-week intervals for the first 6 weeks and then monthly for 1 year. Adjuvants included cimetidine, indomethacin and CYA. A total of 136 patients with advanced disease were treated over 5 years. Although uncontrolled, the 5-year survival rate for these patients was 25% compared with 9% for historical controls. Longer survival was significantly associated with both increased DTH and IgM response although, as with other groups using serology as a marker of response (Livingston *et al.*, 1994), this IgM response was short lived. Morton's group have recently transfected melanoma cell lines with the interleukin-2 cDNA with consequent decreased tumorigenicity (see below for principle, Uchiyama *et al.*, 1993) and hope to put this combination into clinical trial. A similar approach is that of Livingston *et al.*, 1994). Patients with stage III melanoma received vaccines of tumor cells expressing predominantly GM-2, a monosialoganglioside. This was given with BCG and randomized against BCG alone. A non-significant trend towards improved survival was seen, and was associated with the development of an antibody response. Overall survival was associated with GM-2 expression. Table 8.1 lists most trials performed with chemically and virally altered autologous and allogeneic cellular vaccines.

8.4 GENETICALLY MODIFIED TUMOR VACCINES

Advances in cell and molecular biology have enabled investigators to manipulate the immunogenicity of tumor cells. It became clear that cytokines such as IL-2 were able to stimulate an anti-tumor response. IL-2 was therefore given systemically, initially to animals and subsequently in large human trials, to mimick the cellular response (Rosenberg *et al.*, 1987). In a few cases dramatic complete responses and cures were seen, suggesting that immune disequilibration of tumor anergy had occurred. This was, however, at great cost both in terms of materials and in patient morbidity and

Table 8.1 Virus and hapten-altered vaccines

Reference	*Tumour*	*Vaccine*	*Adjuvant*	*Stage (nos)*	*Result*
Bohle *et al.* (1985)	Colorectal	Autologous	NDV	Resected Dukes' B/C (23)	
Hersey *et al.* (1987)*	Melanoma	Allogeneic	Vaccinia CYA	Stage I/II (80)	Survival advantage over historical controls
Berd *et al.* (1990)	Melanoma	Autologous	BCG DNP CYA	Stage II (41)	Improved disease-free survival over historical controls
Hanna *et al.* (1993)	Colorectal	Autologous	BCG	Resected Dukes' B/C (80)	
Livingstone (1994)	Melanoma	Allogeneic	*Salmonella* BCG CYA	Resected stage III (58)	
Ulsperger *et al.* (1993)*	Colorectal	Allogeneic	*Vibrio cholerae* Neuraminidase Mitomycin C	Resected Dukes' C (212)	
Berd *et al.* (1990)	Melanoma	Autologous	BCG DNP	Metastatic (46)	9% Partial response
Mitchell *et al.* (1990)	Melanoma	Allogeneic	CYA	Metastatic (106)	20% Overall response
Hoon *et al.* (1990)	Melanoma	Allogeneic	CYA Cimetidine Indomethacin	Metastatic (40)	22% Overall response

* Randomized studies. BCG, Bacille Calmette-Guerin; CYA, cyclophosphamide; DNP, dinitrophenyl benzamine; NDV, Newcastle disease virus

Table 8.2 Transfected cytokines and tumorigenicity

Cytokine	*Reference*	*System*	*Protection*
IL-2	Gansbacher *et al.* (1990a)	CMS-5/Balb/c	–
	Patel *et al.* (1993)	FS29/C57Bl	Poor
	Fearon *et al.* (1990)	CT26/Balb/c	Present
	Tsai *et al.* (1993)	4T07/C57Bl	–
IL-4	Tepper *et al.* (1989)	J558L/Balb/c	–
	Golumbek *et al.* (1991)	Renca/Balb/c	Present
	Patel *et al.* (1993)	FS29/C57Bl	Present
IL-7	Hock *et al.* (1991)	J558L/Balb/c	–
IL-12	Tahara *et al.* (1994)	Bl-6/C57Bl	–
IFN-g	Watanabe *et al.* (1989)	C1300/A/J	Present
	Gansbacher *et al.* (1990b)	CMS-5/Balb/c	–
TNF	Blankenstein *et al.* (1991)	J558L/C57Bl	–
GCSF	Colombo *et al.* (1991)	C26/Balb/c	–
GM-CSF	Dranoff *et al.* (1993)	Various	Present

GCSF, Granulocyte colony stimulating factor; GM-CSF, granulocyte macrophage colony stimulating factor; IFN-g, interferon gamma; IL, interleukin; TNF, tumor necrosis factor

mortality. It became clear that high-dose IL-2 stimulated mainly natural killer (NK) cells which were not the most specific nor effective killer cell. Local delivery of cytokine, being less toxic and more logical, was proposed following the demonstration of immune modulation following the introduction of cytokine-secreting tumor cells to animals. Fearon *et al.* (1990) used a chemically induced murine tumor CT26 in BALB/c mice to demonstrate a paracrine effect of locally secreted IL-2. They transfected tumor cells *in vitro* with a plasmid vector containing IL-2 cDNA and subsequently returned these tumor cells to mice. They found that transfected tumors were nearly all rejected, although some tumors progressed when a higher cell inoculum was used. Mice which had rejected tumors were challenged with parental untransfected tumor 2 weeks after rejection. Most were protected from tumor although this 'memory' did not persist to 5 weeks. This rejection did not require CD4+ cells and was tumor-specific.

Tumors have now been engineered to secrete a number of cytokines and demonstrated to provide variable degrees of protection upon re-challenge with the parental tumor (Table 8.2). We have examined the use of both IL-2 and IL-4 in the transplantable murine tumor FS29, a tumor with known low immunogenicity (Patel *et al.*, 1993). Cytokine-secreting tumors were rejected and the best protection after rejection of IL-4 expressing cells (Figure 8.2). Interestingly, those cells expressing IL-2 were lost rapidly, indicating stronger selection pressure against them. This would suggest that low levels of IL-4 encourages specific tumor killing, and concurs with *in vitro* selection methods for growing tumor-specific CTL rather than lymphokine activated killer (LAK) cells. Despite these promising results,

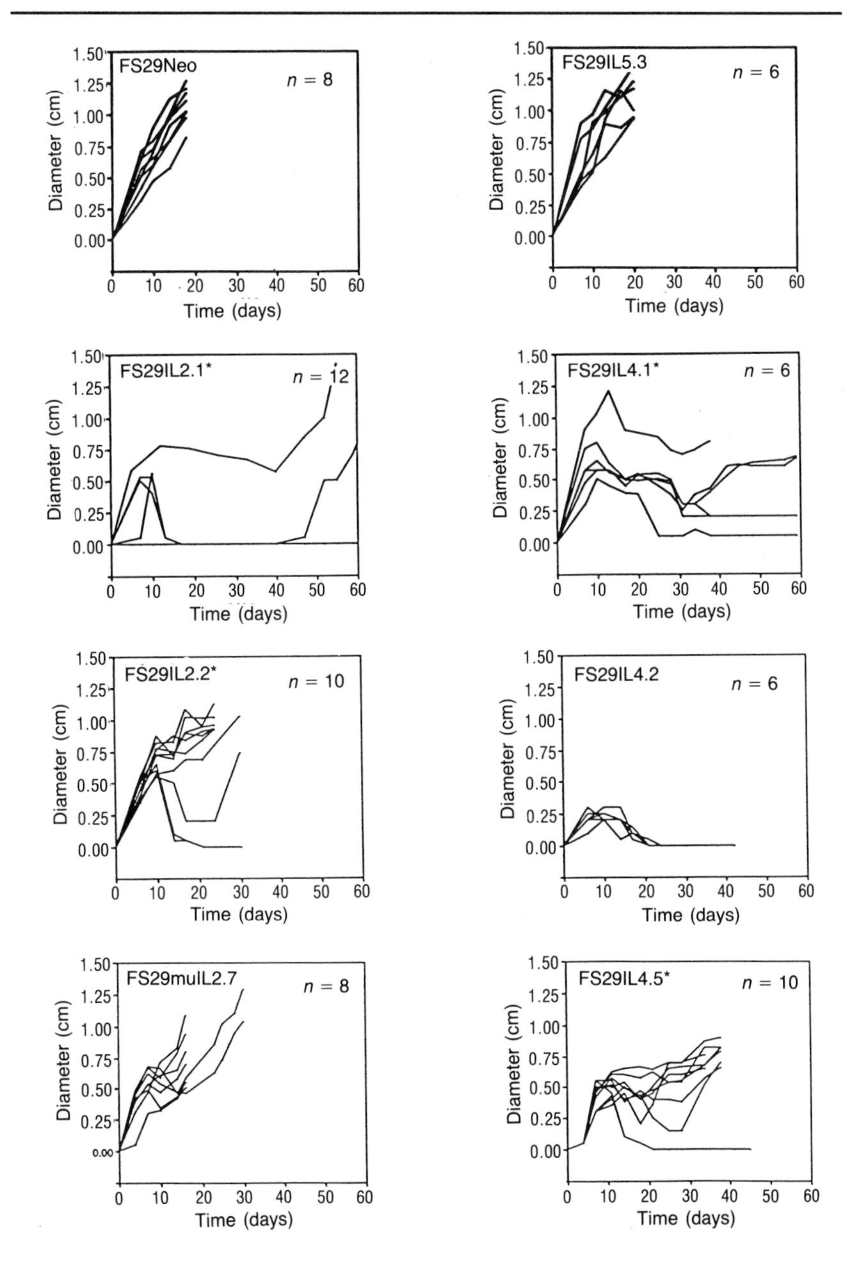

Figure 8.2 Growth of cytokine-secreting tumor cells. Exponentially growing cells (10^6) were injected subcutaneously into C57Bl mice. Tumor diameter was measured twice weekly. Growth for each tumor in a typical experiment is shown. $^*P<0.005$ when tumor size at 16 days compared to FS29Neo (tumor infected with neomycin resistance gene alone) by calculating the standard error of the difference between the means.

several difficulties in the interpretation of animal data concerning cytokine-secreting tumor vaccines remain:

1. Efficacy against established metastatic disease has been shown in only one animal study (Porgador *et al.*, 1993). In phase I trials, most patients will have metastatic disease.
2. Current genetically modified vaccines may provide only limited or variable protection. Blankenstein's group have shown that in some murine models, cytokine modified vaccine is equivalent to vaccination with BCG cell lysate (Hock *et al.*, 1993). Transfection of some murine models may also enhance rather than abrogate tumorigenicity (Qin *et al.*, 1993).
3. Unequivocal evidence of efficacy of cytokine has been shown in only one study. Dranoff *et al.* (1993) showed increased protection of irradiated granulocyte macrophage colony stimulating factor (GM-CSF)-secreting cells over irradiated parental cells. A possible mechanism of action would be intratumoral migration of macrophages leading to increased antigen presentation: both CD4+ and CD8+ cells are required for efficacy. Other studies have not described an unequivocal advantage of irradiated modified vaccine over irradiated parental vaccine, partly because of the poor protection provided by irradiated cells. Conversely, Khazaie *et al.* (1994) have shown that protection is closely associated with the persistence of live tumor at very low levels in the marrow; thus efficacy of vaccines previously described may be due to residual live tumor vaccine rather than to secreted cytokine.

8.5 HUMAN TRIALS USING TUMOR VACCINES

Tumor cells modified to secrete cytokine are being used as tumor vaccines in human gene therapy trials (Table 8.3). The recombinant DNA advisory committee (RAC) in the USA and the genetic therapy advisory committee in the UK (GTAC) have emphasized safety requirements, in particular adequate animal data demonstrating safety and lack of germ-line alteration. Current gene therapy trials in the USA and Europe (Osanto *et al.*, 1993) include modified tumors secreting IL-2, IL-4 and IFN-γ in melanoma, renal cell cancer and neuroblastoma. All trials, once approved, are published in the journal, *Human Gene Therapy*. In conjunction with the Royal Marsden Hospital, we hope to commence a phase I trial using autologous melanoma engineered to secrete IL-2 by retroviral infection (Figure 8.3). Tumor specimens will be resected from patients with metastatic melanoma and established *in vitro*. Once established, cell lines will be multiply infected with replication incompetent retrovirus containing IL-2 cDNA. When the transfected tumor tissue is shown to secrete IL-2, cell samples will be irradiated and returned subcutaneously. The emphasis will be on safety and feasibility.

8.6 TUMOR ANTIGENS AND CO-STIMULATION

In 1991, Boon and co-workers (van der Bruggen *et al.*, 1991) isolated the gene encoding a human melanoma cell surface antigen recognized by specific cytotoxic T lymphocytes. The gene coded for an antigen called MAGE1 which is expressed on 40% of melanomas, many breast and lung tumors and was absent in normal tissue apart from testis. It is presented to CTL by MHC alleles A1 or B45. Reversal of tumor anergy may be possible by presenting MAGE1 by HLA-A1 or HLA-B45 together with an accessory molecule such as B7. Boon and co-workers have proposed a human gene therapy trial using an allogeneic cell line expressing both MAGE1 and HLA-A1 as tumor vaccine (Marchand *et al.*, 1993). If patients are HLA-A1 positive, polymerase chain reaction (PCR) is performed on PBL to determine MAGE1 carriage. If present, the patient will be eligible for vaccine, although only 10% of all melanoma patients will qualify; inclusion of patients who are HLA-B45 positive, an HLA association recently identified, will increase the proportion of patients eligible for treatment. In addition to MAGE, a product of the tyrosinase gene has been identified as an antigen recognized by CTLs, present in both melanocyte and tumor (Brichard *et al.*, 1993). Further use of tyrosinase as a target for gene therapy will depend on strategies to differentiate between tyrosinase on normal tissue and tumor.

Townsend and Allison (1993) have demonstrated the need for co-stimulation of helper lymphocytes by antigen-presenting cells with both MHC II in conjunction with antigen and a co-stimulatory molecule such as B7. Such molecules are typically not expressed by epithelial tumors and presentation of tumor antigen to T cell without B7 could lead to anergy rather than immunity. The expression of B7 in tumor cells increases anti-tumor immunity (Chen *et al.*, 1994) but only in relatively immunogenic tumors: mostly those expressing MHC II. Blocking the interaction with CTLA4 or CD28, natural ligands for B7, abrogates this response. In non-immunogenic tumors, transfection of tumor with another cytokine or stimulation of antigen expression with interferon or IL-7 will also bypass tumor resistance. Similar strategies for manipulating antigen presentation as a mechanism for overturning anergy were demonstrated originally in 1985. MHC I was shown to be expressed in transfected murine sarcoma with associated rejection of inoculated tumor (Tanaka *et al.*, 1985). This effect concurs with data showing concordance between tumorigenicity and loss of MHC expression on tumors (Uyttenhove *et al.*, 1983). Manipulation of MHC expression represents a further avenue for immune intervention.

8.7 FUTURE PROSPECTS

The clinical prospects for tumor vaccines are balanced between virally or hapten-altered cell lysates which provide 20 years of clinical experience and the recent developments in genetically modified tumor vaccine. The latter

Table 8.3 Current and pending human gene therapy trials

Reference/Principal investigator	*Place of study*	*Outline of trial*	*Date(s) of approval (NIH/RAC)*
Rosenberg *et al.* (1992b)	National Institutes of Health	Immunization of cancer patients using autologous cancer cells modified by insertion of the gene for tumor necrosis factor (TNF)	15/10/91, 7/10/91
Rosenberg *et al.* (1992a)	National Institutes of Health	Immunization of cancer patients using autologous cancer cells modified by insertion of the gene for interleukin-2 (IL-2)	15/10/91, 7/10/91
Nabel *et al.* (1992)	University of Michigan	Immunotherapy of malignancy by *in vivo* gene transfer into tumors	17/4/92, 10/2/92
Brenner *et al.* (1992)	St Jude's Children's Research Hospital	Neuroblastoma cells for treatment of relapsed refractory neuroblastoma	14/8/92, 1/6/92
Gansbacher *et al.* (1992a)	Memorial Sloan Kettering	Immunization with HLA-A2 matched allogeneic melanoma cells that secrete interleukin-2 in patients with metastatic melanoma	14/8/92, 2/6/92
Gansbacher *et al.* (1992b)	Memorial Sloan Kettering	Immunization with interleukin-2-secreting allogeneic HLA-A2 matched renal cell carcinoma cells in patients with advanced renal cell carcinoma	14/8/92, 2/6/92
Lotze *et al.* (1994)	University of Pittsburgh	Gene therapy of cancer: a pilot study of IL-4 gene modified antitumor vaccine	5/2/93, 15/9/93

Table 8.3 *Continued*

Reference/Principal investigator	*Place of study*	*Outline of trial*	*Date(s) of approval (NIH/RAC)*
Simons*	Johns Hopkins Oncology Center	Phase I study of non-replicating autologous tumor cell injections using cells prepared with or without granulocyte-macrophage colony stimulating factor gene transduction in patients with metastatic renal cell cancer	–, 1/3/93
Seigler	Duke University Medical Center	A phase I trial of human gamma interferon transduced autologous tumor cells in patients with disseminated malignant melanoma	3/9/93, 7/6/93
Nabel *et al.* (1994)	University of Michigan	Immunotherapy for cancer by direct gene transfer into tumors	3/9/93, 7/6/93
Cassileth*	Miami Veterans Hospital	Phase I study of transfected cancer cells expressing the interleukin-2 gene product in limited stage small cell lung cancer	–, 9/9/93
Das Gupta*	University of Illinois	Immunization of malignant melanoma patients with interleukin-2 secreting melanoma cells expressing defined allogeneic histocompatibility antigens	–, 10/9/93
Economou*	University of California	Genetically engineered autologous tumor vaccines producing interleukin-2 for the treatment of metastatic melanoma	–, 10/9/93

* NIH approval pending 3/3/94

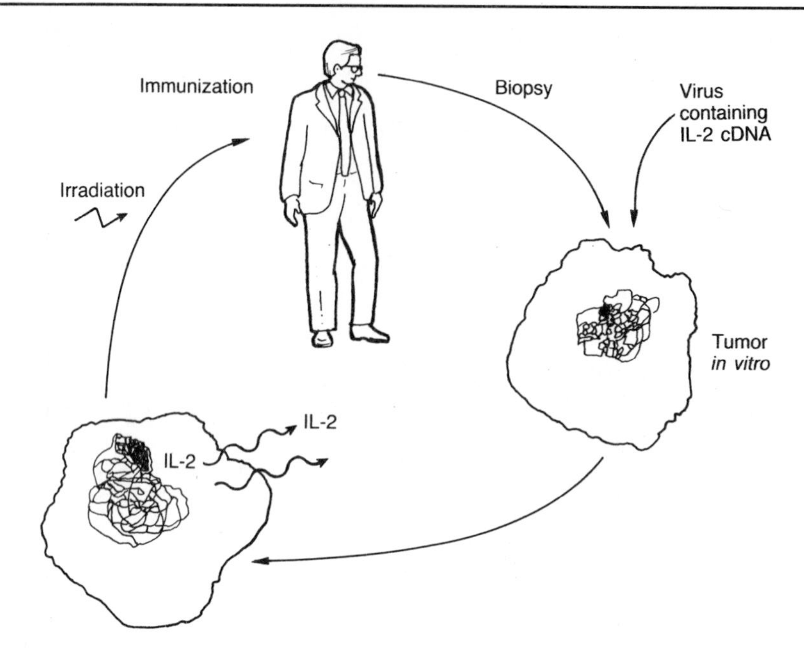

Figure 8.3 Proposed gene therapy for melanoma protocol from the Institute of Cancer Research and Royal Marsden Hospital, 1993.

affords logical scientific rational, supported by elegant *in vitro* and *in vivo* demonstration of cellular anti-tumor immunity. The former has been demonstrated to elicit a humoral response but the cellular responses have not, with few exceptions, been well characterized. A better understanding of this cellular response will be needed for further development. Genetically modified vaccines are in their infancy: as more experience in clinical usage is gained, more sophisticated vaccines demonstrating greater experimental potential will replace them. Single cytokine-secreting tumor vaccines are unlikely to demonstrate dramatic responses. Increasingly sophisticated vaccines secreting more than one cytokine, vaccines transduced with artificial chromosomes or tumor vaccines altered by whole-cell inoculation (Guo *et al.*, 1994) may be more promising. For both approaches, the need for randomized trials following stage I trials is essential, as only in such trials will the intrinsic variability in immunogeneic tumors and enthusiasm among eager physicians be appropriately controlled.

REFERENCES

Berd, D., Maguire, H.C., McCue, P. and Mastrelangelo, M.J. (1990) Active specific immunotherapy for melanoma with an autologous tumour cell vaccine: clinical and immunological results in 64 patients. *J. Clin. Oncol.*, 8, 1858–67.

Blankenstein, T., Qin, Z., Urbela, K. *et al.* (1991) Tumour suppression after tumour cell targeted tumour necrosis factor alpha gene transfer. *J. Exp. Med.*, **173**, 1047–52.

Bohle, W., Schlag, P., Liebrich, W. *et al.* (1990) Post-operative active specific immunisation in colorectal cancer patients with virus modified autologous tumour vaccines: first clinical results with tumour cell vaccines modified with live but avirulent Newcastle disease virus. *Cancer*, **66**, 1517–23.

Boon, T., Snick, J.V., Pel, A.V., Uyttenhove, C. and Marchand, M. (1980) Immunogenic variants obtained by mutagenesis of mouse mastocytoma P815. II. T lymphocyte mediated cytolysis. *J. Exp. Med.*, **152**, 1184–93.

Brenner, M.K. *et al.* (1992) Phase I study of cytokine gene modified autologous neuroblastoma cells for treatment of relapsed refractory neuroblastoma. *Human Gene Ther.*, **3**, 665– 76.

Brichard, V., Pel, A.V., Wolfe, T. *et al.* (1993) The tyrosinase gene encodes an antigen recognised by autologous cytolytic T lymphocytes on HLA-A2 melanomas. *J. Exp. Med.*, **178**, 489–95.

Cassel, W.A., Murray, D.R. and Phillips, H.S. (1983) A phase II study on the postsurgical management of stage II–III malignant melanoma with a newcastle disease virus oncolysate. *Cancer*, **52**, 856–60.

Chen, L., McGowan, P., Ashe, S. *et al.* (1994) Tumor immunogenicity determines the effect of B7 costimulation on T cell-mediated tumor immunity. *J. Exp. Med.*, **179**, 523–32.

Colombo, M.P., Ferrari, G., Stoppacciaro, A. *et al.* (1991) Granulocyte colony stimulating factor gene transfer suppresses tumorigenicity of a murine adenocarcinoma *in vivo*. *J. Exp. Med.*, **173**, 889–97.

de Waal-Malefyt, R., Yssel, H., Roncarlo, M.-G., Spits, H. and de Vries, J. (1992) Interleukin 10. *Curr. Opin. Immunol.*, **4**, 314–20.

Dranoff, D., Jaffee, E., Lazenby, A. *et al.* (1993) Vaccination with irradiated tumour cells engineered to secrete murine GM-CSF stimulates potent, specific and long lasting anti-tumour immunity. *Proc. Natl Acad. Sci. USA*, **90**, 3539–53.

Fearon, E., Pardoll, D., Itaya, T. *et al.* (1990) Interleukin-2 production by tumour cells bypasses T helper function in the generation of anti-tumour response. *Cell*, **60**, 397–403.

Gansbacher, B., Zier, K., Daniels, B., Cronin, K., Bannerji, R. and Gilboa, E. (1990a) Interleukin-2 gene transfer into tumour cells abrogates tumourgenicity and induces protective immunity. *J. Exp. Med.*, **172**, 1217–24.

Gansbacher, B., Bannerji, R., Daniels, B., Zier, K., Cronin, K., and Gilboa, E. (1990b) Retroviral vector mediated g-interferon gene transfer into tumour cells generates potent and long lasting antitumour immunity. *Cancer Res.*, **50**, 7820–5.

Gansbacher, B. *et al.* (1992a) A pilot study of immunisation with HLA-A2 matched allogeneic melanoma cells that secrete interleukin-2 in patients with metastatic melanoma. *Human Gene Ther.*, **3**, 677–90.

Gansbacher, B., *et al.* (1992b) A pilot study of immunisation with interleukin secreting allogeneic HLA-A2 matched renal cell carcinoma cells in patients with advanced renal cell cancer. *Human Gene Ther.*, **3**, 691–703.

Gastl, G.A., Abrams, J.S., Nanus, D.M. *et al.* (1993) Interleukin 10 production by human carcinoma cell lines and its relationship to interleukin 6 expression. *Int. J. Cancer*, **55**, 96–101.

Golumbek, P., Lazenby, A., Levitsky, H. *et al.* (1991) Treatment of established renal cell cancer by tumour cells engineered to secrete interleukin-4. *Science*, **254**, 713–17.

Guo, Y., Wu, M., Chen, H. *et al.* (1994) Effective tumour vaccine generated by fusion of hepatoma cells with activated B cells. *Science*, **263**, 518–20.

Hanna, M.G., Ransom, J.H., Pomato, N. *et al.* (1993) Active specific immunotherapy of colorectal carcinoma with an autologous tumour cell/Bacillus Calmette-Guerin vaccine. *Ann. N.Y. Acad. Sci.*, **690**, 135–46.

Hellstrom, I. and Hellstrom, K.E. (1993) Tumour immunology: an overview. *Ann. N.Y. Acad. Sci.*, **690**, 24–33

Hersey, P., Edwards, A., Coates, A., Shaw, H., McCarthy, W.H. and Milton, G.W. (1987) Evidence that treatment with vaccinia melanoma lysates (VMCL) may improve survival of patients with stage II melanoma. *Cancer Immunol. Immunother.*, **25**, 257–65.

Hock, H., Dorsch, T., Diamanstein, T. and Blankenstein, T. (1991) Interleukin 7 induces CD4+ T cell dependent tumour rejection. *J. Exp. Med.*, **174**, 1291–8.

Hock, H., Dorsch, M., Kunzendorf, U. *et al.* (1993) Vaccinations with tumour cells genetically engineered to produce different cytokines: effectively not superior to classical adjuvant. *Cancer Res.*, **53**, 714–16.

Hoon, D.S.B., Foshag, L.G., Nizze, A.J., Bohman, R. and Morton, D.L. (1990) Suppressor cell activity in a randomised trial of patients receiving active specific immunotherapy with low dosages of cyclophosphamide. *Cancer Res.*, **50**, 5358–64.

Kan-Mitchell, J., Huang, X.-Q., Steinman, L. *et al.* (1993) Clonal analysis of *in vitro* activated CD8+ T lymphocytes from melanoma patient responsive to active specific immunotherapy. *Cancer Immunol. Immunother.*, **37**, 15–25.

Khazaie, K., Prifti, S., Beckhove, P. *et al.* (1994) Persistence of dormant tumour cells in the bone marrow of tumour cell vaccinated mice correlates with long term immunological protection. *Proc. Natl Acad. Sci. USA*, **91**, 7430–4.

Livingston, P.O., Wong, G.Y., Adluri, S. *et al.* (1994) Improved survival in Stage III melanoma patients with GM2 antibodies: a randomized trial of adjuvant vaccination with GM2 ganglioside. *J. Clin. Oncol.*, **12**, 1036–44.

Lotz, M.T., Rubin, J.T., Carty, S. *et al.* (1994) Gene therapy of cancer: a pilot study of IL-4-gene-modified fibroblasts admixed with autologous tumor to elicit an immune response. *Human Gene Ther.*, **5**, 41–55.

Marchand, M., Brasseur, F., van der Bruggen, P., Coulie, P. and Boon, T. (1993) Perspective for immunization of HLA-A1 patients carrying a malignant melanoma expressing gene, MAGE-1. *Dermatology*, **186**, 278–80.

Mitchell, M.S., Harel, W., Kempf, R.A. *et al.* (1990) Active specific immunotherapy for melanoma. *J. Clin. Oncol.*, **8**, 856–69.

Mitchell, M.S., Harel, W., Kan-Mitchell, J. *et al.* (1993) Active specific immunotherapy of melanoma with allogeneic cell lysates. *Ann. N.Y. Acad. Sci.*, **690**, 153–66.

Morton, D.L., Hoon, D.S.M., Nizze, J.A. *et al.* (1993) Polyvalent melanoma vaccine improves survival of patients with metastatic melanoma. *Ann. N.Y. Acad. Sci.*, **690**, 120–34.

Mukherji, B., Guha, A., Chakraborty, N.G. *et al.* (1989) Clonal analysis of cytotoxic and regulatory T cell responses against human melanoma. *J. Exp. Med.*, **169**, 1961–76.

Nabel G., *et al.* (1992) Immunotherapy of malignancy by *in vivo* gene transfer into tumours. *Human Gene Ther.*, **3**, 399–410.

Nabel G.J., Chang, A.E., Nabel, E.G. *et al.* (1994) Immunotherapy for cancer by direct gene transfer into tumours. *Human Gene Ther.*, **5**(1), 57–77.

Osanto, S., Brouwenstyn, N., Vaessen, N. *et al.* (1993) Immunisation with interleukin-2 transfected melanoma cells. *Human Gene Ther.*, **4**, 323 –30.

Patel, P.M., Flemming, C.L., Russell, S.J. *et al.* (1993) Comparison of the potential therapeutic effects of interleukin 2 or interleukin 4 secretion by a single tumour. *Br. J. Cancer*, **68**(2), 295–302.

Porgador, A., Gansbacher, B., Bannerji, R. *et al.* (1993) Anti-metastatic vaccination of tumour bearing mice with IL-2 gene inserted tumour cells. *Int. J. Cancer*, **53**, 471–7.

Qin, Z., Kruger-Krasagakes, S., Kunzendorf, U., Hock, H., Diamantstein, T. and Blankenstein, T. (1993) Expression of tumour necrosis factor by different tumour cell lines results in tumour suppression or augmented metastasis. *J. Exp. Med.*, **178**, 355–60.

Rosenberg, S.A., Lotze, M.T., Muul, L.M. *et al.* (1987) A progress report on the treatment of 157 patients with advanced cancer using lymphokine activated killer cells and interleukin-2 or high dose interleukin-2 alone. *N. Engl. J. Med.*, **316**, 889–97.

Rosenberg, S.A. *et al.* (1992a) Immunisation of cancer patients using autologous cancer cells modified by the insertion of the gene for interleukin-2. *Human Gene Ther.*, **3**, 75–90.

Rosenberg, S.A. *et al.* (1992b) Immunisation of cancer patients using autologous cancer cells modified by the gene for tumour necrosis factor. *Human Gene Ther.*, **3**, 57–73.

Tahara, H., Zeh, H.J., Storkus, W.J. *et al.* (1994) Fibroblasts engineered to secrete interleukin-12 can suppress tumour growth and induce antitumour immunity to a murine melanoma *in vivo*. *Cancer Res.*, **54**, 182–9.

Tanaka, K., Isselbacher, K., Khoury, G. and Jay, G. (1985) Reversal of oncogenesis by the expression of a major histocompatability complex class I gene. *Science*, **228**, 26–30.

Tepper, R., Pattengale, P. and Leder, P. (1989) Murine interleukin-4 displays potent anti-tumour activity *in vitro.Cell*, **57**, 503–12.

Townsend, S.E. and Allison, J.P. (1993) Tumour rejection after direct costimulation of CD8+ T cells by B7-transfected melanoma cells. *Science*, **259**, 368–70.

Tsai, S.C., Gansbacher, B., Tait, L., Miller, F.R. and Heppner, G.H. (1993) Induction of antitumour immunity by interleukin 2 gene transduced mouse mammary tumour cells versus transduced mammary stromal fibroblasts. *J. Natl Cancer Inst.*, **85**, 546–53.

Uchiyama, A., Hoon, D.S.B., Morisaki, T., Kaneda, Y., Yuzuki, D.H. and Morton, D.L. (1993) Transfection of interleukin-2 gene into human melanoma cells augments cellular immune response. *Cancer Res.*, **53**, 949–52.

Ulsperger, E., Rainer, H., Locker, G. *et al.* (1993) Adjuvant vaccination in colorectal carcinoma. *Ann. N.Y. Acad. Sci.*, **690**, 360–3.

Uyttenhove, C., Maryanski, J. and Boon, T. (1983) Escape of mouse mastocytoma P815 after nearly complete rejection is due to antigen loss variants rather than immunosuppression. *J. Exp. Med.*, **157**, 1040–52.

van der Bruggen, P., Traversari, P., Chomez, P. *et al.* (1991) A gene encoding an antigen recognised by cytotoxic T lymphocytes on human melanoma. *Science*, **254**, 1643–8.

Watanabe, Y., Kuribayashi, K., Miyatake, S. *et al.* (1989) Exogenous expression of mouse interferon gamma cDNA in mouse neuroblastoma C1300 cells results in reduced tumourigenicity by augmented anti-tumour activity. *Proc. Natl Acad. Sci. USA*, **86**, 9456–60.

Weiskirch, L.M., Bar-Dagan, Y. and Mokyr, M.B. (1994) Transforming growth factor-beta-mediated down-regulation of antitumour cytotoxicity of spleen cells from MOPC-315 tumour bearing mice engaged in tumour eradication following low dose melphalan therapy. *Cancer Immunol. Immunother.*, **38**, 215–24.

9

Inherited immunodeficiencies

CHRISTINE KINNON

9.1 INTRODUCTION

The inherited immunodeficiency disorders arise as consequences of naturally occurring mutations in the genes which encode various components of the immune system. Increasingly, the genes responsible for these disorders are being identified (Table 9.1). This opens up the possibility for the design of more effective strategies for treating these disorders, including somatic gene therapy.

9.1.1 Elucidation of the defective genes

Table 9.1 summarizes the major inherited immunodeficiencies, dividing them into X-linked and autosomal recessive patterns of inheritance. The defective genes and their chromosomal locations, where known, are indicated. The chromosomal localization information has been important for the identification of the genes in many cases. Such information is necessary for the successful deployment of the positional cloning strategy. This strategy involves the detailed genetic and physical mapping of the gene locus. Genetic linkage studies of affected families and the construction of long-range physical maps of the relevant chromosomal regions, using techniques such as pulsed field gel electrophoresis, serve to reduce the extent of DNA over which the gene can be contained. Once this is reduced to a manageable size, usually about 2Mb or a size which is small enough to be contained in yeast artificial chromosomes (YACs) and cosmids, cDNA selection or exon trapping techniques can be used to identify potential coding sequences. The

Molecular and Cell Biology of Human Gene Therapeutics
Edited by George Dickson
Published in 1995 by Chapman & Hall. ISBN 0 412 62550 4

Table 9.1 Inherited immunodeficiencies

Disease and pattern of inheritance	*Gene defect*	*Chromosomal location*
X-linked disorders		
Severe combined immunodeficiency	Common gamma chain of IL2R, IL4R & IL7R	Xq13.1
Chronic granulomatous disease	gp91-*phox*	Xp21.1
X-linked agammaglobulinemia	*btk*	Xq22
X-linked agammaglobulinemia with hyper IgM	CD40 ligand	Xq26
Properdin deficiency	Properdin	Xq11.23–21.1
X-linked lymphoproliferative syndrome	?	Xq25
Wiskott–Aldrich syndrome	?	Xp11
Autosomal recessive disorders		
Severe combined immunodeficiency	Adenosine deaminase	20q13.11
	Purine nucleoside phosphorylase	14q13.1
	Others	?
MHC class II deficiency	RFX	19p13?
	CIITA	?
Chronic granulomatous disease	p47-*phox*	7q11.23
	p67-*phox*	1q25
	p22-*phox*	16q24
Leukocyte adhesion deficiency	CD18	21q22.3
Ig heavy chain deficiency	Ig heavy chain constant regions	14q23
Ig light (kappa) chain deficiency	Ig kappa light chain	2p11?
IgA deficiency	?	6p21?
Common variable immunodeficiency	?	6p21?
Ataxia telangiectasia	?	11q23
Di–George syndrome	?	22q11
Chediak–Higashi syndrome	?	?

identification of the causative genes has in many cases relied heavily on the development and use of such sophisticated molecular biological techniques.

The positional cloning strategy has been particularly important in the elucidation of the X-linked genes, with the first successful use of this strategy resulting in the identification of the gp91-*phox* gene as being responsible for X-linked chronic granulomatous disease (CGD) (Royer-Pokera *et al.*, 1986). gp91-*phox* was found to be a component of the NADPH oxidase of neutrophils (reviewed in Casimir and Teahan, 1994).

More recently, the positional cloning approach was successfully employed to isolate the *Btk* gene, identified as causing X-linked agammaglobulinemia (XLA) (Vetrie *et al.*, 1993). At the same time, the equivalent murine gene was isolated independently and suggested to be a candidate for the XLA gene based on its chromosomal localization (Tsukada *et al.*, 1993). Similarly, the CD40 ligand (CD40L) and the common γ chain of the interleukin (IL) -2, -4 and -7 receptors were suggested as being candidates for X-linked agammaglobulinemia with hyper IgM (HIGMX1) (Allen *et al.*, 1993a; Aruffo *et al.*, 1993; DiSanto *et al.*, 1993; Korthäuer *et al.*, 1993; Ramesh *et al.*, 1993) and X-linked severe combined immunodeficiency (SCIDX1) (Noguchi *et al.*, 1993b; Puck *et al.*, 1993), respectively.

In some cases the genes for the autosomal recessive disorders have been identified, most notably perhaps, for the autosomal recessive forms of CGD and increasingly for the various forms of severe combined immunodeficiency (SCID), but in many cases they have yet to be elucidated. Problems hindering their characterization are that affected families are much rarer than are those with X-linked conditions and the autosomes are less well mapped than the X chromosome. Perhaps elucidation of the defective genes for the remaining disorders will rely more heavily on candidate genes being suggested.

9.1.2 Diagnosis and treatment

The precise mapping of the genes causing the various immunodeficiency diseases has provided opportunities for the prevention of these diseases. Most often this has been through the identification of carrier females in affected families and first trimester prenatal diagnosis. This approach has been particularly successful for the X-linked immunodeficiencies where it has been possible to use assays based on X-inactivation analysis and gene tracking. X-inactivation analysis relies on the determination of unilateral patterns of X-inactivation in particular cell lineages, those where the expression of an X-linked gene is essential for the development of that cell type. This has been most useful in cases of mothers of single affected males where there has been no previous family history of the disease. Gene tracking is dependent on the use of closely linked polymorphic probes.

Where the genes are now identified it is possible to diagnose the diseases and assess carrier status unambiguously for 'at-risk' females by direct mutation analysis or through the use of intragenic polymorphisms. The identification of the causative genes should lead to an improved understanding of the nature of these disorders and, in time, to improved management and treatment. The currently available treatments remain far from adequate and, in many cases, these diseases are still fatal. Perhaps the most exciting prospect in this area is treatment by somatic gene therapy.

9.1.3 Requirements for gene therapy

For a number of reasons the inherited immunodeficiencies make particularly good candidate disorders for treatment by somatic gene therapy. As discussed already, they are caused by single gene defects and the genes in many cases are now cloned (Table 9.1). In many cases expression of the relevant genes is restricted to cells of the hematopoietic lineage. This provides a relatively accessible target organ for treatment, the peripheral blood or bone marrow. In many cases these disorders are already treatable by bone marrow transplantation. Correction of the defect and reconstitution of a functioning immune system may only require low levels of expression of the relevant protein in some of these situations. This means that although the techniques available at present are far from perfect, it may still be acceptable to attempt this type of treatment in cases where the genetically modified cells may have a selective advantage over the unmodified cells and where there is often no satisfactory alternative treatment.

In this chapter, the nature of some of these disorders is discussed and what is currently known about the defective genes and/or their protein products. In addition, the possibilities for improved counselling and treatment, particularly by somatic gene therapy, are also discussed.

9.2 SEVERE COMBINED IMMUNODEFICIENCY

The SCIDs comprise a syndrome that is characterized by absent cell-mediated and humoral immunity. They encompass a wide variety of different diseases, all with similar clinical characteristics but with a variety of underlying causes.

In many forms of SCID genetic counselling and prenatal diagnosis are possible. In the remainder, fetal blood sampling and lymphocyte subset analysis can be performed at 18–20 weeks' gestation, providing accurate prenatal prediction in most cases.

All forms of SCID are usually fatal by the first or second year of life if not treated successfully by allogeneic bone marrow transplantation. The success rate varies depending on the nature of the donor (reviewed in Fischer *et al.*, 1990). A 90% success rate can be generally achieved using a matched donor, usually a sibling, the situation in about one-third of cases. This is reduced dramatically to a 50 to 60% success rate using a haplo-identical donor, such as a parent (about two-thirds of all cases). Consequently, since matched donors are not always available, there is still a real need to develop a generally applicable and consistent therapy for SCID.

9.2.1 Autosomal recessive severe combined immunodeficiency

About one-third of cases of SCID are caused by defects in autosomal genes. Most of these are caused by defects in adenosine deaminase (ADA) (less than 15% of total SCIDs) and purine nucleoside phosphorylase (PNP) (about 4% of total SCIDs), enzymes involved in purine metabolism. These disorders are indistinguishable phenotypically from other forms of SCID and can be diagnosed directly by analyzing enzyme levels in the blood. There are additional rare autosomal recessive forms of SCID which can involve defects in various components of the T-cell receptor and T-cell activation pathways and which are less well understood (reviewed in Arnaiz-Villena *et al.*, 1992; see also Arpaia *et al.*, 1994).

Adenosine deaminase deficiency

Although other treatments, such as administration of intravenous polyethylene glycol conjugated bovine ADA (PEG-ADA), can be used to treat ADA-deficient SCID, this disorder has become the prime target immunodeficiency disorder for treatment by somatic gene therapy techniques. So far this is the only inherited immunodeficiency for which clinical trials have been undertaken using these techniques (reviewed in Bordignon, 1993; Cournoyer and Caskey, 1993).

From patient studies it is clear that only low levels of enzyme expression are sufficient to reconstitute normal immune function (Hirschhorn, 1990). There is no requirement for the regulated expression of the gene since it is expressed in all cells of the body but only appears to be critical for the normal function of the T cells. Defects in ADA have been recognized as being causative of SCID for over 20 years and the ADA gene was cloned in 1983. Thus, for over 10 years ADA-deficient SCID has been a candidate disorder for this form of treatment. Many basic experiments have been performed, including murine and primate studies, which have demonstrated the efficacy of such a procedure (van Beusechem *et al.*, 1992; Bordignon, 1993; Cournoyer and Caskey, 1993).

Several ADA-deficient SCID patients have now been treated at different centers in the USA and Europe with retroviruses containing a functional ADA gene. The aim is to correct their defects by reconstituting expression in peripheral T cells. This has been attempted directly by introducing the gene into peripheral blood T cells expanded with interleukin 2 (IL-2) *in vitro* – known as tumor-infiltrating (TIL) cells. The first clinical application of somatic gene therapy for ADA-SCID was for the retrovirus-mediated transduction of TIL cells and a second protocol involved transduction of both TIL cells and T-cell depleted bone marrow (Bordignon, 1993; Cournoyer and Caskey, 1993). In more recent protocols 'stem' cell or progenitor cell populations have been used, derived from either the peripheral blood or bone marrow. The advantages of using precursor cells is that they should provide a renewable long-term source of mature T cells. The approach using more mature cells is technically easier but does not effect a long-term cure, requiring the treatment to be repeated at regular intervals. Although the stem cell approach is more difficult, success will mean that the treatment may only need to be performed once. So far, the results of these trials have not been published, but anecdotal evidence suggests cautious optimism for the future of this fast-moving field.

Enzyme replacement, using PEG-ADA to increase stability and half-life and decrease immunogenicity, has also been used to treat this disorder. This has been reported as successful both alone and used in combination with the TIL cell gene therapy protocol, although efficacy appears to be reduced under long-term treatment conditions because of problems of immunogenicity.

Purine nucleoside phosphorylase deficiency

In the same way that ADA-deficient SCID is a prime target for treatment by somatic gene therapy so too is PNP-deficient SCID (Markert, 1991). Very little progress has been made along these lines; however, metabolic correction of PNP-deficient murine lymphoid cells has been achieved by retroviral-mediated gene transfer. Most likely this is a less attractive disorder to researchers because of its relative rarity, only 33 patients worldwide having been reported, 29 of which have died.

MHC class II deficiency (bare lymphocyte syndrome)

Major histocompatibility complex (MHC) class II deficiency is characterized by the presence of normal levels of T and B cells, but these cells have apparently little functional activity since they fail to make cellular or humoral responses to foreign antigens. This means that it is possible to classify this disorder as an autosomal recessive form of SCID. All bone marrow-derived cells in affected individuals fail to express HLA class II antigens (DR, DP and DQ), and this property forms the basis of the diagnosis of the disease.

This lack of expression is the consequence of a failure to synthesize the α and β chain proteins. The defect is not in the class II genes themselves and does not appear to be located within the MHC locus. The defect appears to be in the regulation of expression of the class II genes. There is a specific defect in the binding of a protein factor, RFX, to the highly conserved X box of the HLA class II promoter in some of these patients (Steimle *et al.*, 1993). In a second group of patients which do not display this defect, mutations have been found in another protein, CIITA, which acts as an MHC transactivator (Steimle *et al.*, 1993). This novel regulatory gene is essential for the constitutive expression of all MHC class II genes in B lymphocytes. Thus, two types of HLA class II regulatory mutants can now be distinguished at the molecular level which involve different sets of patients. So far only the CIITA gene has been cloned and transfection experiments have demon-

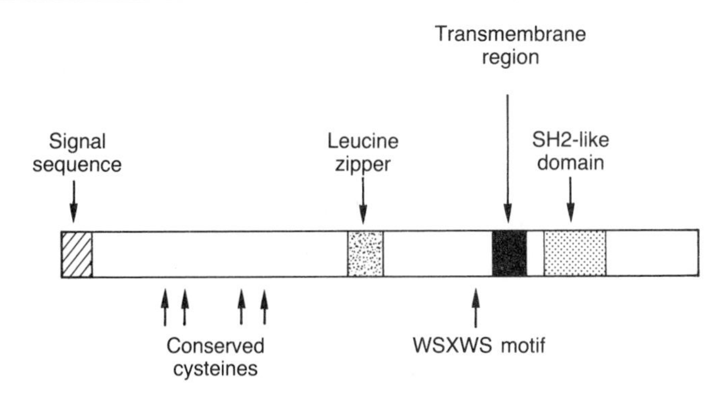

Figure 9.1 A schematic representation of the IL-2R γ chain, highlighting the functional domains.

strated full complementation of the deficiency in MHC class II expression in cell lines from the relevant group of patients. This raises the possibility of treating some patients with this disease by somatic gene therapy in the future.

9.2.2 X-linked severe combined immunodeficiency

The X-linked form of SCID, responsible for about two-thirds of all cases, was recently shown to be caused by defects in the IL-2 receptor γ chain gene (IL-2RG) (Noguchi *et al.*, 1993b; Puck *et al.*, 1993). The IL-2RG is a transmembrane glycoprotein with a number of structural motifs characteristic of cytokine receptor superfamily members (Figure 9.1). Originally identified as a component of the high and intermediate affinity IL-2 receptor that is required to achieve full ligand binding affinity and internalization, the γ chain is expressed constitutively in hematolymphoid cells and is now known to be a component of additional cytokine receptors, including the IL-4 and IL-7 receptors (IL-4R and IL-7R; Kondo *et al.*, 1993; Noguchi *et al.*, 1993a; Russell *et al.*, 1993). It is postulated that in each of these receptors the unique chain recognizes the specific cytokine, and thus confers the signaling specificity, while the common γ chain transduces the signal via its cytoplasmic domain. All three receptors are involved at different stages of T and B cell growth and maturation and the combined effect of disrupting their expression may augment the immune defect. In addition, this could explain the apparent discrepancy between the much milder phenotypes observed in naturally occurring IL-2-deficient humans and IL-2 and IL-4-deficient mice engineered using homologous recombination strategies (Kondo *et al.*, 1993; Noguchi *et al.*, 1993a; Russell *et al.*, 1993).

The identification of mutations in this gene will have important implications for the families involved. However, the majority of SCIDX1 families have not yet been investigated and in many families a typical X-linked pedigree is absent meaning that it is still difficult to diagnose the X-linked form of this disease with certainty, and to differentiate it from one of the autosomal recessive forms. The most effective way to confirm X linkage in families with no previous history is through X-inactivation analysis of T cells from potential female carriers (Goodship *et al.*, 1989). This can provide about a 90% accuracy of diagnosis in about 90% of cases, using highly polymorphic dinucleotide repeat locus-specific probes.

No restriction length polymorphisms (RFLPs) have been found to be associated with the IL-2RG cDNA but single strand conformation polymorphisms (SSCPs) have been detected within both introns 1 and 2 of the gene which should aid in carrier assessment and prenatal diagnosis in some families (Noguchi *et al.*, 1993b). In addition, the recent characterization of multiple, highly polymorphic dinucleotide repeat markers which flank the SCIDX1 locus means that genetic linkage analysis can be performed in most families who have been diagnosed as SCIDX1 with over 99% accuracy (Markiewicz *et al.*, 1993).

The IL-2R, IL-4R and IL-7R are all widely studied cell surface molecules which means that a great deal is already understood concerning their function and biology. Given the severity of this disorder and the relative abundance of patients, SCIDX1 will undoubtedly become a prime target disorder for treatment by somatic gene therapy techniques. A canine model of SCIDX1 already exists, and undoubtedly mice lacking a γ chain will soon be engineered, both of which will be useful as animal models for genetic reconstitution studies. The γ chain protein is expressed constitutively in lymphoid cells, suggesting that the precise control and regulation of expression of the gene will not be required for successful reconstitution of immune function.

9.3 CHRONIC GRANULOMATOUS DISEASE

Chronic granulomatous disease (CGD) is a group of several disorders characterized by defective phagocyte function. Phagocytic cells in children with CGD ingest organisms normally but fail to kill them as a result of defects in the superoxide-producing pathway. This results in recurrent and chronic bacterial and fungal infections in affected individuals.

Treatment with prophylactic antibiotics, especially trimethoprim and sulfamethoxazole, may partially control the frequency of infections. There is now evidence that treatment with γ-interferon improves patient phagocyte function. In addition, in a small number of cases bone marrow transplantation using matched donors has been successful in curing the disease (Casimir and Teahan, 1994; Seger and Ezekowitz, 1994).

All forms of the disease affect the various components of the NADPH oxidase system of phagocytic cells (Figure 9.2). The NADPH oxidase is now

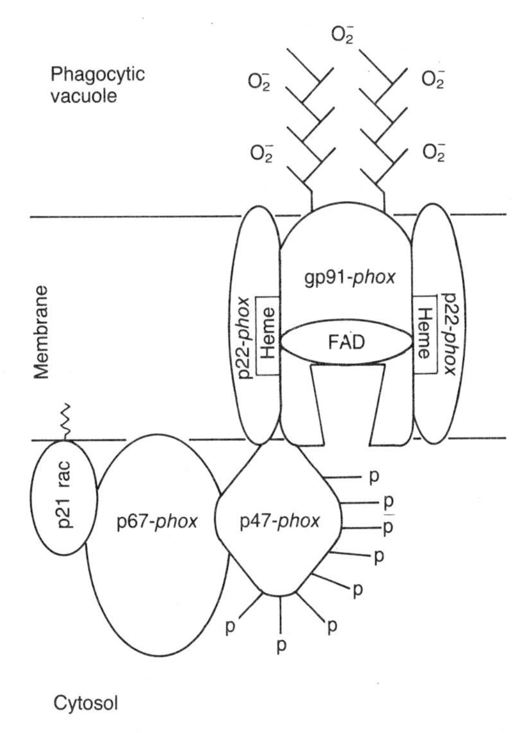

Figure 9.2 A schematic representation of the activated NADPH oxidase of the neutrophils showing the stoichiometry of the complex, the components defective in CGD, p22-*phox*, p47-*phox*, p67-*phox* and gp91-*phox*, p21 rac and the heme, FAD and NADPH binding sites.

known to consist of four specific components, two of which are cytosolic, p47-*phox* and p67-*phox*, and two are the membrane-bound flavocytochrome b_{-245} components, gp91-*phox* and p22-*phox*. Defects in any of these components give rise to CGD, with over 90% of all cases caused by defects in gp91-*phox* and p47-*phox*.

Genetic analysis can be used for unambiguous carrier determination in about 60% of families affected by the X-linked form of this disease (about two-thirds of all cases). A small percentage (about 15%) of patients have deletions within or involving the gp91-*phox* gene, which can be detected by cytogenetic analysis or altered patterns on Southern blot analysis. Although the point mutations in many families have now been identified (reviewed in Casimir and Teahan, 1994), carrier determination and prenatal diagnosis have relied largely on the identification of RFLPs. There are two intragenic RFLPs which should be informative in about 45% of families, allowing reliable carrier detection and first-trimester prenatal diagnosis (Pelham *et al.*, 1990). The majority (over 85%) of all families with defects in p47-*phox* could be diagnosed by virtue of an identical gene abnormality consisting of a dinucleotide deletion at a GTGT tandem repeat at the first intron–exon boundary (Casimir and Teahan, 1994). In families where genetic analysis is not possible, carrier detection can be performed using the nitroblue tetrazolium (NBT) test with more accurate diagnosis depending on investigation of the relevant proteins presence or absence in the neutrophils. Prenatal diagnosis can be performed using the NBT test on fetal blood samples.

Gene therapy experimentation in this area, to date has made extensive use of retroviral vectors to introduce cDNA sequences into the target cell population. Retroviral vectors containing full-length cDNAs for p22-*phox* (Porter *et al.*, 1994), p47-*phox* (Thrasher *et al.*, 1992) and gp91-*phox* (Porter *et al.*, 1993) have been employed to transduce CGD patient lymphoid cell lines and demonstrated significant functional correction of the NADPH oxidase expression and activity. Recently it has been reported that peripheral blood stem cells from p47-*phox*-deficient patients transduced with a p47-*phox*-containing retrovirus could mature into neutrophils and monocytes producing significant levels of superoxide (Sekhsaria *et al.*, 1993). Such experiments must be viewed as the first steps in the process of developing the tools for the treatment of CGD by somatic gene therapy.

9.4 LEUKOCYTE ADHESION DEFICIENCY

Leukocyte adhesion deficiency (LAD) is a rare autosomal recessive disorder that leads to delayed detachment of the umbilical cord and to recurrent severe infections due to impaired leukocyte function (Fischer and Lisowska-Grospierre, 1988). The disease leads to an inability of phagocytic cells to adhere to endothelial cells and subsequently to migrate to sites of infection.

At present, the disease can be diagnosed as early as around 20 weeks of gestation by investigating expression of adhesion molecules on fetal polymorphonuclear and mononuclear cells. It is generally managed by treating infectious episodes with antibiotics.

The disorder is caused by defects in the β chain (CD18) of the adhesion molecules LFA-1, Mac-1 (CR3), and p150,95, which results in their reduced or absent expression. Each of these molecules is a heterodimer consisting of a different α chain (CD11a, CD11b or CD11c) and a shared common β chain (CD18). The CD18 gene has been cloned and heterogeneous mutations described which alter the expression and/or structure of this component. The severity of the disease seems to correlate with the level of expression of these molecules. Biochemical and genetic studies have defined five different types of LAD, including severe and moderate phenotypes.

Studies in somatic cell hybrids suggested that the introduction of a functional copy of CD18 into cells derived from LAD patients reconstituted CD11 and CD18 expression. In a similar way to that described for CGD, the genetic and functional abnormalities in an EBV-transformed B cell line could be corrected by retrovirus-mediated transduction of a functional CD18 gene (Wilson *et al.*, 1990). As discussed above, the clinical severity in LAD is correlated with level of expression and thus significant clinical improvement may be achieved through less than full genetic reconstitution. In addition, some success in curing LAD has been achieved with matched bone marrow transplantation and there is a canine model of LAD which may be useful for testing genetic reconstitution strategies. All these observations suggest that LAD may be a good model system for the further development of somatic gene therapy techniques.

9.5 X-LINKED AGAMMAGLOBULINEMIA

XLA is characterized by agammaglobulinemia involving all immunoglobulin subclasses and is the direct result of a lack of mature B cells. Pre-B cells are present at normal levels in the bone marrow and the defect appears to involve a block in their differentiation.

The treatment is usually a life-time of regular intravenous immunoglobulin infusions. This significantly reduces the episodes of infection and with good management the life expectancy of most boys is extended to a near normal span. Despite this however, the treatment is not a long-term cure and some of these patients succumb to the consequences of bacterial infection and infection with rare neurotropic viruses.

The gene for this disorder was identified as *Btk*, which encodes a novel tyrosine kinase expressed in B cells (Kinnon *et al.*, 1993; Tsukada *et al.*, 1993; Vetrie *et al.*, 1993). The predicted amino acid sequence of Btk is highly homologous to members of the src-related non-receptor tyrosine kinase family, which includes many tyrosine kinases thought to be involved in signal transduction in hematopoietic cells. Based on homology, the Btk protein can be subdivided into a number of functional domains (Figure 9.3).

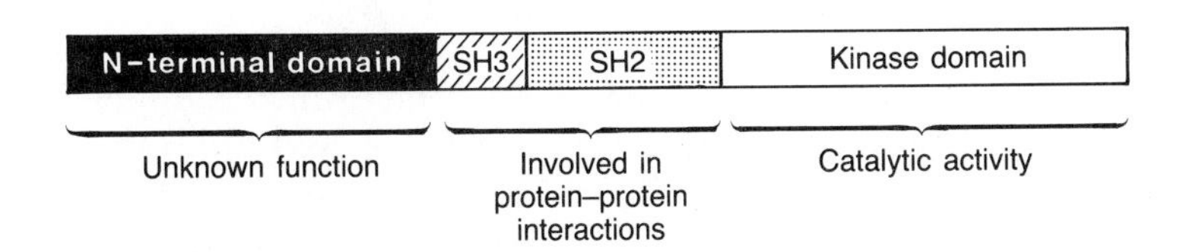

Figure 9.3 A schematic representation of Btk protein, highlighting the functional domains.

The identification of deletions within *Btk* in some patients (about 15%) indicates that the disorder is likely to arise as a loss-of-function of the protein (Vetrie *et al.*, 1993). The further investigation of such mutations, particularly point mutations resulting in amino acid substitutions in this protein, may shed light on its function in B cell development. Furthermore, this could help elucidate the functioning of other tyrosine kinases where mutations more usually result in a gain-of-function, for example, in diseases where constitutive activation of the kinase domain arising from point mutations are associated with an oncogenic potential.

The identification of mutations in *Btk* in XLA patients has allowed unambiguous assessment of carrier status in female relatives of affected boys (Lovering *et al.*, 1994). An SSCP polymorphism has been detected with a heterozygosity of about 35% which may be useful for carrier determination in some families (Bradley *et al.*, 1994). Genetic mapping of the XLA locus has revealed a number of conventional RFLP and dinucleotide repeat loci that flank the *Btk* locus (Kinnon *et al.*, 1993) and accurate genetic counselling on this basis should now be available and for most families. Carrier determination is also possible by analysis of X inactivation patterns in B cells of potential female carriers.

With an improved understanding of the role of Btk in XLA and in normal immune function it will be possible to develop new treatments, including somatic gene therapy. From what we know already, it is likely that reconstitution of Btk activity, even at low levels, in B cells at the correct stage of development should be sufficient to confer a growth advantage to those cells. However, before such a treatment can be envisaged, it will be necessary to investigate the effects of over-expression of Btk, as it is possible that it may have a potential for oncogenesis. The generation of a transgenic mouse over-expressing the Btk protein will provide more definitive evidence for a lack of oncogenic potential for Btk. Reconstitution of a complete immune system in an immunodeficient mouse with a mutation in Btk, the *xid* mouse strain (reviewed in Kinnon *et al.*, 1993), would be a useful first step to the development of such techniques for eventual clinical use.

9.6 X-LINKED AGAMMAGLOBULINEMIA WITH HYPER IgM SYNDROME

HIGMX1 is characterized by agammaglobulinemia which is distinguished from XLA by normal or aberrantly high levels of IgM antibodies, with absent or very low IgG and IgA. The IgM antibodies may in some cases be monomeric and may lack function. B cells are present at normal levels in this disease.

Although diagnosis is relatively straightforward, some cases are misdiagnosed as XLA. In addition to the elevated levels of IgM, affected boys are prone to infections more typical of T-cell deficiencies, such as *Pneumocystis carinii* and cryptosporidial diarrhea. As with XLA, the treatment is by regular immunoglobulin infusion.

Until recently the HIGMX1 locus was mapped only crudely, largely because of the rarity of families suffering from this disorder, but has now been shown to map close to the HPRT locus in Xq26 (Padayachee *et al.*, 1992). At around the same time the gene for CD40L, a cell surface marker of activated T cells, was isolated in several laboratories and, with the knowledge that it maps near to the HPRT locus, CD40L became a very good candidate for the HIGMX1 gene. A total of 14 HIGMX1 patients from different families have now been analyzed and found to have mutations in their CD40L genes (Allen *et al.*, 1993a; Aruffo *et al.*, 1993; DiSanto *et al.*, 1993; Korthäuer *et al.*, 1993; Ramesh *et al.*, 1993). As with the other disorders, these mutations are varied, unique to individual families and will provide a basis for carrier determination and prenatal diagnosis. A dinucleotide repeat polymorphism has also been identified in the 3′ untranslated region of the CD40L gene (Allen *et al.*, 1993b; DiSanto *et al.*, 1993). The heterozygosity, reported as being between 70 to 80%, depending on the ethnic origin of the population studied, suggests that this should be useful in ascertaining carrier status (Allen *et al.*, 1993b). Furthermore, the genomic structure of this gene has now been elucidated and will be useful for the identification of additional mutations and for genetic counselling in this disease. Already this information has been applied to prenatal diagnosis of HIGMX1 using polymerase chain reaction (PCR) amplification of genomic DNA derived from fetal samples (Villa *et al.*, 1994).

It has been shown that CD40L can be expressed in transfected human B cell lines and that such cells can be used to induce isotype switching to IgE production in the presence of IL4 (Allen *et al.*, 1993a). The reconstitution of expression of CD40L in patient peripheral T cells and the demonstration of their ability to induce switch recombination in B cells *in vitro* must be the next step towards the development of somatic gene therapy for HIGMX1. The main problem that can be envisaged is that CD40L is usually only expressed in activated T cells and deregulated expression may lead to as yet unforeseen problems. The isolation of genomic clones allows study of the genetic control elements of CD40L expression (Villa *et al.*, 1994). It can be foreseen that such elements will be incorporated into constructs designed to allow more specific control of expression.

9.7 PROPERDIN DEFICIENCY

Properdin is a serum protein that belongs to the alternative pathway of complement activation; it is involved in the stabilization of the enzyme complex $C3b_2Bb$. Absence or deficiency of properdin is associated with severe, often fatal pyogenic infections, often with meningococci.

The properdin protein has been purified and found to be a multimeric glycoprotein with a subunit molecular weight of 56 kDa. The gene has been cloned and it is not difficult to envisage that gene therapy for this disorder would be a relatively simple task since the protein is secreted and would not require tissue-specific expression. Most likely it has not stimulated much attention in this area because of the relatively few patients affected by the disorder.

9.8 FUTURE PROSPECTS

It is likely with the current rate of progress continuing that the next few years will see the elucidation of the genes involved in the remaining primary immunodeficiency disorders described in Table 9.1. Indeed, it is possible to envisage that additional genes with specific roles in immune function will be identified as playing a part in more rare sporadic immunodeficiencies. For example, genes involved in rare chromosomal breakpoints and translocations which result in immune dysfunction. Some of these disorders will become targets for treatment by the same kind of somatic gene therapy protocols described here.

In addition, other disorders caused by single gene defects will become targets for this kind of treatment. Although most of the first clinical protocols for somatic gene therapy have targeted the congenital disorders of the hematolymphoid system and have had hematopoietic stem cells as their target tissue of choice, these techniques can be applied to many other congenital disorders. This includes not only a variety of hematological disorders such as the hemophilias, but also those caused by defects in secreted proteins normally present in the blood but not necessarily of hematolymphoid origin or acting on these cells (Table 9.2). It is possible to envisage the blood used as a gene delivery system for other target organs.

As discussed in this chapter, gene therapy experimentation to date in the area of primary immunodeficiencies has made extensive use of retroviral vectors to introduce DNA sequences into the target cells. Increasingly attempts are being made to transduce purified stem cell populations derived from bone marrow and peripheral blood. However, these populations are still rather heterogeneous in nature. It is likely that stem cell purification procedures must be improved to give higher yields of stem cells, rather than the presently obtained mixture of progenitors and stem cells, before this approach can be completely successful.

A major disadvantage affecting the use of retroviral vectors is that they have a very low efficiency of integration in cells not actively in cycle, including stem cells. One particularly attractive gene transduction system for potential use in clinical protocols is based on adeno-associated virus (AAV). AAV is an endogenous human virus with many properties which suggest it could be utilized as an ideal vector for stem cell transduction. Perhaps most significantly, cell division is thought not to be a prerequisite for AAV infection (Lebkowski *et al.*, 1988; Philip *et al.*, 1994). Further improvements to retroviral vector systems are likely to be in the form of refinements rather than radical in nature.

Table 9.2 Candidate disorders, in addition to the inherited immunodeficiencies, for hematolymphoid stem cell somatic gene therapy

Disease	*Gene defect*
Hemophilias	
Factor IX deficiency	Factor IX
Factor VIII deficiency	Factor VIII
Anemias	
β-thalassemia	β-globin
Sickle cell anemia	β-globin
Lysosomal storage disorders	
Gaucher's disease	Glucocerebrosidase
Hurler's syndrome	Iduronidase
Mucopolysaccharidosis type VII	β-glucuronidase
Lipid metabolism disorders	
Hypercholesterolemia	Low density lipoprotein (LDL) receptor
Emphysema	
α_1-antitrypsin deficiency	α_1-antitrypsin
Urea cycle disorders	
Hyperammonemia	Ornithine transcarbamylase (OTC)
Citrullinemia	Argininosuccinate synthetase
Other metabolic disorders	
Phenylketonuria	Phenylalanine hydroxylase
Hypoxanthine-guanine phosphoribosyl transferase (HGPT) deficiency	HGPT

In summary, substantial progress has been made along the pathway of establishing somatic gene therapy as a viable treatment and potential cure for many of the inherited immunodeficiencies. The next few years should see the development and refinement of the techniques necessary for successful treatment by gene therapy. Indeed, in the near future we should see the fruition of much of this work in the development and application of clinical protocols for several of these diseases.

ACKNOWLEDGEMENTS

I would like to thank colleagues, particularly Dr Colin Casimir, and members of my laboratory for helpful discussions, and the Wellcome Trust, the Medical Research Council (UK) and Action Research for their support.

REFERENCES

Allen, R.C., Armitage, R.J., Conley, M.E. *et al.* (1993a) CD40 ligand gene defects responsible for X-linked hyper IgM syndrome. *Science*, **259**, 990–3.

Allen, R.C., Spriggs, M.K. and Belmont, J.W. (1993b) Dinucleotide repeat polymorphism in the human CD40 ligand gene. *Hum. Mol. Genet.*, **2**, 828.

Arnaiz-Villena, A., Timón, M., Rodriguez-Gallego, C. *et al.* (1992) Human T-cell activation deficiencies. *Immunology Today*, **13**, 259–65.

Arpaia, E., Shahar, M., Dadi, H., Cohen, A. and Roifman, C.M. (1994) Defective T cell receptor signalling and $CD8^+$ thymic selection in humans lacking Zap-70 kinase. *Cell*, **76**, 947–58.

Aruffo, A., Farrington, M., Hollenbaugh, D. *et al.* (1993) The CD40 ligand, gp39, is defective in activated T cells from patients with X-linked hyper-IgM syndrome. *Cell*, **72**, 291–300.

Bordignon, C. (1993) Progress toward the clinical application of somatic gene therapy. *Curr. Opin. Hematol.*, **1**, 246–51.

Bradley, L.D., Sweatman, A.K., Lovering, R.C. *et al.* (1994) Mutation detection in the X-linked agammaglobulinemia gene, *BTK*, using single strand conformation polymorphism analysis. *Hum. Mol. Genet.*, **3**, 79–83.

Casimir, C.M. and Teahan, C.G. (1994) The respiratory burst of neutrophils and its deficiency, in *Immunopharmacology of Neutrophils*, (eds P.G. Hellewell and T.J. Williams), Academic Press, London, pp. 27–54.

Cournoyer, D. and Caskey, C.T. (1993) Gene therapy of the immune system. *Annu. Rev. Immunol.*, **11**, 297–329.

DiSanto, J.P., Bonnefoy, J.Y., Gauchat, J.F. *et al.* (1993) CD40 ligand mutations in X- linked immunodeficiency with hyper IgM. *Nature*, **361**, 541–3.

Fischer, A. and Lisowska-Grospierre, B. (1988) Leukocyte adhesion deficiency: molecular basis and functional consequences. *Immunodeficiency Rev.*, **1**, 39–54.

Fischer, A., Landais, P., Friedrich, W. *et al.* (1990) European experience of bone-marrow transplantation for severe combined immunodeficiency. *Lancet*, **336**, 850–4.

Goodship, J., Levinsky, R.J. and Malcolm, S. (1989) Linkage of PGK1 to X-linked severe combined immunodeficiency (IMD4) allows predictive testing in families with no surviving male. *Hum. Genet.*, **84**, 11–14.

Hirschhorn, R. (1990) Adenosine deaminase deficiency. *Immunodeficiency Rev.*, **2**, 175–98.

Kinnon, C., Hinshelwood, S., Levinsky, R.J. and Lovering, R.C. (1993) X-linked agammaglobulinemia – gene cloning and future prospects. *Immunology Today*, **14**, 554–8.

Kondo, M., Takeshita, T., Ishii, N. *et al.* (1993) Sharing of the interleukin-2 (IL-2) receptor γ chain between receptors for IL-2 and IL-4. *Science*, **262**, 1874–7.

Korthäuer, U., Graf, D., Mages, H.W. *et al.* (1993) Defective expression of T-cell CD40 ligand causes X-linked immunodeficiency with hyper-IgM. *Nature*, **361**, 539–41.

Lebkowski, J.S., McNally, M.M., Okarma, T.B. and Lerch, L.B. (1988) Adeno-associated virus: A vector system for efficient introduction of DNA into a variety of mammalian cell types. *Mol. Cell Biol.*, **8**, 3988–96.

Lovering, R.C., Sweatman, A., Genet, S.A. *et al.* (1994) Identification of deletions in the *btk* gene allows unambiguous assessment of carrier status in families with X-linked agammaglobulinaemia. *Hum. Genet.*, **94**, 77–9.

Markert, M.L. (1991) Purine nucleoside phosphorylase deficiency. *Immunodeficiency Rev.*, **3**, 45–81.

Markiewicz, S., DiSanto, J.P., Chelly, J. *et al.* (1993) Fine mapping of the human SCIDX1 locus at Xq12–13.1 *Hum. Mol. Genet.*, **2**, 651–4.

Noguchi, M., Nakamura, Y., Russell, S.M. *et al.* (1993a) Interleukin-2 receptor γ chain: a functional component of the interleukin-7 receptor. *Science*, **262**, 1877–80.

Noguehi, M., Yi, H., Rosenblatt, H.M. *et al.* (1993b) Interleukin-2 receptor γ chain mutation results in X-linked severe combined immunodeficiency. *Cell*, **73**, 147–57.

Padayachee, M., Feighery, C., Finn, A. *et al.* (1992) Mapping of the X-linked form of hyper IgM syndrome (HIGM1) to Xq26 by close linkage to HPRT. *Genomics*, **14**, 551–3.

Pelham, A., O'Reilly, M.-A., Malcolm, S. *et al.* (1990) RFLP and deletion analysis for X-linked chronic granulomatous disease using the cDNA probe: potential for improved prenatal diagnosis and carrier determination. *Blood*, **76**, 820–4.

Philip, R., Brunette, E., Kilinski, L. *et al.* (1994) Efficient and sustained gene expression in primary T lymphocytes and primary and cultured tumor cells mediated by adeno-associated virus plasmid DNA complexed to cationic liposomes. *Mol. Cell Biol.*, **14**, 2411–18.

Porter, C.D., Parkar, M.H., Levinsky, R.J. *et al.* (1993) X-linked chronic granulomatous disease: correction of NADPH oxidase defect by retrovirus-mediated expression of gp91-*phox*. *Blood*, **82**, 2196–202.

Porter, C.D., Parkar, M., Verhoeven, A.J. *et al.* (1994) p22-*phox*-deficient chronic granulomatous disease: identification of a biosynthetic intermediate of gp91-*phox* and reconstitution by retrovirus-mediated expression. *Blood*, **84**, 2767–75.

Puck, J.M., Deschênes, S.M., Porter, J.C. *et al.* (1993) The interleukin-2 receptor γ chain maps to Xq13.1 and is mutated in X-linked severe combined immunodeficiency, SCIDX1. *Hum. Mol. Genet.*, **2**, 1099–104.

Ramesh, N., Fuleihan, R., Ramesh, V. *et al.* (1993) Deletions in the ligand for CD40 in X-linked immunoglobulin deficiency with normal or elevated IgM (HIGMX-1). *Int. Immunol.*, **5**,769–73.

Royer-Pokera, B., Kunkel, L.M., Monaco, A.P. *et al.* (1986) Cloning the gene for an inherited human disorder – chronic granulomatous disease – the basis of its chromosomal location. *Nature*, **322**, 32–8.

Russell, S.M., Keegan, A.D., Harada, N. *et al.* (1993) Interleukin-2 receptor γ chain: a functional component of the interleukin-4 receptor. *Science*, **262**, 1880–3.

Seger, R.A. and Ezekowitz, R.A.B. (1994) Treatment of chronic granulomatous disease. *Immunodeficiency*, **5**, 113–30.

Sekhsaria, S., Gallin, J.I., Linton, G.F. *et al.* (1993) Peripheral blood progenitors as a target for genetic correction of p47*phox*-deficient chronic granulomatous disease. *Proc. Natl Acad. Sci. USA*, **90**, 7446–50.

Steimle, V., Otten, L.A., Zufferey, M. and Mach, B. (1993) Complementation of an MHC class II transactivator mutated in hereditary MHC class II deficiency (or bare lymphocyte syndrome). *Cell*, **75**, 135–46.

Thrasher, A., Chetty, M., Casimir, C. and Segal, A.W. (1992) Restoration of superoxide generation to a chronic granulomatous disease-derived B-cell line by retrovirus mediated gene transfer. *Blood*, **80**, 1125–9.
Tsukada, S., Saffron, D.C., Rawlings, D.J. *et al.* (1993) Deficient expression of a B cell cytoplasmic tyrosine kinase in a human X-linked agammaglobulinemia. *Cell*, **72**, 279–90.
van Beusechem, V.W., Kukler, A., Heidt, P. J. *et al.* (1992) Long-term expression of human adenosine deaminase in rhesus monkeys transplanted with retrovirus-infected bone marrow cells. *Proc. Natl Acad. Sci. USA*, **89**, 7640–4.
Vetrie, D., Vorechovsky, I., Sideras, P. *et al.* (1993) The gene involved in X-linked agammaglobulinaemia is a member of the *src* family of protein-tyrosine kinases. *Nature*, **361**, 226–33.
Villa, A., Notarangelo, L.D., DiSanto, J.P. *et al.* (1994) Organization of the human CD40L gene: implications for molecular defects in x chromosome-linked hyper-IgM syndrome and prenatal diagnosis. *Proc. Natl Acad. Sci. USA*, **91**, 2110–14.
Wilson, J.M., Ping, A.J., Kraus, J.C. *et al.* (1990) Correction of CD18-deficient lymphocytes by retrovirus-mediated gene transfer. *Science*, **248**, 1413–16.

10

Anti-viral strategies

MARINEE K.L. CHUAH, THIERRY VANDENDRIESSCHE
and RICHARD A. MORGAN

10.1 INTRODUCTION

A variety of viral diseases that are refractory to vaccination or therapy are potentially amenable for treatment by gene therapy. Human gene therapy can be defined as the introduction of new genetic material into cells of an individual with resulting therapeutic benefit to the individual (Morgan and Anderson, 1993). Gene therapy for viral diseases requires the introduction of anti-viral genes into cells, to prevent or inhibit viral gene expression or function and consequently limit viral replication and pathogenesis.

Anti-viral gene therapy offers new opportunities for the intervention of viral replication at the molecular level such as the possibility to target essential viral proteins or conserved cis-acting regulatory sequences. A combination of different anti-viral genes can be simultaneously introduced into cells to target multiple stages in the viral life cycle. In addition to this intracellular intervention, gene therapy may be employed to intervene with viral spread at the extracellular level. This could be achieved by sustained expression *in vivo* of a secreted anti-viral protein at high local concentrations or by stimulation of an anti-viral immune response.

It is beyond the scope of this review to discuss in detail all human pathogenic viruses that are potential candidates for treatment by gene therapy. Instead, the emphasis will be on the underlying principles of gene therapy for viral diseases and therefore only the most typical viruses will be presented. Since most efforts in developing anti-viral gene therapy strategies were aimed at inhibiting HIV, anti-HIV gene therapy will be discussed more

Molecular and Cell Biology of Human Gene Therapeutics
Edited by George Dickson
Published in 1995 by Chapman & Hall. ISBN 0 412 62550 4

extensively and the emerging concepts from these HIV studies can be extrapolated to other viral diseases.

10.2 ANTI-VIRAL GENE THERAPY STRATEGIES

To implement an efficacious gene therapy strategy for viral infection, it is critical to efficiently deliver the anti-viral genes to the appropriate target cell. Various methods of gene transfer, including viral vectors based on retrovirus, adeno-associated virus and adenovirus as well as non-viral mediated gene transfer techniques, are being evaluated as gene-delivery systems. A detailed description of these gene transfer methods has been presented in other chapters and will therefore not be discussed here.

Viruses can be inhibited by developing strategies that target any of the viral proteins or their respective cis-acting regulatory elements. Anti-viral gene therapy strategies can be either protein-based (trans-dominant viral proteins, cellular proteins, gene vaccines and toxic 'suicide' proteins), RNA-based (antisense, RNA decoys and ribozymes) or DNA-based (antisense oligonucleotides).

10.2.1 Trans-dominant proteins

Viral regulatory or structural proteins containing dominant-negative (or trans-dominant) mutations can inhibit viral replication. By definition, such mutants not only lack intrinsic wild-type activity but also inhibit the function of their cognate wild-type protein *in trans*. Inhibition may occur because the mutant competes for an essential substrate or cofactor that is available in limiting amounts; or, for proteins that form multimeric complexes, the mutant may associate with wild-type monomers to form an inactive mixed multimer (Figure 10.1). A potential drawback in the use of trans-dominant viral proteins is their possible immunogenicity when expressed by the transduced cells. The engineered cells may consequently induce an immune response that might result in their own destruction. This may diminish the efficacy of anti-viral gene therapy using trans-dominant proteins. The use of non-viral, cellular proteins might overcome this drawback.

10.2.2 Anti-viral cellular proteins

Proteins derived from normal cellular genes have been identified that exhibit anti-viral activity. These anti-viral activities may act through the inhibition of virus binding to cells, by binding directly to the viral regulatory/structural proteins, or indirectly by inducing or repressing cellular factors that in turn

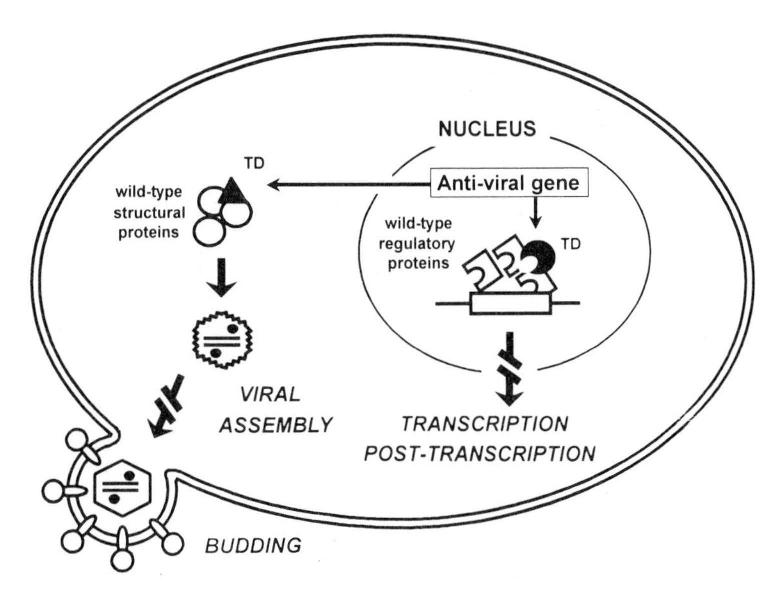

Figure 10.1 Anti-viral trans-dominant (TD) mutant proteins. Trans-dominant mutant forms of either structural or regulatory proteins can interfere with the normal functioning of their wild-type counterparts and thereby inhibit viral transcription, post-transcriptional events and/or viral assembly. This inhibition is most likely due to the formation of inactive mixed multimeric complexes between the mutant and wild-type proteins.

influence viral gene expression (Figure 10.2). Since 'self' proteins are non-antigenic, cells engineered with these cellular anti-viral genes may not be eliminated by the recipient's immune system. This is an obvious advantage as compared with the use of genes encoding potentially immunogenic, trans-dominant viral proteins for gene therapy.

10.2.3 Suicide genes

Instead of protecting a cell from viral infection by blocking viral replication, reduced viral spread could also be achieved by selectively killing the infected cell (Figure 10.3). This approach requires the conditional virus-dependent induction of the expression of a 'suicide' gene, that either directly or indirectly causes cell death upon infection with the virus. Suicide genes can be designed to express toxic proteins under the control of the viral promoter. Examples of suicide genes approaches include engineering cells with either a diphtheria toxin A-chain (*DT–A*) gene or the herpes simplex virus thymidine kinase (*HSV–tk*) gene. HSV–tk can mediate cell death when

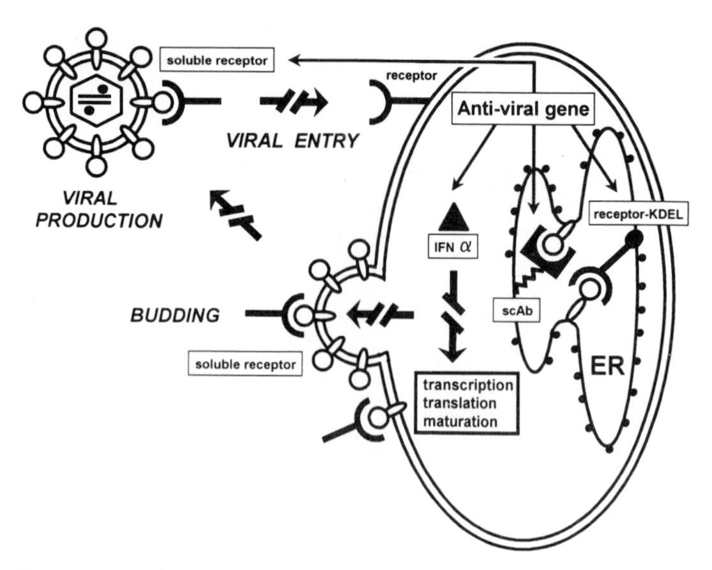

Figure 10.2 Anti-viral strategies based on cellular proteins. The use of an endoplasmic reticulum (ER) retention signal (KDEL) coupled to a viral receptor or a single-chain monoclonal antibody (scAb) can retain essential viral proteins within the endoplasmic reticulum. This retention can prevent maturation and transport of essential viral proteins to the cell membrane. Envelope processing or viral maturation can also be inhibited. Additionally, IFN-α2 (IFN α) may be used to prevent transcription, translation and/or viral maturation as well.

growing cells are exposed to anti-herpetic nucleoside analogs such as ganciclovir, and this prodrug is metabolized by HSV–tk to a toxic analog (HSV–tk has a much greater affinity for ganciclovir than do mammalian tk enzymes). Alternatively, a potent immunogen, such as the influenza H5 HA antigen, could be used as a suicide gene that would eliminate the infected cells through the induction of a cytocidal immune response.

10.2.4 Gene vaccines

Gene therapy is now being actively investigated as a method for optimizing immunization and vaccination strategies. 'Gene vaccines', by definition, involve the transfer into cells of genes that encode the native viral proteins (Figure 10.3). These cells need not be the target cells that are normally infected by the virus. The rationale of these gene vaccines is to deliver the intracellularly synthesized viral antigenic peptides to the MHC class I antigen presentation pathway in an attempt to induce a potent virus-specific CD8+ T-cell-mediated immune response.

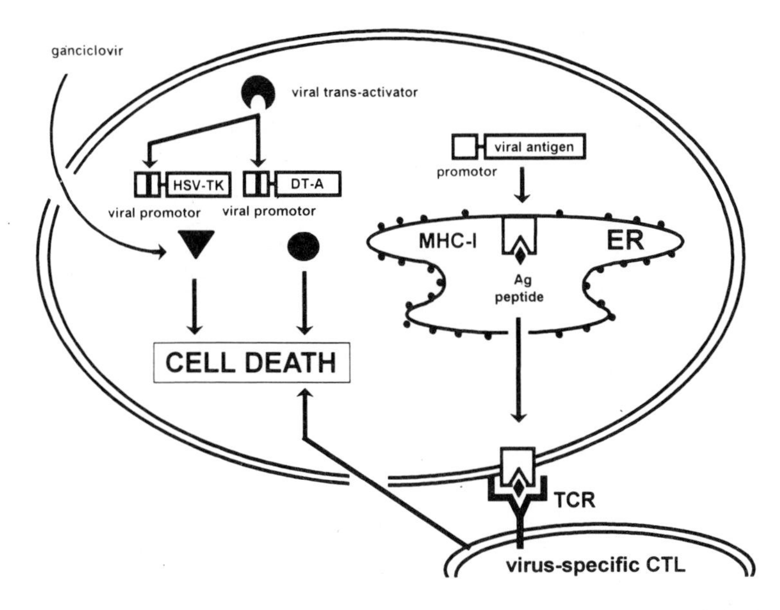

Figure 10.3 Anti-viral suicide genes and gene vaccines. Expression of the herpes simplex virus thymidine kinase (*HSV–tk*) gene, or the diphtheria toxin A-chain (*DT–A*) gene, directly kills viral infected cells. Unlike DT–A, HSV–tk requires administration of ganciclovir to mediate cell death. Alternatively, a strong immunogen may indirectly cause cell death by the induction of a virus-specific MHC class I-restricted cytotoxic T-lymphocyte (CTL) response and may constitute the basis of genetic vaccination. TCR, T-cell receptor.

10.2.5 Antisense RNA and DNA

Antisense RNA transcripts can be designed that target various regions of the viral genome and inhibit viral expression by antisense hybridization (Figure 10.4). The mechanism of antisense-mediated inhibition of gene expression is not well defined but it has been assumed that the formation of RNA–RNA duplexes either triggers degradation of the duplex by RNase or prevents subsequent translation. Similarly, antisense synthetic oligonucleotides or oligonucleotide analogs with modified backbones can also be engineered to inhibit viral gene expression (Tonkinson and Stein, 1993). The uptake of synthetic oligonucleotides from the extracellular milieu is generally not efficient and inhibition is usually transient. Moreover, the use of synthetic antisense oligonucleotides is frequently associated with non-specific inhibition of gene expression. Intracellular expression of antisense genes may be the method of choice for inhibiting viral gene expression by antisense. A general advantage of antisense RNA (and also ribozymes, RNA decoys and DNA oligonucleotides) for anti-viral gene therapy is their lack of

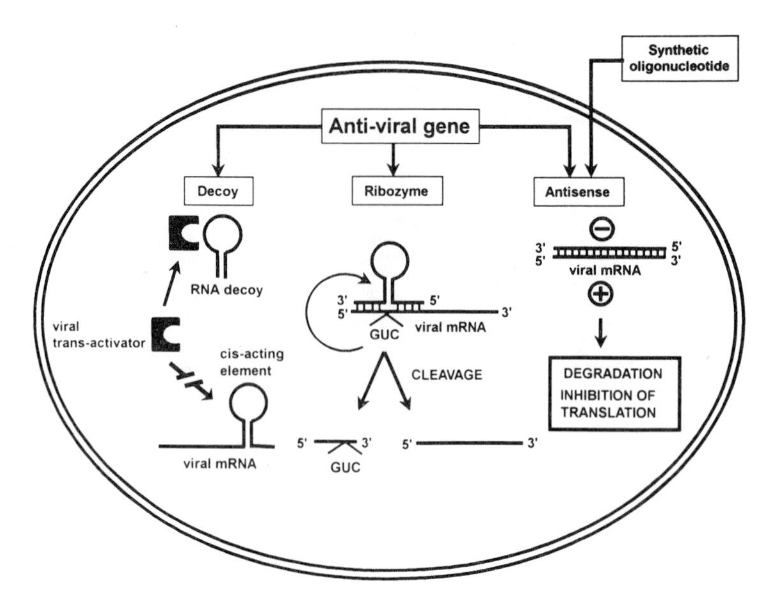

Figure 10.4 Antisense RNA/DNA, ribozymes, and RNA decoys. Anti-viral genes can be expressed in the form of RNA to inhibit viral production. These anti-viral RNA genes consist of RNA decoys, ribozymes, and antisense RNA. Overexpression of short RNA molecules corresponding to viral cis-acting elements prevent binding of viral protein to the cognate sequences found on viral mRNA transcripts. Ribozymes act by hybridizing specifically to the complementary viral RNA target and functionally inactivating it by irreversible cleavage at GUG, GUA, or GUU sequences. The ribozyme can subsequently go through another round of cleavage. Intracellular expressed antisense RNA transcripts or synthetic DNA oligonucleotides hybridize to their complementary viral RNA targets and inhibit viral expression either by RNaseH-mediated degradation or inhibition of translation.

immunogenicity. Consequently, cells that have been engineered by either of these approaches will not be eliminated by the immune system of the recipient.

10.2.6 RNA decoys

Overexpression of short RNA molecules corresponding to critical cis-acting regulatory elements can been used as decoys for viral trans-activating proteins preventing binding of these trans-activators to their corresponding cis-acting elements in the viral genome. Besides their therapeutic potential as decoys for viral trans-activating proteins, RNA molecules can also be obtained that directly inhibit essential viral proteins. There is some question as to whether or not RNA decoy strategies will be as benign to cell physiology as antisense RNAs. The sequestration of cellular factors by RNA decoys raises the concern whether overexpression of RNA transcripts may have deleterious effects on cell viability or function.

10.2.7 Ribozymes

Ribozymes are catalytic RNA molecules that hybridize specifically to a complementary RNA target and functionally inactivate it by cleaving the phosphodiester backbone at a specified location (Figure 10.4). Cleavage of the target RNA occurs after a GUC, GUU or GUA sequence and involves hydrolysis of a 3′–5′ phosphodiester bond. Because the catalytic RNA is not consumed during the cleavage reaction a large number of target molecules can be cleaved. Alternatively, ribozymes targeting different regions of the viral genome could be used in the same vector. The development of ribozymes that co-localize in the same subcellular compartment as their viral target may further increase their effectiveness as anti-viral agents.

10.3 GENE THERAPY FOR VIRAL DISEASES

10.3.1 Human immunodeficiency virus

Human immunodeficiency virus (HIV-1) establishes a chronic infection that results in a progressive depletion of CD4+ lymphocytes (and the destruction of lymphoid–organ architecture), leading to a severe immune deficiency and ultimately death. Since no vaccine or treatment is available for this acquired immune deficiency syndrome (AIDS), various gene therapy strategies were developed in an attempt to combat this devastating disease. One of the strategies involves trans-dominant proteins (Feinberg and Trono, 1992).

The Rev protein of HIV facilitates the extra-nuclear transport of unspliced and singly-spliced HIV-1 mRNA and therefore represents an important target for anti-HIV gene therapy. A trans-dominant Rev mutant protein with two adjacent point-mutations in the activator domain (M10) was shown to diminish production of HIV-1 either when co-transfected with a proviral DNA molecular clone (Malim *et al.*, 1989) or upon stable retroviral-mediated transduction of a human T-cell line (Malim *et al.*, 1992). We have shown that a trans-dominant Rev mutant with a single point-mutation in the activator domain inhibited HIV-1 production not only in T-cell lines but also in primary human PBL (Morgan *et al.*, 1994). Trans-dominant Rev molecules are one of the most potent anti-HIV-1 gene therapies currently available. It is therefore not surprising that a phase I clinical protocol has recently been approved by the Recombinant DNA Advisory Committee (RAC) to assess trans-dominant Rev gene therapy in HIV-1 infected patients (see below). Other dominant-negative mutants of

HIV-1 proteins (Feinberg and Trono, 1992) such as Gag, Tat and Env or the SIV Vpx protein (Matsuda *et al.*, 1993) have been produced and shown to inhibit HIV-1 infection.

Besides trans-dominant proteins, anti-HIV gene therapy strategies based on cellular proteins have also been developed. The CD4 protein is the cell surface receptor mediating entry of HIV-1 into cells and several groups have shown that soluble forms of CD4 (sCD4) can bind to gp120 and thereby prevent viral entry into cells. *In vivo* administration of sCD4 has been attempted with disappointing clinical results (Schooley *et al.*, 1990). Effective inhibition of HIV-1 infection, may require continuous expression of high levels of these soluble receptors, and this may not be possible by standard drug delivery methods. To achieve this goal, we have constructed several retroviral vectors that express the truncated, soluble CD4 receptor (Morgan *et al.*, 1990) which led to HIV-1 inhibition in transduced T-cells.

A slightly different CD4-based anti-HIV-1 gene therapy approach has recently been developed (Buonocore and Rose, 1993). A mutated soluble CD4 molecule that contained a specific endoplasmic reticulum (ER) retention signal (Ser-Glu-Lys-Asp-Glu-Leu or SEKDEL), blocked secretion of gp120 and surface expression of gp120/41, when expressed either by a vaccinia virus expression system or by retroviral vectors. By blocking the transport of the HIV-1 glycoprotein, this retained CD4 molecule (CD4-KDEL) prevented maturation of infectious HIV-1 in transduced cells. Since mutations that decrease the affinity of CD4 for gp120 as much as 500-fold have little effect on the ability of CD4-KDEL to retain gp120 in the ER, it is likely that CD4-KDEL may effectively inhibit HIV-1.

A single-chain antibody, derived from a human monoclonal antibody that recognizes the CD4 binding region of HIV-1 gp120, has been designed for intracellular expression in eukaryotic cells as an HIV-1 gene therapy (Marasco *et al.*, 1993). The single-chain antibody is composed of an immunoglobulin heavy-chain leader sequence, for targeting to the ER, and heavy- and light-chain variable regions that are joined by a flexible interchain linker. The antibody is stably expressed and retained in the ER, probably by virtue of its ability to bind to the ER-localized BiP protein. This single-chain antibody binds to the envelope glycoprotein made in the same cell and inhibits env protein maturation and function as judged by inhibition of gp160 cleavage to gp120/gp41 and reduction of syncytia formation. The virus particles released from the antibody-producing HIV-1-infected cells proved more than 1000 times less infectious than ordinary HIV-1.

Another cellular protein that inhibits HIV *in vitro* is human interferon (IFN-α2). However, the inhibition is moderate and reversible upon IFN-α2 removal. To achieve prolonged high concentrations of IFN-α2 and to assess the feasibility of IFN-α2 gene therapy, retroviral vectors expressing IFN-α2 were transduced into T-cell lines that were subsequently challenged with

HIV-1 (Bednarik *et al.*, 1989). IFN-α2 was conditionally expressed under the control of the HIV-1-LTR and therefore responsive to Tat transactivation. The replication of HIV-1 in the transduced cell lines was significantly inhibited, probably as a consequence of transcriptional inhibition. However, effects of IFN-α2 on other steps in the HIV-1 life cycle could not be excluded, such as on assembly and maturation. In our own studies, we transduced PBL with retroviral vectors expressing a Tat and Rev-dependent INF-α2 construct, and observed almost no HIV-1 production versus control engineered PBL (unpublished observation). Since interferons are pleiotropic molecules, it will be important to exclude possible side effects in using IFN-α2 for gene therapy.

The conditional HIV-1-dependent induction of the expression of a suicide gene can lead to the selective killing of the HIV-1 infected cells. In uninfected cells, suicide gene expression is at low level, but can be greatly induced by the Tat protein following infection by HIV-1. In principle, suicide gene expression will lead to cell death before HIV-1 structural protein synthesis, thus preventing viral spread. Suicide gene approaches for anti-HIV gene therapy include HIV-dependent expression of HSV–tk and DT–A or of a potent immunogen such as the influenza H5 HA antigen (Buchschacher and Panganiban, 1992). Using these approaches, inhibition of HIV-1 production in cell lines was reported using stable (DNA-mediated) transfection (Caruso and Klatzman, 1992), and retroviral transduction (Harrisson *et al.*, 1992).

The use of gene vaccines is another strategy that can be employed to inhibit HIV. The therapeutic rationale for this approach is that induction of augmented CTL response to HIV-1 antigens, such as Env and/or Rev may lead to the elimination of HIV-1 infected cells. A reduction in viral burden may further result in positive therapeutic benefits. Cells transduced with a retroviral vector encoding the HIV-1-IIIB env gene, were capable of inducing an effective HIV-1-specific cellular and humoral immune response in mice, cross-reactive to peptides derived from the V3 hypervariable region of other isolates (Warner *et al.*, 1991; Jolly *et al.*, 1992). These observations underscore the potential of gene transfer for the generation of a potent anti-HIV-1 immune response that will hopefully provide heterologous protection against different polymorphic HIV-1 isolates. A phase I clinical trial for the evaluation of anti-HIV-1 gene vaccines in humans has recently been approved by the RAC. In these approved phase I clinical safety trials, HIV-1-infected individuals will either have fibroblasts removed for *ex vivo* engineering, or vectors will be directly injected. *In vivo* administration will hopefully stimulate immune responses to the HIV-1 Env and Rev gene products.

There have been many attempts to inhibit HIV-1 by intracellular antisense RNA expression and only the most typical examples will be presented here. An extended comparative analysis was performed to assess the efficacy of various antisense RNA targeted at 10 different regions of the HIV-1 genome (Rittner and Sczakiel, 1991). The most inhibitory antisense RNA expression plasmids were targeted against a 1-kb region within the Gag open reading frame and against a 562 bp region covering the *Tat* and *Rev* genes. In general, antisense strategies should be targeted at regions or genes within the HIV-1 genome that are essential for HIV-1 gene expression and conserved among divergent HIV-1 strains. Based on these criteria, the TAR-region of HIV-1 (because it mediates Tat trans-activation) is an appropriate candidate for antisense gene therapy. Moreover, the TAR element is present on all HIV-1 transcripts, including the RNA genome. An antisense complementary to bases at +13 to +75 within the 5′ untranslated region of HIV-1 (encompassing the TAR-region) was cloned into an AAV-based vector and transduced in human hematopoietic and non-hematopoietic cell lines (Chatterjee *et al.*, 1992). Cell lines expressing this antisense RNA showed up to 95% inhibition of gene expression directed by the HIV-1 long terminal repeat and greater than 99% reduction in infectious HIV-1 production. We have also shown that this antisense TAR can effectively inhibit HIV-1 in retrovirally transduced T-cell lines (Chuah *et al.*, 1994).

Overexpression of critical cis-acting HIV-1 regulatory elements, such as TAR and RRE, have been used as decoys for viral trans-activating proteins Tat and Rev, respectively. Overexpression of TAR decoys, expressed by retroviral vectors in CEM-SS cells, prevented Tat-mediated trans-activation and HIV-1 viral production and greatly reduced the amount of HIV-1 produced (Sullenger *et al.*, 1990). Overexpression of TAR in retroviral vectors had been achieved either by expressing the decoy from a strong constitutive pol III initiator $tRNA_i^{Met}$ promoter in a double copy configuration or by expressing a multimerized TAR decoy (containing up to 50 TAR repeats) from a Tat-inducible HIV-1-LTR (Lisziewicz *et al.*, 1993). Similarly, overexpression of RRE decoys by retroviral vectors inhibits HIV-1 most likely by preventing binding of Rev to the normal RRE sequences, found in singly-spliced and non-spliced HIV-1 mRNA (Lee *et al.*, 1992).

Finally, ribozymes have also been employed as an anti-HIV gene therapy strategy. CD4+ HeLa cells transfected with a vector expressing a hammerhead ribozyme targeted to HIV-1 Gag transcripts showed significant protection against a subsequent HIV-1 challenge (Sarver *et al.*, 1990). Another target for ribozyme-mediated anti-HIV-1 gene therapy is the 5′ leader sequence present on all HIV-1 transcripts. The leader sequence is essential for reverse transcription, trans-activation, 5′ capping and translation. In two separate studies the HIV-1 5′ leader sequence was targeted by either retroviral transduction of a hammerhead ribozyme into T-cell lines (Weerasinghe *et al.*, 1991) or transfection of a hairpin ribozyme expression vector into HeLa cells (Yu *et al.*, 1993). In both cases, an inhibition of HIV-1 production was observed. A clinical protocol has been approved by the RAC that is based on these studies and involves retroviral vector-mediated gene transfer of the anti-leader ribozyme into CD4+ T-cells in HIV-1 infected individuals. One potential limitation of the use of ribozymes is the high mutability of HIV-1 that could render the HIV-1 RNA resistant to the ribozyme. Multitarget-ribozymes have recently been developed that cleave at up to nine highly conserved HIV-1 Env RNA regions and effectively inhibit replication of different HIV-1 isolates (Chen *et al.*, 1992). The development of ribozymes that co-localize in the same subcellular compartment as their HIV-1 target may enhance their efficacy (Sullenger and Cech, 1993).

Presently, it is not clear which gene therapy strategy is most effective in inhibiting HIV-1, since side-by-side comparative studies have not been reported. Therefore, we are currently comparing the efficacy of at least seven different anti-HIV constructs for their ability to inhibit HIV-1. In addition, some of the gene therapy approaches described here, such as the expression of soluble receptors, intracellular antibodies, IFN and toxic 'suicide' proteins, have only been evaluated in the context of anti-HIV gene therapy. Viral diseases, other than HIV-1 infection, are also potentially amenable to treatment by these novel gene therapy strategies (see below). This is particularly relevant in the case of IFN gene therapy, since this cytokine possesses general anti-viral properties acting on several groups of viruses including herpesvirus, papovavirus, myxovirus, paramyxovirus, rhabdovirus, togavirus, picornavirus, hepadnavirus and reovirus.

10.3.2 Human T-cell lymphotropic virus

Human T-cell lymphotropic virus (HTLV-I) is a pathogenic human retrovirus that causes adult T-cell leukemia (ATL) and has also been linked to progressive demyelinating neuropathy. Although HTLV-I is a type C oncogenic retrovirus, while HIV-1 is a prototypical lentivirus, both of these viruses utilize similar strategies to regulate their gene expression. Consequently, similar gene therapy strategies can be employed to inhibit HTLV-I gene expression. Since the Rex protein of HTLV is essential for HTLV-I replication and since it is functionally and structurally homologous to the HIV-1 Rev protein, it is likewise an appropriate candidate target gene for inhibition by gene therapy. Trans-dominant Rex mutant proteins have recently been generated that inhibit their cognate wild-type Rex protein function (Bogerd and Greene, 1993). Presently it is not known whether anti-HTLV-I gene therapy based on trans-dominant Rex proteins can inhibit or prevent the development of ATL and/or the HTLV-I associated myelopathy.

The HTLV-I Tax protein is also essential for viral replication and transactivates HTLV-I transcription. In addition, Tax is implicated in the induction of IL-2 secretion and its corresponding IL-2 receptor, which may possibly contribute to T-cell transformation by stimulating cellular proliferation in an autocrine fashion. Transduction of HTLV-I-infected umbilical cord mononuclear cells with retroviral vectors expressing antisense Tax RNA inhibited HTLV-I replication and virus-mediated immortalization (von Ruden and Gilboa, 1989). However, this inhibition was not absolute since the onset of IL-2-independent T-cell growth was only delayed and not prevented. Hence, because of the dominant phenotype of HTLV-I-induced malignant cell transformation, anti-HTLV-I gene therapy will require extremely high efficiencies of *in vivo* gene transfer to prevent ATL. This represents a serious obstacle, not only for HTLV-I induced leukemia in particular, but also for all virally-induced malignancies.

10.3.3 Human papillomavirus

Human papillomavirus (HPV) is a DNA virus that causes a spectrum of benign and malignant diseases. Despite the widespread occurrence of HPV-infection as a sexually transmitted disease and its association with malignancy, no specific anti-viral agents for HPV are available. Many different strains of HPV have been identified. HPV-1 appears to be principally responsible for ordinary skin warts, HPV-6 for genital warts and about 90% of all oral and cervical carcinomas are associated with persistent infections by specific types of human papillomaviruses, for instance HPV-16 and HPV-18. Current treatment of genital warts is non-specific and consists primarily of surgical or chemical removal of the wart. These therapies are not curative since the viral DNA episome still persists in the basal epithelium of latently infected patients. Transfection of the DNA of HPV-16 or HPV-18 into primary human keratinocytes gives rise to immortalized keratinocyte cell lines. These immortalizing functions are mediated by the HPV E6 and E7 proteins, which also seem to regulate the mitotic activity of various cervical cancer cell lines. Endogenously expressed antisense RNA directed against these viral E6 and E7 oncoproteins inhibited cell growth as well as tumorigenicity of human cervical carcinoma cells in nude mice (von Knebel Doeberitz *et al.*, 1992). In addition, inhibition of cell growth of HPV-18-positive human oral and cervical carcinoma cell lines was also observed with either antisense oligonucleotides or antisense RNA (Steele *et al*, 1993). Similarly, a phosphorothioate antisense oligonucleotide targeted to the translation initiation of both HPV-6 and HPV-11 E2 trans-activator mRNA was shown to inhibit E2-dependent trans-activation (Cowsert *et al.*, 1993). In addition to antisense *per se*, HPV-specific ribozymes have also been designed that specifically cleave cottontail rabbit papillomavirus E7 RNA *in vitro* (Wisotzkey *et al.*, 1993) but it is not known whether these ribozymes effectively inhibit HPV-induced cellular transformation. These data underscore the potential of antisense RNA or DNA and ribozymes as anti-HPV gene therapy strategies. Furthermore, since HPV is epitheliotropic, problems of tissue accessibility and delivery of anti-HPV genes is greatly reduced. However, it remains to be seen whether antisense can inhibit tumor progression when administered to precancerous or cancerous HPV-positive lesions in humans. Gene therapy for papilloma-associated carcinomas will require highly efficient *in vivo* gene transfer which will be extremely difficult to achieve. An alternative gene therapy approach that had proven effective for brain tumors (Culver *et al.*, 1992), that could also potentially be applied to HPV-associated carcinomas, consists of transducing the carcinoma cells *in vivo* with retroviral vectors expressing a suicide gene, such as *HSV–tk*. This could be achieved by intra-tumor inoculation of irradiated, retroviral vector producer cells. The retroviral vector particles would preferentially stably transduce the rapidly proliferating carcinoma cells. Subsequent administration of ganciclovir may result not only in the elimination of the transduced carcinoma cells expressing HSV–tk, but also of their neighboring, non-transduced cancer cells. The mechanism of this bystander toxicity is not fully understood but may involve intercellular transport of toxic metabolites through gap junctions.

10.3.4 Hepatitis B virus

There are an estimated 300 million carriers of hepatitis B virus (HBV), making it a major cause of chronic hepatitis, cirrhosis and hepatocellular carcinoma. At present, vaccination for HBV is the sole protective measure available against HBV infection. Although IFN-α treatment may in some cases be beneficial, no satisfactory medical treatment for chronic HBV infection is currently available. Antisense oligonucleotides inhibited either human HBV in hepatocellular carcinoma cells (Goodarzi *et al.*, 1990) or duck HBV in primary duck hepatocytes *in vitro* as well as in duck HBV-infected Pekin ducks *in vivo* (Offensperger *et al.*, 1993). The most effective antisense oligodeoxynucleotide was directed against the gene encoding the surface antigen (pre-S) and resulted in a complete inhibition of viral replication and gene expression *in vitro* and *in vivo*. However, slow reactivation of HBV replication followed after cessation of the antisense gene therapy. To augment the specificity of gene transfer to hepatocytes, anti-HBV chemically-modified antisense molecules were coupled to the asialoglycoprotein (AGP) via a poly-L-lysine polycationic bridge. These antisense–AGP complexes were targetable to hepatocytes via AGP receptors present on those cells and significantly inhibited HBV gene expression and replication in HBV transfected HepG2 cell lines (Wu and Wu, 1992). These strategies require continuous administration of the antisense molecules and therefore vector-based approaches permitting continuous and stable expression of the anti-HBV genes will be more suitable for the treatment of chronic HBV infection by gene therapy.

10.3.5 Influenza virus

The influenza virus is a human pathogen that leads to significant morbidity and sometimes mortality in infected individuals and frequently causes devastating epidemics throughout the world. Neutralizing antibodies can be generated against influenza virus, and form the basis of the current vaccination strategies. However, since these antibodies recognize the highly polymorphic influenza envelope proteins they are generally strain-specific and do not cross-protect against heterologous influenza strains. This polymorphism accounts for the recurrence of influenza epidemics. In contrast to antibodies, cytotoxic T lymphocytes (CTLs) specific for conserved viral antigens can respond to different strains of virus. The generation of such CTLs requires endogenous expression of the antigen. To overcome the limitation of current influenza vaccines, genes encoding conserved influenza proteins, such as the nucleoprotein, were introduced by direct intramuscular DNA injection into mice (Ulmer *et al.*, 1993). This gene vaccine resulted in the generation of nucleoprotein-specific CTLs and protection from a subsequent challenge with a heterologous strain of influenza A virus, as measured by decreased viral lung titers, inhibition of mass loss and increased survival. This gene therapy-based immunization scheme is an important step forward in the prevention of influenza and is also being evaluated for other polymorphic viruses, such as HIV. Influenza virus replication and viral protein production was also inhibited *in vitro* in a MDCK epithelial cell-line, with chemically modified oligonucleotides (Zerial *et al.*, 1987; Kabanov *et al.*, 1990). These two strategies may ultimately prove effective as an influenza gene therapy for humans.

10.3.6 Epstein–Barr virus

Epstein–Barr virus (EBV) is a major pathogen in human and causes lymphoproliferative disease in immune-deficient humans. In addition it is associated with Burkitt's lymphoma, Hodgkin's disease, and nasopharyngeal carcinoma. It has also been linked to other diseases such as chronic mononucleosis, chronic fatigue syndrome and mononucleosis-like syndrome as well as oral hairy leukoplakia in patients with AIDS. EBV efficiently establishes a non-replicative infection in primary B lymphocytes and expresses six nuclear proteins (EBNAs) (LMP1, LMP2A and LMP2B) and two small RNAs (EBERs). Though its replication can be effectively inhibited by available antiviral drugs, latent EBV infection is unaffected by any of these conventional antiviral agents. The episome is the molecular basis for latent infection and therefore its replication is the primary target for inhibition by antisense oligonucleotides. The antisense strategy focuses on blocking the synthesis of a single protein, the EBNA1 protein which is the sole protein needed for replication of EBV episomes and secondarily, synthesis of other

EBV latent products. Antisense phosphorothioated oligodeoxynucleotides with or without sequence specificity complementary to the EBV genes have been found to be potent inhibitors of EBV replication in cell culture (Yao *et al.*, 1993). Other antisense oligomers that target to several sites within the EBNA1 open reading frame had also been tested (Pagano *et al.*, 1992). Though antisense seems promising for inhibiting latent EBV infection further work is needed to prove efficacy and specificity of the antisense effect.

10.3.7 Herpes simplex virus

The clinical manifestations of herpes simplex virus (HSV)-infection are very diverse and range from relatively benign skin and genital lesions to potentially lethal encephalitis, neonatal herpes or disseminated progressive disease in immunocompromised hosts. HSV infection may also be a cofactor in cervical carcinoma. Effective chemotherapy with viral DNA synthesis inhibitors such as acyclovir and vidarabine is available to treat the majority of cases. However, drug-resistant HSV mutant strains have arisen in patients, causing a more severe and progressive mucocutaneous disease. Gene therapy may serve as an alternative in those patients that are refractory to treatment by chemotherapy. Various groups have now reported that HSV-1 gene expression and replication can be inhibited by chemically modified antisense deoxyoligonucleotides targeted at various regions in the HSV genome (Tonkinson and Stein, 1993). The efficacy of antisense HSV oligonucleotides has been demonstrated *in vitro* as well as in animal models. Some antisense oligonucleotides were actively inhibiting acyclovir-resistant, thymidine kinase deletion mutant strains of HSV-1 *in vitro*. In addition, dominant-negative inhibitory mutant proteins were derived from various HSV-1 proteins such as the HSV regulatory proteins ICP0, ICP4 ICP8 and ICP27 and were shown to inhibit viral gene expression and replication (Shepard *et al.*, 1990; Weber *et al.*, 1992). The ICP4 trans-dominant proteins were engaged in non-functional complexes with their wild-type counterparts, hereby altering the conformation of their trans-activating domain. The current antisense and trans-dominant based gene therapy strategies may prove useful for the treatment of some of the life-threatening HSV infections.

10.3.8 Cytomegalovirus

Cytomegalovirus (CMV) is a ubiquitous herpes-virus which causes mild or subclinical disease in immunocompetent adults but which may lead to severe morbidity or mortality in neonates or immunocompromised individuals. Mortality due to CMV pneumonia can be as high as 40% in bone marrow transplant recipients. Disseminated CMV infection is common in

AIDS patients and is often associated with gastroenteritis and sight-threatening chorioretinitis. Ganciclovir and foscarnet have been used beneficially for the treatment of CMV infections. However, in immunocompromised individuals, such as AIDS patients, disease invariably recurs after cessation of treatment, necessitating prolonged maintenance therapy. Drug toxicity and the emergence of resistant virus strains associated with long-term therapy have limited the effectiveness of these compounds and demonstrated the need for new treatment strategies. Synthetic oligonucleotides represent a novel alternative to currently available antiviral drugs. Antisense was targeted to some genes essential for production of infectious virions. The production of infectious CMV requires the DNA polymerase gene and the major immediate–early (IE) transcriptional unit of CMV which encodes several proteins responsible for regulation of viral gene expression. Phosphorothioate oligonucleotides complementary to the major immediate–early regions 1 and 2 (IE1 and IE2) were evaluated (Azad *et al.*, 1993). Oligonucleotides complementary to the IE2 region exhibited the most potent antiviral activity. One of the oligonucleotides was at least 30-fold more potent than the anti-herpetic drug ganciclovir. Other oligonucleotides complementary to IE1 coding region had been shown to inhibit CMV infection as well (Bryant and Sinclair, 1993).

10.4 PROSPECTS FOR ANTI-VIRAL GENE THERAPY

A large variety of anti-viral gene therapy strategies have been developed, each with their own advantages and disadvantages. At this point, it is difficult to evaluate which anti-viral gene therapy strategy is most effective. It is not known whether absolute resistance to viral infection can be achieved by any single anti-viral gene therapy strategy, particularly in the case of a high infectious viral dose. It might be possible to achieve better protection in such a situation by combining multiple anti-viral genes into the same vector system (Morgan *et al.*, 1994). However, for some viral diseases, such as AIDS, it is not clear what *in vitro* challenge dose of virus mimics the true *in vivo* condition that varies greatly depending on the clinical stage and the anatomic location (Pantaleo *et al.*, 1993). Similarly, anti-viral gene therapy for virus-associated malignancies, such as HPV-associated cervical carcinoma, EBV-associated Burkitt's lymphoma or HTLV-I associated T-cell lymphomas, may only be effective in the early stages of the disease since other genes contribute to tumor progression. The anti-viral gene therapy strategies that were described were evaluated *in vitro* and it will be useful to test either with appropriate animal models or in the context of clinical trials, whether viral resistance can be achieved *in vivo*. Six anti-viral gene therapy protocols have been reviewed and approved by the Recombinant DNA Advisory Committee (RAC) (as of December 1993). One protocol involves the marking of EBV-specific CTL with retroviral vectors to follow their fate and persistence *in vivo* (Heslop *et al.*, 1994). Five protocols are confined to the evaluation of anti-HIV gene therapy strategies or complications associated with HIV-infection. These protocols involve: (i) marking of HIV-specific CTL with a retroviral vector expressing a marker/suicide gene (HygR/HSV–TK) (Greenberg, 1992). The *HSV–tk* gene is a built-in safety feature that, if unexpected CTL growth or an unexpectedly vigorous CTL response is observed in patients, the engineered cells can be ablated by administration of ganciclovir to the patients; (ii) marking of syngeneic T-cells with retroviral vectors in an HIV-1-uninfected twin in order to follow their fate in the HIV-1-infected sibling upon adoptive transfer (Walker, 1993); (iii) retroviral-mediated gene transfer of trans-dominant Rev (M10) in CD4+ T-cells of an HIV-infected patient; (iv) retroviral vector-mediated gene transfer of an anti-HIV leader sequence ribozyme in CD4+ T-cells of an HIV-1 infected individual; and (v) anti-HIV gene vaccine with retroviral vectors expressing Rev or Env in an HIV-1 infected individual. A major technical hurdle that still needs to be overcome for all anti-viral gene therapy strategies is to achieve high-efficiency stable gene transfer, preferably by direct *in vivo* gene transfer and hereby minimizing *ex vivo* manipulation of patients' cells. Only then can gene therapy be made accessible on a wider scale.

ACKNOWLEDGEMENTS

The authors acknowledge Ms. Theresa Lumsden for secretarial assistance in the preparation of this manuscript. Thierry VandenDriessche is a Research Fellow of the Belgian National Fund for Scientific Research (N.F.W.O.-Kom Op Tegen Kanker Actie).

REFERENCES

Azad, R.F., Driver, V.B., Tanaka, K. *et al.* (1993) Antiviral activity of a phosphorothioate oligonucleotide complementary to RNA of the human cytomegalovirus major immediate-early region. *Antimicrob. Agents Chemother.*, 37, 1945.

Bednarik, D.P., Mosca, J.D., Raj, N.B.K. *et al.* (1989) Inhibition of human immunodeficiency virus (HIV) replication by HIV-trans-activated α_2-interferon. *Proc. Natl Acad. Sci. USA*, **86**, 4958.

Bogerd, H. and Greene, W.C. (1993) Dominant negative mutants of human T-cell leukemia virus type I rex and human immunodeficiency virus type 1 Rev fail to multimerize *in vivo*. *J. Virol.*, **67**, 2496.

Bryant, L.A. and Sinclair, J.H. (1993) Inhibition of human cytomegalovirus major immediate early gene expression by antisense RNA expression vectors. *J. Virol.*, **74**, 1965.

Buchschacher, G.L. Jr and Panganiban, T. (1992) Human immunodeficiency virus vectors for inducible expression of foreign genes. *J. Virol.*, **66**, 2731.

Buonocore, L. and Rose, J.K. (1993) Blockade of human immunodeficiency virus type 1 production in $CD4^+$ T cells by an intracellular CD4 expressed under control of the viral long terminal repeat. *Proc. Natl Acad. Sci. USA*, **90**, 2695.

Caruso, M. and Klatzman, D. (1992) Selective killing of $CD4^+$ cells harboring a human immunodeficiency virus-inducible suicide gene prevents viral spread in an infected cell population. *Proc. Natl Acad. Sci. USA*, **89**, 182.

Chatterjee, S., Johnson, P.R. and Wong, K.K. (1992) Dual-target inhibition of HIV-1 *in vitro* by means of an adeno-associated virus antisense vector. *Science*, **258**, 1485.

Chen, C.J., Banerjea, A.C., Harrison, G.G. *et al.* (1992) Multitarget-ribozyme directed to cleave at up to nine highly conserved HIV-1 env RNA regions inhibits HIV-1 replication-potential effectiveness against most presently sequenced HIV-1 isolates. *Nucleic Acids Res.*, **20**, 4581.

Chuah, M.K.L., VandenDriessche, T. and Morgan, R.A. (1994) Inhibition of human immunodeficiency virus type-1 by retroviral vectors expressing antisense-TAR. *Human Gene Ther.*, **5**, 1467.

Cowsert, L.M., Fox, M.C., Zon, G. *et al.* (1993) *In vitro* evaluation of phosphorothioate oligonucleotides targeted to the E2 mRNA of papillomavirus: potential treatment for genital warts. *Antimicrob. Agents Chemother.*, **37**, 171.

Culver, K.W., Ram, Z., Wallbridge, S. *et al.* (1992) *In vivo* gene transfer with retroviral vector-producer cells for treatment of experimental brain tumors. *Science*, **256**, 1550.

Feinberg, M.B. and Trono, D. (1992) Intracellular immunization: trans-dominant mutants of HIV gene products as tools for the study and interruption of viral replication. *AIDS Res. Hum. Retroviruses*, **8**, 1013.

Goodarzi, G., Gross, S.C., Tewari, A. *et al.* (1990) Antisense oligodeoxyribonucleotides inhibit the expression of the gene for hepatitis B virus surface antigen. *J. Gen. Virol.*, **71**, 3021.

Greenberg, P.D. (1992) Phase I study of cellular adoptive immunotherapy using genetically modified $CD8^+$ HIV-specific T cells for HIV seropositive patients undergoing allogeneic bone marrow transplant. *Human Gene Ther.*, **3**, 319.

Harrisson, G.S., Long, C.J., Curiel, T. *et al.* (1992) Inhibition of human immunodeficiency virus-1 production resulting from transduction with a retrovirus containing an HIV-regulated diphtheria toxin A chain gene. *Human Gene Ther.*, **3**, 461.

Heslop, H.E., Brenner, M.K., Rooney, C. *et al.* (1994) Administration of neomycin resistance gene marked EBV specific cytotoxic T lymphocytes to recipients of mismatched-related or phenotypically similar unrelated donor marrow grafts. *Human Gene Ther.*, **5**, 381.

Jolly, D., Chada, S., Townsend, K. *et al.* (1992) CTL cross reactivity between HIV strains. *AIDS Res. Hum. Retroviruses*, **8**, 1369.

Kabanov, A.V., Vinogradov, S.V., Ovcharenko, A.V. *et al.* (1990) A new class of antivirals: antisense oligonucleotides combined with a hydrophobic substituent effectively inhibit influenza virus reproduction and synthesis of virus-specific proteins in MDCK cells. *FEBS Lett.*, **259**, 327.

Lee, T.C., Sullenger, B.A., Gallardo, H.F. *et al.* (1992) Overexpression of RRE-derived sequences inhibits HIV-1 replication in CEM cells. *New Biol.*, **4**, 66.

Lisziewicz, J., Sun, D., Smythe, J. *et al.* (1993) Inhibition of human immunodeficiency virus type l replication by regulated expression of a polymeric Tat activation response RNA decoy as a strategy for gene therapy in AIDS. *Proc. Natl Acad. Sci. USA*, **90**, 8000.

Malim, M.H., Bohnlein, S., Hauber, J. *et al.* (1989) Functional dissection of the HIV-1 rev trans-activator-derivation of a trans-dominant repressor of rev function. *Cell*, **58**, 205.

Malim, M.H., Freimuth, W.W., Liu, J. *et al.* (1992) Stable expression of trans-dominant rev protein in human T cells inhibits human immunodeficiency virus replication. *J. Exp. Med.*, **176**, 1197.

Marasco, W.A., Haseltine, W.A. and Chen, S.Y. (1993) Design, intracellular expression, and activity of a human anti-human immunodeficiency virus type 1 gp120 single-chain antibody. *Proc. Natl Acad. Sci. USA*, **90**, 7889.

Matsuda, Z., Yu, X., Yu, Q.-C. *et al.* (1993) A virion-specific inhibitory molecule with therapeutic potential for human immunodeficiency virus type 1. *Proc. Natl Acad. Sci. USA*, **90**, 3544.

Morgan, R.A. and Anderson, W.F. (1993) Human gene therapy. *Annu. Rev. Biochem.*, **62**, 191.

Morgan, R.A., Looney, D.J., Muenchau, D.D. *et al.* (1990) Retroviral vectors expressing soluble CD4: a potential gene therapy for AIDS. *AIDS Res. Hum. Retroviruses*, **6**, 183.

Morgan, R.A., Ragheb, J., Chuah, M. *et al.* (1994) Development of retroviral vectors that optimize the inhibition of HIV-1 at multiple points of the viral life cycle. *J. Cell Biochem.*, (**18** Suppl A), 243.

Offensperger, W.B., Offensperger, S., Walter, E. *et al.* (1993) *In vivo* inhibition of duck hepatitis B virus replication and gene expression by phosphorothioate modified antisense oligodeoxynucleotides. *EMBO J.*, **12**, 1257.

Pagano, P.S., Jimenez, G., Sung, N.S. *et al.* (1992) Epstein–Barr viral latency and cell immortalization as targets for antisense oligomers. *Ann. N.Y. Acad. Sci.*, **660**, 107.

Pantaleo, G., Graziosi, C., Demarest, J.F. *et al.* (1993) HIV infection is active and progressive in lymphoid tissue during the clinically latent stage of disease. *Nature*, **362**, 355.

Rittner, K. and Sczakiel, J. (1991) Identification and analysis of antisense RNA target regions of the human immunodeficiency virus type 1. *Nucleic Acids Res.*, **19**, 1421.

Sarver, N., Cantin, E.M., Chang, P.S. *et al.* (1990) Ribozymes as potential anti-HIV-1 therapeutic agents. *Science*, **247**, 1222.

Schooley, R.T., Merigan, T.C., Gaut, P. *et al.* (1990) Recombinant soluble CD4 therapy in patients with the acquired immunodeficiency syndrome (AIDS) and AIDS-related complex. *Ann. Intern. Med.*, **112**, 247.

Shepard, A.A., Tolentino, P. and DeLuca, N.A. (1990) *trans*-Dominant inhibition of herpes simplex virus transcriptional regulatory protein ICP4 by heterodimer formation. *J. Virol.*, **64**, 3916.

Steele, C., Cowsert, L.M. and Shillitoe, E.J. (1993) Effects of human papillomavirus type 18-specific antisense oligonucleotides on the transformed phenotype of human carcinoma cell lines. *Cancer Res.*, **53**, 2330.

Sullenger, A., Gallardo, H.F., Ungers, G.E. *et al.* (1990) Overexpression of TAR sequences renders cells resistant to human immunodeficiency virus replication. *Cell*, **63**, 601.

Sullenger, B.A. and Cech, T.R. (1993) Tethering ribozymes to a retroviral packaging signal for destruction of viral RNA. *Science*, **262**, 1566.

Tonkinson, J.L. and Stein, C.A. (1993) Anitsense nucleic acids – prospects for antiviral intervention. *Antiviral Chem. Chemother.*, **4**, 193.

Ulmer, J.B., Donnelly, J.J., Parker, S.E. *et al.* (1993) Heterologous protection against influenza by injection of DNA encoding a viral protein. *Science*, **259**, 1745.

von Knebel Doeberitz, M., Rittmuller, C., zur Hausen, H. *et al.* (1992) Inhibition of tumorigenicity of cervical cancer cells in nude mice by HPV E6-E7 anti-sense RNA. *Int. J. Cancer*, **51**, 831.

von Ruden, T. and Gilboa, E. (1989) Inhibition of human T-cell leukemia virus type I replication in primary human T-cells that express antisense RNA. *J. Virol.*, **63**, 677.

Walker, R. (1993) A study of the safety and survival of the adoptive transfer of genetically marked syngeneic lymphocytes in HIV-infected identical twins. *Human Gene Ther.*, **4**, 659.

Warner, J.F., Anderson, C.G., Laube, L. *et al.* (1991) Induction of HIV-specific CTL and antibody responses in mice using retroviral vector-transduced cells. *AIDS Res. Hum. Retroviruses*, **7**, 645.

Weber, P.C., Kenny, J.J. and Wigdahl, B. (1992) Antiviral properties of a dominant negative mutant of the herpes simplex virus type 1 regulatory protein ICP0. *J. Gen. Virol.*, **73**, 2955.

Weerasinghe, M., Liem, S.E., Asad, S. *et al.* (1991) Resistance to human immunodeficiency virus type 1 (HIV-1) infection in human $CD4^+$ lymphocyte-derived cell lines conferred by using retroviral vectors expressing an HIV-1 RNA-specific ribozyme. *J. Virol.*, **65**, 5531.

Wisotzkey, J.D., Krizenoskas, A., DiAngelo, S. *et al.* (1993) Cleavage of cottontail rabbit papillomavirus E7 RNA with an anti-E7 ribozyme. *Biochem. Biophys. Res. Commun.*, **192**, 833.

Wu, G.Y. and Wu, C.H. (1992) Specific inhibition of hepatitis B viral gene expression *in vitro* by targeted antisense oligonucleotides. *J. Biol. Chem.*, **267**, 12436.

Yao, G.-Q., Grill, S., Egan, W. *et al.* (1993) Potent inhibition of Epstein–Barr virus by phosphorothioate oligodeoxynucleotides without sequence specification. *Antimicrob. Agents Chemother.*, **37**, 1420.

Yu, M., Ojwang, J., Yamada, O. *et al.* (1993) A hairpin ribozyme inhibits expression of diverse strains of human immunodeficiency virus type 1. *Proc. Natl Acad. Sci. USA*, **90**, 6340.

Zerial, A., Thuong, N.T. and Helene, C. (1987) Selective inhibition of the cytopathic effect of type A influenza viruses by oligodeoxynucleotides covalently linked to an intercalating agent. *Nucleic Acids Res.*, **15**, 9909.

11

AIDS and HIV infection

CLAY SMITH and BRUCE A. SULLENGER

11.1 INTRODUCTION

Infection with the human immunodeficiency virus (HIV) is a paradigm for serious viral diseases which may be treatable with gene therapy. The biology of the virus and the pathophysiology of infection with HIV are rapidly being clarified and a number of potential targets for gene therapy interventions have been identified. HIV infects primarily the hematopoietic system which is relatively easy to obtain, manipulate, assess and transplant. Current pharmacologic, immunologic, and biotherapeutic approaches to treating HIV infection have been disappointing, propelling the development of alternative treatment approaches. This chapter will review the features of HIV biology relevant to gene therapy, HIV gene therapy inhibition strategies and gene delivery systems, preclinical models for evaluating the safety and efficacy of gene therapy approaches to HIV infection, and the initial potential clinical applications of this technology.

11.2 BIOLOGY OF HIV INFECTION

11.2.1 Clinical manifestations of HIV

HIV is transmitted by exposure to bodily fluids including blood, semen, and saliva of infected individuals (Fauci, 1988). Within several weeks of inoculation by HIV, at least 50% of individuals who progress to a sustained infection develop an acute syndrome characterized by fever, malaise, adenopathy and rash. This syndrome lasts several weeks and then spontaneously resolves coincident with the development of antibody responses to HIV.

Molecular and Cell Biology of Human Gene Therapeutics
Edited by George Dickson
Published in 1995 by Chapman & Hall. ISBN 0 412 62550 4

Subsequently, an asymptomatic phase develops which lasts months to years and is characterized by low levels of viremia and steadily declining peripheral blood CD4+ T-cell concentrations. In some individuals, as HIV infection progresses, a syndrome of persistent generalized adenopathy develops. At the most advanced stages of HIV infection, opportunistic infections, malignancies and generalized constitutional symptoms which are diagnostic of the acquired immune deficiency syndrome (AIDS) develop. Progression from infection with HIV to AIDS may eventually occur in close to 100% of patients, and the life expectancy once a diagnosis of AIDS is made is 1–2 years (Chin, 1991). In the USA, epidemiological surveys indicate that over 1 million persons are currently infected with HIV and over 180 000 now have AIDS; worldwide, at least ten million people are currently infected with HIV.

11.2.2 Life cycle of the HIV

HIV is an RNA virus in the lentivirus family of non-oncogenic retroviruses. The HIV virion consists of a core composed of two copies of single-stranded genomic length RNA complexed with several HIV encoded core proteins. The core contains a reverse transcriptase, protease, integrase, p24 and other structural proteins, all encoded by the HIV *gag* and *pol* genes. This core is encapsulated in an envelope composed of the gp120 and gp41 glycoproteins encoded for by the HIV *env* gene (Haseltine, 1992). At the time of HIV infection, the virion binds to its receptor, the CD4 cell surface protein, and enters the cell, possibly by direct fusion (Figure 11.1). Once inside the cell, the HIV RNA genome is reverse transcribed via a complex series of events into a double-stranded DNA molecule (Panganabian and Fiore, 1988). Reverse transcription is primed by a ^{Lys} tRNA which had bound to a primer binding site (PBS) located in the U5 region of the HIV RNA long terminal repeat (LTR) at the time of virion assembly (Haseltine, 1992). The viral DNA is then integrated into the host chromosomal DNA via the activity of the core integrase. The integrated HIV proviral genome is organized as depicted in Figure 11.2.

Following integration, the HIV provirus remains latent until the HIV LTR is transcriptionally activated by one of several mechanisms. These may include concurrent infection with other viruses, stimulation by various cytokines, or induction of immune response. Early viral gene expression is characterized by production of multiply spliced HIV mRNAs encoding the regulatory proteins tat, rev, and nef (Cullen, 1992). Tat is an RNA-binding protein which recognizes a sequence present in the 5′ untranslated leader portion of nascently transcribed HIV RNA termed the Trans Activation Response (TAR) region in conjunction with one or more putative cellular factors. Rev is a second essential HIV RNA binding protein which binds to and multimerizes on a region in HIV RNA termed the Rev Response

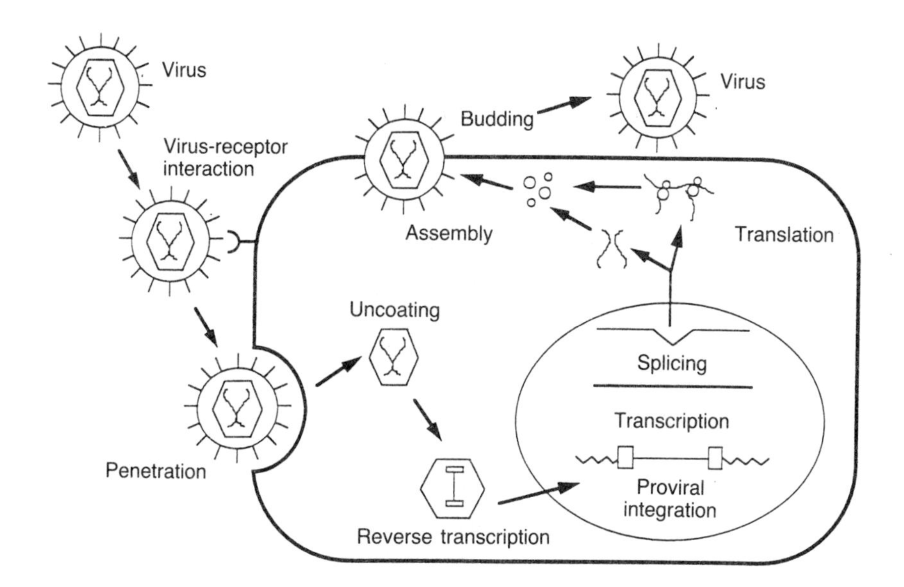

Figure 11.1 The lifecycle of the HIV virus.

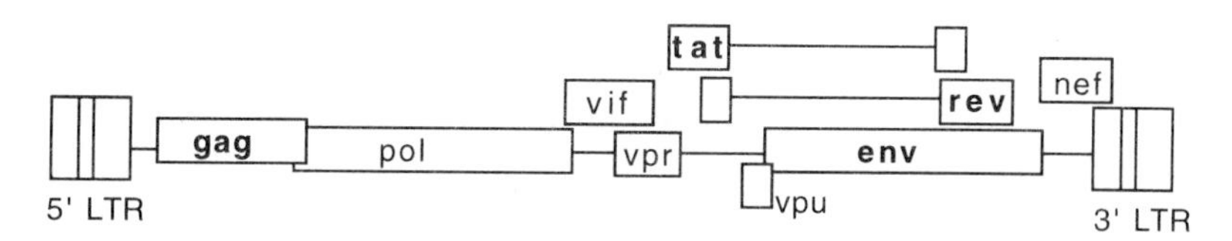

Figure 11.2 Organization of the HIV proviral genome. Coding sequences for the early viral genes *tat*, *rev* and *nef* are shown in grey. Coding sequences for the late viral genes, *gag*, *pol* and *env* are shown in white. The TAR and RRE RNAs are indicated by brackets.

Element (RRE). Binding of tat to TAR results in increased production of HIV mRNAs while binding of rev to RRE is responsible for the transport of singly spliced and unspliced genomic length HIV RNAs from the nucleus (*ibid*). The exact role of nef remains obscure but it appears to enhance HIV infection *in vivo* and is frequently defective in HIV strains which have been adapted to efficient growth in tissue culture (Kestler *et al.*, 1991).

Late viral gene expression occurs after sufficient concentrations of tat and rev have been reached and is characterized by a rapid escalation in the production of singly spliced and unspliced RNAs (Pomerantz *et al.*, 1992). This results in increased production of the envelope and core proteins, as well as genomic RNAs. HIV genomic RNAs are transported from the nucleus to the cell surface and encapsidated into budding viral particles. Formation of the virion core is mediated by recognition of specific packaging sequences(s) present on genomic HIV RNAs which are recognized by the *gag* encoded capsid proteins. The core complex is coated with plasma membrane containing envelope proteins as it buds from the cell. The core proteins are subsequently processed by the viral protease to yield mature virions. A variety of other gene products including vif, vpr, vpu and vpt are expressed in the later stages of the HIV replication cycle as well. These proteins appear to be important in transcriptional control and generation of infectious virions, but the mechanisms of these effects are poorly understood.

11.2.3 Immunopathology of HIV infection

In vivo, HIV specifically infects CD4+ T cells, monocytes/macrophages, and possibly other cells in the nervous system and gastrointestinal tract (Fauci, 1988). HIV does not appear, in most instances, to infect the developmentally primitive totipotent hematopoietic stem cells which continually produce both the lymphoid and myeloid populations (Molina *et al.*, 1990). Numerous abnormalities occur in both the cellular and humoral immune systems following infection with HIV (Miedema, 1992). The acute stages of infection are characterized by transiently depressed or normal numbers of peripheral blood CD4+ T-cells. Subsequently, CD4+ T cells and monocyte/macrophages in the lymph nodes and other lymph tissue appear to be extensively infected by HIV within the first several months following inoculation (Embretson *et al.*, 1992; Pantaleo *et al.*, 1993). Both latent and productive forms of infection are observed in these cells at this time. In addition, dendritic cells in the lymph tissue appear to extensively trap HIV in immune complexes, with or without becoming productively infected themselves (Cameron *et al.*, 1992). Significant and progressive destruction of lymph tissue occurs during the asymptomatic phase of HIV infection resulting in extensive loss of lymphatic dendritic cells and CD4+ T cells (Embretson *et al.*, 1992) and eventually, progressive loss of peripheral blood CD4+ T-cell number (Fauci, 1988). CD4+ T-cell loss results in diminished helper activity for cytotoxic T-cell function and B-cell response to antigen. The dysregulation of B-cell function typically results in polyclonal B-cell activation and frequently, in the development of autoantibodies. Monocyte/macrophage populations also exhibit significant dysfunction in phagocytosis and antigen processing following HIV infection (Poli and Fauci, 1992). These changes contribute to general immune dysfunction and specifically to an increased incidence of infection with encapsulated and intracellular organisms. At the latest stages of HIV infection, the peripheral blood CD4+

T-cell count drops below critical levels, substantial increases in viremia occur, and AIDS develops.

Currently, the mechanisms causing depletion and dysfunction of CD4+ T cells, monocytes and dendritic cells are poorly understood. In addition to direct cytotoxicity due to intracellular accumulation of env, tat or unintegrated HIV provirus (Koga *et al.*, 1990), several non-exclusive etiologies have been proposed including: (i) preferential infection of memory or progenitor CD4+ T-cells which are then eliminated during an amnestic response (Lane *et al.*, 1990); (ii) depletion of CD4+ T cells via recruitment of uninfected cells into syncitia; (iii) autoimmune destruction of uninfected cells presenting epitopes which mimic those on infected cells (Blackburn *et al.*, 1991); (iv) alterations in cysteine/glutathione metabolism following HIV infection (Droge, 1993); (v) secretion of immunomodulatory toxins including tat by HIV infected cells which damages uninfected cells (Benjouad *et al.*, 1993); (vi) disruption of cellular and humoral immune regulatory networks (Yarchoan *et al.*, 1991); (vii) induction of cellular apoptosis (Meyaard *et al.*, 1992); and (viii) destruction of cellular elements critical to ontogeny, such as the thymus.

A critical feature of HIV infection is the substantial strain heterogeneity which occurs in infected populations and individuals. Extensive differences in the cytopathicity and tissue tropism of different strains of HIV appear to be critical to the rate and effects of infection (Schwartz *et al.*, 1989). Individuals appear to be infected with complex mixtures of these strains and subsequently may develop further progressive substrain heterogeneity as the infection progresses. The occurrence and continual development of HIV strains and substrains (or quasispecies) with altered epitopes and altered biologic properties contributes to the resistance of HIV to endogenous immune mechanisms and to exogenously administered chemotherapeutic, biotherapeutic and immunologic interventions. In addition, substantial differences exist between strains isolated from individuals and those adapted to grow *in vitro*.

11.3 CURRENT THERAPY OF HIV INFECTION

Current therapy of HIV infection mainly involves treatment of the infectious and malignant complications of AIDS. These treatment modalities do not affect the underlying pathophysiology of HIV infection. Attempts to attenuate the progression from acute and chronic HIV infection to AIDS have focused primarily on the use of chemotherapeutic and biologic agents which limit viral replication and spread (Purdy and Plaisance, 1991). A limited number of chemotherapeutic agents have demonstrated a modest effect against HIV infection. These agents may have efficacy only at defined stages of infection (i.e. when the peripheral blood CD4+ T-cell count is between 500 and 1000/mm^3), have not proven curative, and are associated with substantial toxicity. Biotherapeutic approaches have been useful in retarding the progression of HIV-associated disease only when CD4 T-cell numbers are relatively preserved and to date, no intervention has consistently improved immune function or been curative.

An alternative approach for the treatment of HIV infection is with immunotherapy. Three general approaches have been described to date: (i) active immune reconstitution with peripheral blood lymphocytes (PBL) from syngeneic donors (Davis *et al.*, 1983); (ii) bone marrow transplantation from syngeneic or allogeneic donors (Saral and Holland, 1993); and (iii) vaccination (Graham *et al.*, 1993). The goal of adoptive transfer of syngeneic peripheral blood lymphocytes for treating HIV infection is to restore a clinically relevant immune repertoire in order to diminish the onset, frequency, and severity of opportunistic infections. This approach can be used at advanced stages of HIV infection when endogenous CD4+ cell numbers are very low. Small clinical trials involving transplantation of peripheral blood lymphocytes between identical twins, with and without concomitant administration of antiretroviral chemotherapy, have resulted in minimal and transient improvements in peripheral blood CD4+ cell number, delayed hypersensitivity response, and *in vitro* lymphocyte function assays. No clinical improvements were noted in studies published to date. The most likely explanation for these observations is that the established HIV infection in the recipient rapidly depleted the transplanted cells due to one or more of the mechanisms outlined above. Another significant limitation to this approach is the small likelihood of an available syngeneic donor.

A potentially more powerful approach to treating HIV infection is by transplantation of allogeneic, syngeneic or autologous hematopoietic stem cells (Holland *et al.*, 1989; Saral and Holland, 1993). Theoretically, the entire immunologic and hematopoietic system could be restored fully following transplantation. A small number of HIV-infected patients with B-cell lymphoma and other diseases have been treated with syngeneic or allogeneic bone marrow transplantation combined with pre-transplant treatment with high doses of anti-retroviral chemotherapeutic agents. In some of these individuals, HIV was eliminated transiently following transplantation, but all eventually succumbed to recurrent HIV infection, recurrent B-cell lymphoma, or complications of transplantation.

Vaccination of HIV seropositive individuals with various HIV encoded proteins or live attenuated virus has been carried out in several clinical trials (Graham *et al.*, 1993). Vaccination may have two applications: first, as prophylaxis for previously uninfected individuals and second, as a means for enhancing anti-HIV immunity in previously infected individuals. In a trial evaluating vaccination of HIV seropositive individuals with gp160, attenuation of the progressive loss of CD4+ cell number in peripheral blood, and enhancement of immune function as measured by delayed hypersensitivity reactions, *in vitro* lymphocyte proliferation, immunoglobulin production, and generation of HIV specific cytotoxic T lymphocyte (CTL) clones was observed (Redfield *et al.*, 1991). These responses were observed only in individuals with relatively preserved CD4+ T-cell numbers and no observations regarding change in clinical status have been published to date.

11.4 STRATEGIES FOR TREATING HIV WITH GENE THERAPY

Gene therapy strategies may prove useful for augmenting adoptive immunotherapy, hematopoietic stem cell transplantation and vaccination approaches to treating HIV infection. Gene therapy strategies for treating HIV infection fall into three general categories based on their goals: (i) inhibition of HIV via intracellular expression of genetic elements which limit HIV expression and replication, previously termed 'intracellular immunization' (Baltimore, 1988); (ii) elimination of HIV infected cells by HIV transactivated toxic molecules; and (iii) enhancement of HIV immune responses. The HIV gene therapy strategies which have been proposed to date are summarized in Table 11.1.

11.4.1 Intracellular immunization strategies

Protection of cells from infection may occur by interfering with a number of aspects of the HIV life cycle (Figure 11.1). This includes inhibition of any of the steps before establishment of the provirus, including receptor binding, reverse transcription and integration; inhibition of expression from an established provirus; and interfering with virion assembly and production. Theoretically, inhibition of proviral establishment would be the most effective intracellular immunization approach since it would preclude any deleterious effects due to HIV expression.

Inhibition of HIV/CD4 interaction

The goal of this strategy is to compete for HIV binding to cellular CD4 with high concentrations of locally produced soluble CD4 (sCD4). Morgan *et al.* (1990) developed a series of murine retroviral vectors which stably secreted high levels of sCD4 from a human HeLa cell line. The sCD4 bound HIV gp120 and protected transduced cells derived from the human T-cell line Sup-T1 from infection with a tissue culture-adapted strain of HIV. In addition sCD4 was also expressed in fibroblasts. Transduced fibroblasts could then act as 'cell factories' and constitutively secrete sCD4 into the circulation. There are several significant drawbacks to this strategy. Natural

Table 11.1 HIV gene therapy strategies

Strategy	*Putative site of action*	*Reference*
Intracellular immunization		
Soluble CD4	Receptor binding	Morgan (1990)
SELEX	Reverse transcriptase (and others)	Tuerk *et al.* (1992)
TAR decoys	Viral RNA transcription/ processing/transport/ stability	Sullenger *et al.* (1990); Lisziewicz *et al.* (1991)
Trans-dominant tat		Green *et al.* (1989)
RRE decoys		Lee *et al.* (1992)
		Bevec *et al.* (1994)
Trans-dominant rev		Malim *et al.* (1989)
Interferon inducible gene		Constantoulakis *et al.* (1993)
Antisense RNA		Sczakiel and Pawlita (1991)
		Chatterjee *et al.* (1992)
Ribozymes		Cech (1988)
Interferons		Bednarik *et al.* (1989)
Trans-dominant gag	Viral assembly	Trono *et al.* (1989)
Trans-dominant env		Buschacher (1992)
		Steffy and Wong-Staal (1993)
vpx		Matsuda *et al.* (1993)
Intracellular CD4,	Binding of rev, env	Bounocore and Rose (1993)
Antibodies		Marasco *et al.* (1993)
Gag–nuclease fusions	Virion disruption	Natsoulis and Boeke (1991)
Conditional toxins	Elimination of HIV-infected cells	Venkatesh (1993)
		Harrison *et al.* (1991)
Augmenting HIV immunity		
Vaccination		Jolly and Warner (1990)
CTL infusion		Riddel *et al.* (1992)

strains of HIV are relatively insensitive to neutralization by sCD4 compared with tissue culture-adapted strains (Orloff *et al.*, 1993) which may obviate this approach unless extremely high concentrations of sCD4 could be achieved in the vicinity of the target cell. In addition, high local concentrations of sCD4 potentially could disrupt normal MHC class II interactions in the target cell and in neighboring immune cells. Finally, HIV may enter cells via non-CD4 dependent mechanisms.

Inhibition of reverse transcription

Tuerk *et al.* (1992) have applied a technique termed Systematic Evolution of Ligands by Exponential Enrichment (SELEX) to isolate specific oligoribonucleotides capable of binding HIV reverse transcriptase with high affinity and inhibiting its activity in *in vitro* cell-free assays. Such inhibitory oligoribonucleotides can potentially bind and inhibit reverse transcriptase and prevent infection. These anti-HIV oligonucleotides may have utility as either exogenously administered pharmacologic agents or as intracellularly expressed RNAs following gene transfer into targets of HIV infection.

Inhibition of HIV expression

Inhibition of tat/TAR

In cells which contain an integrated HIV provirus, it may be possible to inhibit viral gene expression by interfering with the tat/TAR interaction. The potential advantage of this strategy is that inhibition of this interaction could diminish the expression of all HIV RNA species and thus could inhibit production of potentially toxic HIV proteins and infectious virions.

1. **TAR decoys**. Several groups have employed oligoribonucleotides which encode the TAR sequence, termed TAR decoys, to compete for tat binding and inhibit tat dependent HIV transactivation (Sullenger *et al.*, 1990a; Joshi *et al.*, 1991). Sullenger constructed a murine retroviral vector containing the TAR sequence under the transcriptional control of an RNA polymerase III (pol III) human tRNA promoter (Sullenger *et al.*, 1990a, b, 1991). Higher levels of expression of TAR decoys and other oligoribonucleotides may be achieved when using pol III promoters compared with pol II promoters (Sullenger *et al.*, 1990a, b). Selected CEM_{ss} clonal cell lines were found to be more than 95% resistant to challenge with a clonal laboratory isolate of HIV. In addition, these cells were resistant to challenge by simian immunodeficiency virus (SIV), implying that TAR decoys may be effective against a wide variety of HIV species due to sequence conservation of the TAR region. In addition, unselected bulk cultures of TAR decoy-transduced CD4+ T cells appeared to possess a modest and transient survival advantage compared with cells transduced with a mutant non-functional TAR decoy when challenged with a clonal laboratory HIV isolate (Smith *et al.*, 1992). Lisziewicz *et al.* (1991) created an HIV-inducible TAR decoy expression system by constructing vectors with multimerized TAR sequences (polyTAR) expressed under the transcriptional control of an HIV LTR promoter. Selected clones of MOLT-3 cells expressing polyTAR demonstrated prolonged inhibition of HIV-1_{IIIB} (Lisziewicz *et al.*, 1993). Since there is concern that TAR decoys may bind to cellular factors, a potential advantage

of this strategy is that the polyTAR sequences will only be expressed in HIV infected cells since tat is required for expression of the polyTAR transcripts.

2. **Trans-dominant tat mutants.** Variants of proteins which interfere with the normal function of their wild type counterpart are termed dominant negative mutants. Dominant negative mutant proteins exert their effect by forming defective multimers with the wild-type protein or by competing with the wild-type protein for essential cofactors or transport pathways. tat may be an appropriate target for trans-dominant inhibition because it may dimerize before or during binding to TAR, and may interact with several essential cellular cofactors. Several investigators have described a series of tat mutants which trans-dominantly suppressed tat-mediated transactivation of reporter genes (Green *et al.*, 1989; Pearson *et al.*, 1990; Modesti *et al.*, 1991). Several concerns exist regarding the use of tat trans-dominant mutants, including the apparent requirement for high levels of intracellular expression relative to wild-type tat for effectiveness, the possibility that trans-dominant tat as well as all trans-dominant mutant proteins may be immunogenic, and that trans-dominant mutant forms of tat may be toxic to the cells they are designed to protect as well as neighboring cells (Ratner *et al.*, 1987; Benjouad *et al.*, 1993).

Inhibition of rev/RRE

1. **RRE decoys.** As described above, the rev/RRE interaction mediates the switch from early to late gene expression in the HIV life cycle. Lee *et al.* (1992) described an RNA decoy strategy based on intracellular pol II expression of RRE decoys from a retroviral vector. The goal of this strategy is to sequester rev and inhibit it from binding the viral RRE, thus blocking late gene expression and HIV replication. Isolated clones of CEM_{ss} cells expressing RRE decoys inhibited HIV_{SF13} by 95%. Using a similar strategy, Bevec *et al.* (1994) also observed substantial inhibition of HIV-1IIIB replication as well as inhibition of gag expression in HeLa clones and in bulk populations of transduced cells constitutively expressing RRE fused to a neomycin phosphotransferase transcript. A potential advantage of RRE decoys relative to TAR decoys for inhibition of HIV is that creation of an RRE decoy which binds rev but does not sequester cellular factors may be feasible (Cullen, 1992). A disadvantage is that even in the complete absence of rev, expression of multiply and singly spliced RNA can occur, leading to the production of potentially toxic HIV proteins. Also, because rev negatively regulates its own expression, inhibition of rev function by an RRE decoy may only cause HIV to produce more rev until the inhibition is overcome.

2. **Trans-dominant rev mutants.** Rev is an attractive target to inhibit with trans-dominant negative mutants since it appears that rev function requires multimerization. Malim *et al* (1989, 1991, 1992) generated a series of mutant forms of rev which transdominantly inhibited wild type rev function

in vivo. One of these mutants, revM10, contains mutations at the carboxyl terminus which do not affect multimerization or RRE binding but apparently inhibit the binding of rev multimers to cellular factors critical for *in vivo* rev function. Subclones of CEM cells expressing high levels of M10 inhibited a clonal isolate of HIV by >99% while two of three subclones inhibited infection with the heterogeneous HIV-IIIB strain. Lack of effect on the earliest events in HIV expression were inferred from experiments utilizing a CAT reporter gene cloned into the site of *nef* in a full-length HIV expression vector. No effects on NFk-β or tat-mediated LTR activation and no effect on mitogen-induced IL-2 production in transduced Jurkat or EL4 cells was observed, implying that revM10 may not inhibit induction of genes critical to T cells. More recently, other investigators have confirmed these findings by observing inhibition of HIV-IIIB in isolated CEM subclones following intracellular expression of revM10 from a variety of vectors (Bahner *et al.*, 1993; Bevec *et al.*, 1992). Other trans-dominant mutant forms of rev with altered C-terminal regions or an altered nucloelar localization domain have been developed which appear to trans-dominantly inactivate the wild-type rev by forming stable but non-functional multimers (Hope *et al.*, 1992; Kubota *et al.*, 1992).

3. **Interferon-inducible genes**. Constantoulakis *et al.* (1993) found that a cellular protein termed interferon-inducible gene 9–27 binds to RRE and inhibits Rev-dependent RNA expression and viral structural protein production. This protein may be the first of many cellular factors which have the ability to inhibit HIV.

Antisense RNA inhibition

RNA transcripts which can base-pair to target RNA sequences in trans to form duplex RNAs are termed antisense RNAs. Intracellular expression of HIV-specific antisense RNAs may inhibit HIV replication by interfering with RNA processing, transport or translation. The actual mechanism of interference remains obscure, but chemical modification of the duplex RNA, enhanced degradation by RNAseH or other nucleases, or steric hindrance with other cellular factors may be involved. Antisense RNA could potentially render cells resistant to HIV infection by hybridizing to the incoming genomic RNA and blocking reverse transcription. It is unlikely that the HIV genomic RNA is accessible at this point for duplex formation since it is complexed to the core proteins and probably localized in specific subcellular compartments from which antisense RNAs are excluded. Most investigators consequently have focused on inhibiting RNA expressed from the integrated HIV provirus. A number of studies have observed limited, and in most cases, transient inhibition of HIV by vector encoded antisense RNA directed against the highly conserved HIV 5′ leader/*gag* region (Sczakiel and Pawlita, 1991), the tat and rev exons (Sczakiel *et al.*, 1992), various regions of the 5′ end of HIV-1 RNAs including the 18 bp $tRNA^{Lys}$ primer binding site (PBS), the tat, rev, and *vpu* splice acceptor sites, the tat/rev splice donor site, the initiation codons for translation of tat, rev, vpu and env (Rhodes and James, 1990), reverse transcriptase (Meyer *et al*., 1993), and the first coding exon of tat (Lo *et al*., 1992). Chatterjee *et al.* (1992) described development of adeno-associated viral (AAV) vectors expressing antisense RNA which were simultaneously complementary to 5′ untranslated leader sequences and 3′ sequences including the polyadenylation signal of HIV RNAs. AAV vector-transduced, bulk selected, A3.01 cells inhibited HIV-IIIB by more than 99%. These experiments, as well as studies in a variety of other systems using antisense oligonucleotides, indicate that a number of factors need to be addressed in order to develop effective HIV antisense inhibition strategies. These include optimizing the length of the antisense RNA, optimizing intracellular expression, designing antisense sequences to avoid intramolecular base-pairing and non-specific hybridization, optimizing the subcellular trafficking of the antisense RNA so that it can reach its target, and delineating which target sequences yield the greatest biologic effect following inhibition. While no general rules are known for selecting optimal potential antisense target sequences, intramolecular folding of the target sequence as well as interactions of the target sequence with other intracellular molecules may influence their susceptibility to antisense inhibition.

Ribozymes

Certain RNA molecules can adopt three-dimensional conformations which allow them to perform enzymatic reactions. These RNA enzymes, termed ribozymes, contain catalytic cores and active sites which allow them to catalyze cleavage and ligation reactions upon specific substrate RNAs with rate enhancements similar to many protein enzymes (Cech, 1988). Five RNA catalytic motifs have been described which may be useful for trans cleavage of viral target RNAs *in vivo*: RNase P, the hammerhead ribozyme, the hairpin ribozyme, the hepatitis delta virus ribozyme, and the group I intron (Cech, 1988). Of the five motifs, only RNase P cleaves multiple substrate RNAs in trans in nature. This ribozyme recognizes and cleaves tRNA precursors to remove extra sequences from their 5′ ends. Forster and Symons (1987) have demonstrated that *E. coli* RNase P can be directed to cleave non-tRNA sequences *in vitro* by base-pairing an antisense RNA to a target RNA such that the sequence CCA hangs from the end of the duplex, and thus acts as a pre-tRNA substrate. Expression of antisense RNAs, which base-pair to targeted RNAs in this manner, can potentially lead to RNase P-mediated destruction of target RNAs *in vivo*. A major potential advantage of this strategy is that RNase P naturally performs similar reactions continually in cells. However, since the reaction is performed upon pre-tRNAs which probably reside in different nuclear compartments than mRNAs, RNase P may not have access to many targeted RNAs if it is sequestered in intracellular locations essential for pre-tRNA processing.

The most thoroughly characterized ribozyme is the enzymatic form of the *Tetrahymena thermophila* group I intron discovered by Cech and colleagues (1992). This ribozyme recognizes and cleaves a specific hexanucleotide substrate sequence. Because a hexanucleotide sequence is going to be present on many RNAs *in vivo*, use of this ribozyme as an *in vivo* gene inhibitor may be deleterious to cells. The hammerhead (Forster and Symons, 1987), hairpin (Hampel *et al.*, 1990), and hepatitis delta virus ribozymes (Branch and Robertson, 1991) are small cis-cleaving ribozymes which have been found mainly in plant virusoids and satellite RNAs or in a satellite RNA of human hepatitis B virus. Enzymatic versions of each of these ribozymes have been created which are capable of cleaving novel substrates in trans *in vitro* with multiple turnover (Haseloff and Gerlach, 1988; Hampel *et al.*, 1990; Branch and Robertson, 1991). These ribozymes contain guide sequences which allow them to base-pair with and subsequently cleave specific substrate RNAs in trans. Specific substrate sequence requirements are minimal for hammerhead and hairpin ribozymes, and thus allow design of particular ribozymes capable of cleaving a designated RNA by adjusting the length and composition of the ribozymes' guide sequence. Their small size and malleability make these ribozymes very attractive potential gene inhibitors.

Ribozymes posses two theoretical advantages as compared with antisense inhibition strategies: the potential to cleave and directly inactivate target RNAs, and the potential to cleave multiple target sequences catalytically. A number of studies have evaluated the effectiveness of hairpin and hammerhead ribozymes for inhibiting HIV *in vivo*. Sioud and Drlica (1991) constructed a ribozyme targeted to the HIV-1 integrase gene. In cell-free *in vitro* assays, specific cleavage was observed. In addition, following co-transfection of the ribozyme and its target sequence into *E. coli*, loss of target sequence RNA and integrase protein was observed. A control ribozyme lacking an active catalytic center failed to diminish integrase RNA production implying that the inhibitory effect was not due solely to an antisense effect.

Sarver *et al.* (1990) designed two hammerhead type ribozymes targeted to HIV-1 gag transcripts and demonstrated catalytic site specific cleavage with prolonged (>14 hour) incubation in cell-free *in vitro* assays. Subsequently, they constructed plasmids which constituitively expressed these ribozymes and the selectable marker guanosyl-phosphoribosyl transferase (gpt). Selected CD4+ HeLa cells appeared to inhibit HIV by over 95% in short-term cultures relative to controls, and ribozyme-mediated cleavage was inferred from RNA polymerase chain reaction (PCR) analysis of gag transcripts. Other investigators have observed various degrees of inhibition of HIV by intracellularly expressed hammerhead ribozymes directed against the HIV leader sequence (Weerasinghe *et al.*, 1994), the U5 region of the HIV LTR (Dropolic *et al.*, 1992), env (Chen *et al.*, 1992), and the HIV-1 gag, tat, and tat-rev sequences (Taylor and Rossi, 1991). Hairpin-type ribozymes directed against the HIV 5′ leader sequence may also inhibit HIV *in vivo* to varying degrees (Ojwang *et al.*, 1992; Yamada *et al.*, 1994).

While these studies indicate that ribozymes are capable of specifically cleaving HIV RNAs in cell-free *in vitro* assays and may inhibit HIV *in vivo*, direct confirmation of *in vivo* cleavage of HIV transcripts has not been convincingly demonstrated, presumably because cleavage products degrade rapidly inside cells. Likely, additional modifications in ribozyme design will need to be made in order to optimize their enzymatic activity in the complex intracellular milieu. In addition, for ribozymes to be effective, they must be able to localize to the same subcellular compartment as their targets. Sullenger and Cech (1993), in a Moloney murine leukemia virus (MoMuLV) model system, have obtained evidence that the subcellular distribution of ribozymes is critical to their *in vivo* efficacy. An additional concern when using ribozymes as HIV inhibitors is the development of escape mutants. Only one or a small number of nucleotides need to be mutated in a target sequence to abolish a ribozyme's ability to recognize and cleave that sequence. Most likely, vectors encoding ribozymes directed to multiple HIV targets will need to be developed if ribozymes are to be effective HIV inhibitors in patients.

Interferons

HIV-1 replication can be inhibited *in vitro* by exogenously added interferon-α (IFN) (Shirazi and Pitha, 1992). In an effort to achieve high local concentrations of IFN which may be more effective than systematically administered IFN while avoiding the systemic toxicity of IFN, several groups have introduced vectors encoding IFN into CD4+ T cells under the transcriptional control of the HIV-1 LTR. Bednarik *et al.* (1989) constructed tat-transactivated IFN encoding plasmids and retroviral vectors and stably transduced A3.01 cells. Under baseline conditions, no adverse effect on CD4 expression or cellular morphology or growth kinetics was noted. Following infection with HIV-1, over 400 U/ml of a-IF were produced and >95% inhibition of p24 production was observed for 14 days in culture. RNA blot analysis revealed diminished levels of HIV RNA in IFN-transduced cells, suggesting that inhibition of HIV transcription or a reduction in HIV mRNA stability had occurred. Similar effects have also been observed in human promonocytic cells transfected to express interferon-β (Mace *et al.*, 1991).

Inhibition of infectious virion production

Inhibition of virion assembly or virion infectivity would be useful in limiting the spread of HIV. A potential drawback, however, is that toxic HIV gene products produced earlier in HIV replication may accumulate in cells, resulting in direct toxicity or in eliciting an immune response. Therefore, these strategies may be most useful when combined with other HIV inhibition strategies.

Trans-dominant virion proteins

The HIV gag protein is potentially an effective target for trans-dominant negative inhibition because it extensively multimerizes during viral assembly. Trono, Feinberg, and Baltimore developed a number of gag mutants which interfered with production of viral particles in selected HeLa cell clones (Trono *et al.*, 1989). Some gag mutants inhibited viral titer more than p24 production, suggesting that defective virions were being produced. Mutant gag-producing cell lines supported the early stages of HIV replication, indicating that the inhibition occurred at, or following viral assembly. A potential advantage to using trans-dominant negative gag proteins to inhibit HIV replication is that relatively small amounts of the trans-dominant protein may be effective at inhibiting much larger amounts of wild type gag by assembly with it into large defective core complexes.

Steffy and Wong-Staal (1993) constructed an HIV-2 mutant virus containing an in-frame deletion in the CD4 binding domain of gp120 which inhibited HIV viral production. Mechanistic analysis indicated that the mutant gp120 may have inhibited the binding of the wild-type gp120 to CD4 or inhibited post-binding events. In addition, a reduced amount of extracellular HIV-2 gag protein, p26, was observed, indicating that the trans-dominant gp120 may have also interfered with virus assembly or budding. Buschacher and colleagues developed an HIV-1 gp41 mutant, termed 41.2, containing a mis-sense mutation in the hydrophobic amino terminus of the protein. This mutation presumably disrupts gp41 multimerization critical to virion assembly (Buschchacher *et al.*, 1992; Freed *et al.*, 1992). An HIV-based vector (see below) expressing 41.2 conferred partial protection in transduced cells from the cytopathic effects of co-expressed wild-type env and inhibited the spread of wild-type virus following infection with HIV-1. Potential drawbacks to this strategy include the concern that mutant env may be deleterious to cells, the observation that trans-dominant mutant env has been inhibitory only at relatively high molar ratios, and the possibility that the mutant env may be immunogenic.

Vpx is the HIV-2/SIV homolog of the HIV-1 *vpr* gene. Matsuda *et al.* (1993) expressed *Vpx* in T-cell lines and demonstrated that it could be incorporated into HIV-1 virions and cause diminished virion infectivity.

Intracellular traps

Several groups have developed strategies for trapping env intracellularly so that it is unable to assemble onto virions. Bounocore and Rose (1993) modified sCD4 by incorporating into the molecule a peptide sequence which traps the sCD4 in the endoplasmic reticulum and expressed it intracellularly under the control of HIV-1 regulatory elements. HIV env was blocked from transport to the cell surface, and no viral spread was observed in infected cell cultures. Similar strategies involving modification of intracellularly expressed sCD4 to divert gp160 to lysosomes (Lin *et al.*, 1993) and intracellular expression of antibodies to gp120 which are retained in the endoplasmic reticulum (ER) have been described (Marasco *et al.*, 1993) which also appear to inhibit HIV env transport to the cell surface and production of infectious virions. All these strategies suffer from the same concerns discussed above for secreted sCD4. The decreased affinity of natural isolates for sCD4, however, may be overcome by achieving extremely high local concentrations of sCD4 or anti-gp120 within the ER.

gag–nuclease fusion proteins

Natsoulis and Boeke (1991) developed a strategy termed Capsid Targeted Viral Inactivation (CTVI) based on the expression of a chimeric gag–nuclease fusion protein selectively incorporated into a virion. The gag portion of the fusion targets the chimeric molecule to the virion. After budding the nuclease is activated by the elevated calcium concentrations present in the extracellular milieu. Conversely, the nuclease is inactive at intracellular calcium concentrations. As a prelude to developing this approach for inhibition of HIV, CTVI initially was used to inhibit Ty1 virus replication in yeast. Subsequently, CTVI using a fusion of the MoMuLV gag protein and staphylococcal nuclease was used to inhibit MoMuLV virion infectivity in chicken embryo fibroblast cells (Natsoulis and Boeke, 1993). Drawbacks to this approach include potential toxicity due to low levels of leaky intracellular nuclease activity and inability to inhibit virions that are transmitted by cell–cell contact.

In summary, a number of protein and RNA-based intracellular immunization strategies have been developed to inhibit various aspects of the HIV life cycle. Virtually all of these strategies have theoretical or recognized inherent drawbacks and the potential for the outgrowth of insensitive HIV escape mutants exists to a variable degree for each of these strategies. Inhibition strategies which inhibit early events in the HIV lifecycle may be more advantageous as they may potentially reduce toxicity due to HIV expression and less inhibitor may be required to effectively inhibit HIV replication. RNA-based strategies have several advantages relative to protein-based strategies, including decreased immunogenicity, potential for higher levels of expression from pol III promoters, and the ease of construction so that they do not interact with cellular factors. The potential for the generation of escape mutants appears to be less with RNA decoys than with ribozymes; however, the potential for RNA decoys to inhibit cellular factors is a serious concern. Trans-dominant revM10 has been demonstrated reproducibly to be a potent inhibitor of HIV replication and appears to be a very promising molecule for further study. Most likely, no single intracellular immunization strategy will be completely effective in inhibiting HIV, necessitating the use of multiple complementary strategies.

11.4.2 Eliminating the reservoir of HIV-infected cells

An alternative gene therapy approach to intracellular immunization for treating HIV infection is to engineer cells which, once infected by HIV, are selectively eliminated. In all approaches described to date which attempt this strategy, a toxin-encoding gene is introduced into cells under the transcriptional control of the HIV-1 LTR as well as other HIV cis-acting regulatory elements. Such constructs prevent expression of the toxin until the cell is infected with HIV. Advantages to this approach are that only transient expression of small amounts of the conditionally expressed toxin need to occur to eliminate HIV infected cells, and that tat transactivation of the toxin should occur with early viral gene expression before the production of significant amounts of potentially toxic HIV proteins. Harrison *et al* (1992a, b) have described a method for selectively killing HIV infected cells by introducing a vector encoding the diphtheria toxin–A (DT–A) chain under the transcriptional control of the HIV LTR, the cis-acting negative regulatory elements (crs) and RRE (Harrison *et al.*, 1991, 1992a, b). Production of p24 was substantially reduced in H9 cells expressing DT–A. Several groups have described the development of vectors encoding the *HSV–TK* gene under the transcriptional control of the HIV LTR and other HIV regulatory sequences (Venkatesh *et al.*, 1992). The HSV–TK, unlike mammalian TK, can phosphorylate nucleoside analogs such as ganciclovir (GCV) or acyclovir (ACV) to inhibit cellular DNA synthesis and cause cell death. GCV- or ACV-treated transduced cells infected with HIV were selectively eliminated and no non-specific toxicity was noted in transduced cells not infected with HIV. A potential advantage of TK relative to DT–A is that leaky expression of TK may be less toxic to cells. Sun and Baltimore (1989) developed a plasmid vector which expressed the poliovirus 2A protein under the transcriptional control of the HIV-LTR. Poliovirus protein 2A is a protease which is involved in shutting off cellular protein synthesis by inactivation of the translational initiation factor 4F (eIF-4F). Intracellular expression of protein 2A in HeLa cells occurred at low, nonlethal levels until transactivated by tat which resulted in inhibition of cellular mRNA translation. Potential disadvantages of the conditional cytotoxicity strategy include toxicity in uninfected cells due to leaky intracellular expression of toxin, toxicity to neighboring non-transduced cells after cell death, and immunosuppression resulting from widespread death of infected cells which otherwise might have tolerated a latent HIV infection.

11.4.3 Augmenting the immune response to HIV via gene therapy

The immune system plays a dual role in the pathophysiology of HIV infection (Stanley and Fauci, 1993). On one hand, HIV has subverted many of the normal immune processes such as dendritic cell antigen presentation and cytokine mediated T-cell activation to enhance its propagation. On the other hand, substantial evidence indicates that the humoral and cellular immune responses are critical to suppressing HIV infection. For example, in the later stages of HIV infection, patient HIV-specific CD8+ CTL number declines concomitant with increasing viremia, clinical deterioration, and loss of the CD4+ T-cell population. *In vitro* studies have shown that CD8+ CTL, and possibly CD4+ CTL, are effective inhibitors of HIV in culture (Plaeger *et al.*, 1992). These observations imply that enhancement of the HIV-specific CTL response and perhaps antibody response may reduce viral burden and thus slow the progress of HIV infection.

Enhancement of cellular immunity to HIV

Jolly and colleagues have developed a system designed to elicit and enhance cytotoxic T-cell lymphocyte responses to HIV by introducing vectors encoding the HIV gene products env and gag in autologous fibroblasts (Jolly and Warner, 1990; Jolly, 1991). In initial studies, autologous murine fibroblast lines were transduced with a murine retroviral vector encoding the HIV-IIIB env (Warner *et al.*, 1991). gp160 and its cleavage product gp120 were efficiently expressed in these cells and mice immunized with the transduced fibroblasts developed gp120-specific CD8+ CTL and antibodies. More recent studies in primates recapitulated these findings with no apparent toxicity following transplantation of the transduced autologous fibroblasts. Vaccination with naked DNA may elicit anti-HIV cellular and humoral immune responses (Ulmer *et al.*, 1990; Wang *et al.*, 1993). This method of vaccination is significantly less complex than vaccination with genetically modified tissue and determination of its efficacy in non-human primates and humans is currently pending.

Vaccination approaches may have utility both as a therapeutic modality as well as a method for prophylactic vaccination in high-risk populations. The fact that only transient, low-level expression in a small number of cells is necessary to elicit an immune response is an additional advantage of this approach compared with the 'intracellular immunization' approaches described above. Potential disadvantages include eliciting deleterious autoimmune responses and the difficulty in generating immune responses to the wide variety of prevalent HIV strains. In addition, it is unclear whether patients with advanced immunodeficiency will be capable of mounting an effective anti-HIV immune response.

Adoptive transfer of HIV-specific CD8+ CTL

Greenberg and his colleagues have developed an alternative immunotherapy approach to treating HIV infection based on transfusing HIV-infected individuals with *ex vivo* expanded HIV-specific CD8+ CTL (Riddel *et al.*, 1992). These investigators have previously reported clinical responses to infusions of viral-specific CD8+ CTLs in patients with cytomegalovirus infections following bone marrow transplantation (Riddel *et al.*, 1991). Potential drawbacks to this approach for treating HIV infection include the possibility of developing local non-specific cytotoxicity due to inflammatory responses initiated by the CD8+ CTLs in tissues infected with HIV. To address this concern the *HSV–TK* gene was introduced into HIV-specific CD8+ CTL using a retroviral vector in order to elimate transduced CTLs should an inflammatory reaction occur and to follow trafficking of the transfused CTL. Transduced CTLs were eliminated by low concentrations of ganciclovir in *in vitro* cultures as well as in murine models (*ibid*).

11.5 GENE DELIVERY SYSTEMS FOR GENE THERAPY OF HIV INFECTION

For the intracellular immunization strategies described above which rely on prolonged expression of an HIV inhibitor, or the conditional cytotoxicity strategies which rely on prolonged integration of the vector with transient expression of the vector encoded sequences following infection with HIV, vectors capable of stable integration are required. These vectors should ideally be capable of efficiently transducing cells under conditions which do not affect cell physiology and developmental potential. Currently, no vector satisfies these requirements completely. The vectors potentially capable of stable integration which have been used for HIV gene therapy strategies include murine retroviral vectors, adeno-associated viral (AAV) vectors, and HIV vectors. Strategies which aim to elicit an enhanced immune response to HIV can utilize a wider range of vectors for gene delivery since only transient, low-level expression of vector-encoded sequences may be required to elicit effective viral inhibition. A variety of vectors can be effectively used for these purposes.

11.5.1 Murine retroviral vectors

Murine retroviral vectors have been the most commonly used viral vectors for stable gene delivery into mammalian cells (Miller, 1992). The advantages of murine retroviral vectors for use in gene therapy of HIV are that vector containing virions can be safely and efficiently produced in large quantities, there is an accumulating experience with them in human clinical trials, and they have been observed to be capable of stably expressing vector-encoded sequences for periods of greater than 1 year in primate hematopoietic cells (van Beusechem *et al.*, 1992). As described above, a number of studies have utilized retroviral vectors to introduce HIV inhibitory molecules into both immortalized human T-cell lines and primary human peripheral blood T cells. A number of drawbacks exist for using murine retroviral vectors as gene delivery systems for treating HIV infection, however. A major concern is the possibility of introducing a recombinant replication-competent murine retrovirus into an immunocompromised individual, which could result in the development of lymphoma (Donahue *et al.*, 1992). A second related concern is the possibility of pseudotyping the HIV RNA genome into replication-competent amphotropic murine retroviral virions (Spector *et al.*, 1990). This could potentially lead to HIV infection of tissues not typically infectable, with unforeseeable consequences. Other drawbacks to using murine retroviral vectors for HIV gene therapy include a relatively low gene transfer efficiency (1–10%) into either primary human peripheral blood T cells or hematopoietic stem cells capable of sustaining hematopoiesis for long periods of time following transplantation (van Beusechem *et al.*, 1992). In addition, as opposed to *in vitro* studies where durable and consistent vector expression is commonly observed, inconsistent and low-level expression of vector-encoded sequences has been observed in multiple *in vivo* experiments despite persistence of the integrated vector provirus (Williams, 1990; Miller, 1992). An additional drawback to murine retroviral vectors is the apparent requirement of target cells to undergo mitosis for stable retroviral vector integration (Miller *et al.*, 1990). Stimulation of lymphocytes and hematopoietic stem cells to undergo mitosis may be deleterious to their developmental potential and function following transplantation (Dumenil *et al.*, 1989). Finally, because murine retroviral integration is at least semi-random, there is the possibility that integration of the vector may activate an oncogene or inactivate a tumor suppresser gene resulting in malignant transformation.

A number of laboratories are currently attempting to address the drawbacks of murine retroviral vectors so that they are more effective for gene therapy applications. Amphotropic murine retroviral vector packaging lines have recently been constructed which apparently do not produce replication-competent retrovirus (Markowitz *et al.*, 1989). Multiple strategies have been developed to increase the gene transfer efficiency in hematopoietic cells. These include using physical concentration methods and improvements in packaging lines in order to improve retroviral vector titers. In addition, it may be possible to modify the murine retroviral virion env with target cell-specific ligands as well as to use retroviral virions in combination with physical methods such as liposomes or ballistic delivery, in order to further enhance gene transfer efficiency. It may also be possible to modify the murine retroviral vector sequence so that it could be efficiently packaged into HIV virions as a means of promoting the *in vivo* distribution of the inhibitory vector to other cells as they are infected with the pseudotyped HIV. Systematic studies of vector architecture and design have been performed to optimize vector-mediated gene expression in hematopoietic cells *in vivo* (Miller, 1992). Currently, no consistent guidelines exist for designing retroviral vectors with optimal and persistent expression. Simple vectors encoding a single gene and lacking selectable markers, double copy vectors, and vectors employing pol III promoters might be advantageous for these purposes (Hantzopoulos *et al.*, 1989; Sullenger *et al.*, 1 990a, b). In an effort to develop murine retroviral vectors which can integrate stably into non-dividing cells, nuclear localization signals have been added to a component of the retroviral core particle (Bukrinsky *et al.*, 1992). In addition, novel methods for isolating the relevant hematopoietic cells and inducing hematopoietic cells to cycle without prolonged *ex vivo* culturing may help to preserve the biologic function of these cells following transduction (Bregni *et al.*, 1992).

11.5.2 Adeno-associated virus (AAV) vectors

AAV is a non-pathogenic, single-stranded DNA virus of the parvovirus family that stabily integrates as a provirus into a semi-defined region on chromosome 19 (Muzyczka, 1992). The AAV provirus consists of two terminal inverted repeat sequences (ITRs) and internal sequences coding for proteins required for AAV replication including *rep*, *cap* and several *env* genes. AAV by itself is replication incompetent and requires co-infection with adenovirus for propagation. A variety of approaches have been utilized to generate AAV vectors by constructing packaging cells similar to murine retroviral vector packaging lines in order to provide the necessary AAV replication products in trans (Lebkowski *et al.*, 1988). To date, however, co-infection with adenovirus is still required for efficient replication and most packaging systems are transient due to the cytotoxicity of the rep proteins (*ibid*). AAV vectors lacking all the wild-type AAV sequences except the ITRs are able to efficiently transduce mammalian cells, including hematopoietic cells.

Advantages to using AAV as a vector for HIV gene therapy include possibly a lack of a requirement for target cells to be replicating for proviral integration, the ability to transduce hematopoietic cells (LaFace *et al.*, 1988; Muro *et al.*, 1992), and its apparent safety. As described above, AAV vectors have been used to introduce and effectively express HIV antisense sequences into immortalized human T-cell lines (Chatterjee *et al.*, 1992). Potential disadvantages include: the limited insert size range which the vectors can withstand (approximately 2.5–4.7 kb) for efficient virion packaging; the current cumbersome methods for producing AAV virions; the need to produce AAV vector preparations free of contaminating adenovirus which may cause disease in an immunocompromised individual; and the minimal experience with these vectors in preclinical or human clinical trials. Current work is focusing on improving the AAV virion packaging systems and combining AAV with physical methods for improving the gene transfer efficiency in hematopoietic cells.

11.5.3 HIV vectors

HIV itself may be harnessed as a vector system for delivery of HIV inhibitory sequences (Heisig *et al.*, 1989). Potential advantages of this system include the ability to target transduction specifically to the CD4+ cells which are the targets of infection by wild-type HIV and the tolerance of large amounts of cloned genetic material. In addition, integration of HIV vectors may not require cellular proliferation. Originally, HIV vectors were developed which were simple derivatives of the intact HIV genome. Page *et al.* (1990) constructed a series of HIV vectors by introducing drug resistance genes or CAT into a replication-defective HIV genome due to a deletion in *env*, in order to facilitate studies of the early stages of HIV infection. Sodroski and colleagues developed a Cos cell-based packaging system capable of generating helper-free, replication-incompetent, HIV-based vectors (Helseth *et al.*, 1990; Poznansky *et al.*, 1991). In initial studies, a plasmid encoding a non-packageable, full-length HIV genome containing a 19 NT deletion in the 5′ HIV packaging sequence was employed to express the HIV virion structural proteins. This plasmid was co-transfected into Cos cells with a second plasmid encoding a vector containing the HIV LTRs and *gag* region (with the 5′ packaging signal), as well as the NPT gene expressed from an SL3–3 murine leukemia virus LTR. HIV vector virions produced by these packaging cells were capable of transducing Jurkat cells with titers of 5×10^2 to 1×10^4/ml; however, these preparations were contaminated with replication-competent HIV. A second generation of packaging cells with the HIV env and pol virion proteins expressed from two separate plasmids produced titers in Jurkat cells of 1×10^4–1×10^5 per ml, and no replication-competent HIV could be detected in these preparations.

Shimada *et al* (1991) created HIV vector packaging cells by employing a CMV promoter to constitutively express HIV gag (containing a mutated packaging signal), pol and env inside cells. A second plasmid encoding a packagable vector containing the HIV LTRs and full *gag* sequence was introduced into such cells. Transduction rates of the T-cell lines CEM and HPB-ALL with vector-containing virus was equivalent to those seen with a high-titer murine retrovirus as measured by semi-quantitative DNA blotting analysis and G418 resistance assays. In addition, transduction by the HIV vector was inhibited by soluble CD4, implying that gene transfer was specifically dependent on binding of vector to the cellular CD4 receptor. Buschancher and Panganiban (1992) constructed a series of HIV-based vectors with either the hygromycin phosphotransferase gene or the H5 HA gene of influenza virus cloned directly downstream of the 5′ LTR and upstream of the first splice donor site. These vectors were pseudotyped into amphotropic MoMuLV virions following co-transfection with plasmids encoding the amphotropic murine retroviral env and a plasmid encoding an HIV genome with a deletion in the *env* gene. HeLa cells were stabily transduced at low efficiency with the resultant virions. No proviral rearrangements or production of detectable replication-competent HIV was observed, and hygromycin resistance and expression of H5 HA were demonstrated to be tat dependent. In addition, they found that vectors containing 722 bp of the *gag* sequence were more efficiently propagated than vectors containing 50 bp of the *gag* sequence. One proposed application of this system is to elicit an immune response to cells expressing tat-dependent H5 HA so that they may be eliminated. Carroll *et al.* (1992, 1993) have constructed HIV-1 packaging cells by transfection of a plasmid encoding the HIV-1 structural and regulatory genes as well as the hyg^R gene, but lacking the HIV LTRs and packaging signal. HIV structural genes were synthesized at levels equivalent to wild-type infection and no replication-competent virions were produced.

The main potential disadvantages of HIV vectors include the likelihood that they will fail to transduce human hematopoietic stem cells, the possibility of generating replication competent HIV, and the relatively low titers of currently available packaging systems.

11.5.4 Non-viral gene delivery systems

A variety of gene delivery systems for transient delivery of sequences designed to inhibit HIV replication or enhance immunity have been described. These include adenoviral–polylysine–DNA conjugates (Michael *et al.*, 1993) and liposome or direct plasmid DNA injection (Nabel *et al.*, 1992) in addition to the murine retroviral and AAV vector systems described above. It may be possible to modify one or more of these systems to achieve long-term stable gene expression which may obviate many of the concerns discussed above regarding viral vectors.

11.6 PRECLINICAL AND CLINICAL EVALUATION OF GENE THERAPY APPROACHES FOR TREATING HIV INFECTION

Evaluation of the different gene therapy strategies described above includes assessing the efficacy and the safety of the particular approach in a logical progression, similar to the evaluation of anti-retroviral pharmacologic agents or novel cancer therapies. To date, most gene therapy approaches have been characterized *in vitro* in immortalized human cell lines while a minority have been evaluated in the relevant primary cells and in animal models. The most promising approaches will ultimately be evaluated in human clinical trials.

11.6.1 *In vitro* evaluation

To date, initial evaluations of the efficacy and safety of different gene therapy strategies designed to inhibit HIV have been performed in immortalized human lines including H9, CEM and Jurkat as well as HeLa cells. The advantages to performing studies in these cell lines are: the ease in obtaining and maintaining a relatively homogenous population of cells; the ability to obtain large number of cells for biochemical characterization; the ready transduction of these cells with a variety of gene delivery systems; and the typically high level and durable expression of transduced genes in these cells as compared with primary cells. The disadvantages, however, are numerous and profound. Immortalized cell lines are not indicative models for either gene transfer efficiency or expression when compared with primary cells, and they do not recapitulate the functional or developmental aspects of the complex heterogeneous populations of cells that comprise the peripheral blood T cell and hematopoietic stem cell compartment. In addition, *in vitro* cultures do not recapitulate the complex *in vivo* environment and host responses which are critical to the pathophysiology of HIV infection. Furthermore, significant clonal variation exists in the ability of transformed cells to inhibit HIV infection via endogenous mechanisms not associated with the transduced inhibitor. This may significantly confound the interpretation of the effectiveness of gene inhibition strategies (Gruber *et al.*, 1992). Finally, laboratory-adapted heterogeneous strains or clonal isolates of HIV which replicate efficiently in immortalized cells *in vitro* may not reflect the behavior of the complex mixture of HIV species responsible for the pathophysiology of HIV infection in an individual. As a result of these drawbacks, evaluation of HIV gene therapy strategies using immortalized cell lines may be useful only in excluding strategies or gene delivery systems which have fundamental defects. Otherwise, any conclusion from these studies, either positive or negative must be tempered with caution.

Evaluation of HIV gene therapy approaches in primary cells *in vitro* addresses some of these concerns. Evaluation of gene transfer efficiency and the level and durability of vector-mediated gene expression in some of the appropriate cells such as resting and naive T cells and the progeny of hematopoietic stem cells can be performed. Currently, however, no *in vitro* culture system for human hematopoietic stem cells has been rigorously confirmed as being predictive of the developmental potential of stem cells following transplantation.

Evaluation of safety issues can also be performed at a basic level *in vitro*.

Issues such as generation of replication-competent vectors, alterations in cell growth or morphology, alterations in cellular gene expression such as cytokine gene expression, generation of HIV 'escape mutants', overt malignant transformation, and alterations in biologic function due to transduction, such as altered cytotoxicity in CTL assays, can be evaluated. Many of the assays for evaluating these safety issues *in vitro* have limited sensitivity, so again, interpretations must be made cautiously.

11.6.2 Animal models

A potentially much more powerful method for evaluating gene therapy strategies for HIV infection is to use animal models (Koch and Ruprecht, 1992). These provide a more realistic recapitulation of the complex host environmental milieu and immune responses central to the pathophysiology of HIV. In addition, they potentially allow evaluation of the strategy's effectiveness against viral agents which may more closely mimic the biology of the HIV species which are responsible for disease *in vivo* as opposed to HIV laboratory strains. Currently, however, while many potential animal models are available, all suffer from drawbacks which limit their utility.

Many animals including ungulates, cats and mice sustain natural infections with either lentiviruses or oncoviruses which share some phylogenetic and biologic similarities with HIV (*ibid*). In the large animals, these include Maedi-visna virus in sheep, equine infectious anemia virus in horses, bovine leukemia virus, feline leukemia virus, and feline immunodeficiency virus. While these models have had some utility in evaluating anti-retroviral pharmacologic agents and vaccines, in general the substantial differences in the natural history of the diseases caused by these viruses relative to HIV infection, the logistical difficulties in maintaining large animals, as well as the absence of reagents for characterizing hematopoietic cells in these species, limits their use in evaluation of gene therapy approaches.

In mice, several strains of murine leukemia virus which typically contain mixtures of replication competent and defective retroviruses, cause immunodeficiency after infection (Ruprecht *et al.*, 1992). These include Friend and Rauscher viruses, as well as LB-BM5, the etiologic agent of Murine Acquired Immune Deficiency Syndrome. In addition, hematopoietic cells in mice have been extensively characterized and *in vivo* and *in vitro* assays of totipotent hematopoietic stem cells can be readily performed. Despite these advantages, the mechanism of immunodeficiency following infection with the currently characterized murine retroviruses in mice is not sufficiently similar to that following HIV infection in humans to be useful for evaluating gene therapy approaches, other than limited utility in gene therapy vaccination strategies. Transgenic mice which contain the HIV-1 genome have proven useful for studying certain aspects of HIV-1 biology but do not recapitulate actual infection. Several species, including rabbits and chimpanzees, have been infected with HIV itself (Koch and Ruprecht, 1992). Again however, the pathophysiology of the resulting infection does not recapitulate HIV infection, and the hematopoietic systems in these animals have not been well characterized.

Two *in vivo* models currently recapitulate HIV infection sufficiently to be useful for evaluating gene therapy approaches. Infection of macaque monkeys with SIV is currently the gold standard for evaluating anti-retroviral therapies and vaccines in preclinical studies (Letvin, 1992). The genetic sequences and biology of natural isolates of SIV are very similar to natural isolates of HIV, as is the tissue tropism, pathology and natural history of disease following infection. In addition, macaques have been widely used and well characterized as models for hematopoietic stem cell transplantation (van Beusechem *et al.*, 1992). The only HIV primate gene therapy studies reported to date were designed to evaluate the safety and the efficacy of CTL induction following immunization with either transduced fibroblasts or directly injected vectors encoding HIV gene products such as gp160/gp120 (Jolly, 1991). These initial studies were not performed in the setting of SIV infection, thus limiting the usefulness of their conclusions; however, no toxicity was noted and CTLs were efficiently induced. The disadvantages to using SIV infection of macaques as a model to evaluate HIV gene therapy approaches are the logistical difficulties in maintaining primates, the expense and length of time required to perform studies, and subtle but perhaps important differences in the biology of SIV and HIV.

Murine chimeras which contain human hematopoietic cells represent a second potentially useful *in vivo* model for evaluating gene therapy approaches to HIV infection. Several variations on this model have been described. The SCID-hu PBL mouse developed by Mosier and colleagues (1992) is constructed by inoculation of immuno-incompetent mice, such as SCID or xid/nu/beige strains, with large numbers of peripheral blood lymphocytes which then persist at relatively low concentrations for periods of several months in the peritoneum, spleen and peripheral blood of the recipients, depending on the route of inoculation. These mice support HIV infection in the human cells, and features of authentic HIV infection such as transient hypergammaglobulinemia and decline in CD4+ cell concentration is recapitulated. Currently, only initial studies of the efficacy of gene therapy strategies for HIV infection have been performed in this model. Advantages include the efficiency, cost and reproducibility of HIV infection in this model system. Disadvantages are that efficacy and safety issues are evaluated only in mature peripheral blood lymphocytes. Two variations of the Mosier model have been described. Kamel-Reid and Dick (1989) transplanted xid/nu/bg mice with human hematopoietic cells which generated monocytes and their progenitors. This model may allow assessment of gene therapy approaches in monocytes/macrophages. In an alternative model developed originally by Reisner and colleagues (Lubin *et al.*, 1991) and subsequently expanded by Lapidot *et al.* (1992), T-cell proliferation and myelopoiesis has been observed in BalbC mice conditioned with radiation treatment and transplanted with SCID marrow before human hematopoietic cell transplantation. This model may allow evaluation of gene therapy strategies in multiple cell lineages developing from transduced stem cells.

McCune and co-workers (1991) have developed a human/SCID mouse chimera termed the SCID/hu mouse which is transplanted with human fetal lymphoid and hematopoietic organs including thymus, spleen, and liver. This potentially powerful model has been extensively characterized and recapitulates many of the fundamental pathophysiologic effects observed in authentic HIV infection, including infection of lymphoid organs. In addition, this model supports infection with natural isolates of HIV as opposed to laboratory clones. The SCID/hu mouse also appears to support multilineage human hematopoiesis long term and thus may be valuable in evaluating transduced human hematopoietic stem cells and their progeny. Disadvantages to this model include its technical complexity, the use of fetal tissue which may not reflect accurately the biology of adult hematopoietic tissue, and the absence of authentic CNS disease.

In addition to models for assessing the efficacy of alternative HIV inhibition strategies and gene delivery systems, these models can be used to evaluate safety issues such as development of malignancies, development of replication-competent or psuedotyped vectors, and any pharmacologic toxicity by performing histologic examination of transplanted mice. In addition, overt deleterious effects on immune function due to the therapy may be apparent by comparison with control animals.

11.6.3 Clinical trials

Traditionally, clinical trials of pharmacologic agents proceed logically from Phase I dose escalation/toxicity trials to Phase II uncontrolled efficacy trials, to Phase III double-blind comparisons with established agents. Phase I and II trials are typically performed in patient populations who have failed to respond to treatment with established agents. Evaluation of gene therapy approaches to HIV infection can be approached in a similar fashion. Initial Phase I and II toxicity and efficacy trials may be performed in HIV-infected individuals who have a limited life expectancy despite being treated with currently available pharmacologic agents. Such individuals may include patients with a CD4+ cell count $<200/mm^3$ and Kaposi's sarcoma with either an elevated β2-microglobulin or neopterin, patients with an AIDS-defining illness, and CD4+ cell counts $<50/mm^3$, or patients with an HIV-associated malignancy such as B-cell lymphoma (Krown *et al.*, 1991). HIV-infected individuals who do not have these prognostic features may survive for >10 years with currently available supportive care and medical therapies and perhaps are not appropriate candidates for initial evaluation of gene therapy approaches while the toxicities are unknown. The caveat to performing clinical trials in HIV-infected individuals with advanced stages of disease is that failure to achieve efficacy in such individuals does not preclude potential efficacy in patients during earlier stages of the disease. In addition, since the natural history of HIV encompasses a wide spectrum of manifestations, application of any conclusions about efficacy and toxicity obtained from trials of selected patients must be made with caution to other patient groups.

The initial clinical applications of gene therapy for the treatment of AIDS fall into four general categories: (i) restoration of some degree of general immune function by transfusion with transduced peripheral blood lymphocytes; (ii) restoration of the entire hematopoietic and immune systems with transplantation of transduced hematopoietic stem cells; (iii) attempts to limit the reservoir of HIV infection by transplanting hematopoietic stem cells transduced with HIV activated toxins; and (iv) augmentation of HIV-specific immune function by either vaccination or adoptive transfer of HIV-specific CTL.

Peripheral blood lymphocyte transplantation

In HIV-infected individuals, the number and severity of opportunistic infections increases dramatically as the CD4+ T-cell number in the peripheral blood first falls below 100 cells/mm^3. Currently it is not clear whether this is solely a direct consequence of CD4+ PBL depletion. The overall decline in immune function and accumulating HIV viremia are also probably important determinants. It may be possible however, to diminish the frequency and severity of opportunistic infections by infusing transduced CD4+ peripheral blood T cells in sufficient number so that the total functional peripheral blood CD4+ T-cell count is raised. Such an increase in T-cell count can potentially be accomplished by either transplantation of syngeneic PBL derived from a healthy donor or *ex vivo* expansion of autologous PBL before reinfusion. Gene transfer of HIV inhibitory genetic elements into such cells could allow one to follow the trafficking and persistence of infused cells as well as to directly protect the infused T cells from the deleterious effects of HIV. For this approach to be clinically useful, transduction of a wide immunologic repertoire of peripheral blood T cells is essential. Furthermore, these cells must function and traffick appropriately following infusion. Several recent clinical trials utilizing infusions of transduced peripheral blood T cells indicate that this may be feasible. Several clinical trials involving retroviral vector gene transfer of the human adenosine deaminase (ADA) gene into ADA-deficient peripheral blood lymphocytes from patients with severe combined immune deficiency (SCID) revealed that the transduced lymphocytes could persist for several months following infusion, with no apparent toxicity and possibly, with some associated clinical benefit (Blaese *et al.*, 1993; Bordignon *et al.*, 1993). In another clinical trial involving retroviral-mediated gene transfer of a NPT-encoding vector into autologous tumor-infiltrating lymphocytes (TILs) followed by reinfusion into patients with advanced melanoma, vector sequences and NPT production could be detected in lymphocytes derived from recipients for periods of several months, albeit at low levels (Rosenberg *et al.*, 1990).

Currently three clinical trials have been approved by the Recombinant DNA Advisory Committee (RAC) which involve gene transfer into bulk populations of PBL (Yu *et al.*, 1994). A study has been initiated at the National Institutes of Health using infusions of syngeneic T cells which have been transduced with a murine retroviral vector encoding the NPT gene into HIV-infected individuals. This marking study will allow the determination of the infusion size and number required to raise the peripheral blood T-cell count, the persistence and trafficking of transduced cells following transplantation, and will evaluate safety issues of murine retroviral gene transfer in this population. The main safety issues to be evaluated include: the development of vector-related malignancy; the development of replication-competent murine retroviruses; the development of pseudotyped vectors in HIV; pharmacologic toxicity; and alterations in immune functions. This study will provide a foundation for applying gene therapy strategies which inhibit HIV replication. Two other studies have been approved by the RAC which involve transduction of autologous CD4+ PBL with revM10 or with an anti-HIV hairpin ribozyme. In both trials, autologous PBL will be transduced with murine retroviral vectors and expanded *ex vivo* in the presence of anti-retroviral agents before reinfusion. The main endpoints will be evaluation of toxicity and the persistence of the treated cells following infusions.

A number of concerns exist when using infusions of transduced autologous or syngeneic PBL for treating HIV infection. Mitogenic stimulation of previously infected autologous PBLs during gene transfer and *ex vivo* expansion may increase the production of infectious HIV from these cells so that an increase in viral burden may be created in the recipient. *Ex vivo* expansion of CD4+ T cells from syngeneic donors may render them more susceptible to HIV infection than unstimulated cells. In both cases, spread of HIV may actually be promoted rather than inhibited in the recipient. Mitogenic stimulation of T cells may also adversely affect their function and persistence following transplantation, particularly if performed for extended lengths of time. If transplantation of transduced autologous *ex vivo* expanded T cells is to be useful, then safe methods for culturing and transduction of the cells or alternative vectors have to be developed that do not result in increased HIV production or increased susceptibility to infection. Initial studies indicate that the addition of anti-retroviral agents including nevaripine and AZT may effectively inhibit HIV propagation during transduction and expansion (Nabel *et al.*, 1992). If T cells can be expanded for reinfusion without increasing the viral burden or susceptibility to infection, issues such as cell dose, optimal *ex vivo* expansion strategy, and timing of administration of infusions will need to be optimized. Transfusion of transduced peripheral blood T cells will have little effect on other critical cell populations involved in HIV infection, including monocytes, dendritic cells and CNS cells. Finally, since much of the critical damage to the immune system may be occurring in lymphoid organs such as lymph nodes or in antigen-presenting cells such as dendritic cells, it is unclear whether raising the peripheral blood CD4+ T-cell concentration will be beneficial to a patient.

Hematopoietic stem cell transplantation

Transplantation of totipotent hematopoietic stem cells which have been transduced with HIV inhibitory vectors has many obvious potential advantages relative to transplantation of transduced peripheral blood lymphocytes. Totipotent hematopoietic stem cells generate multilineage progeny, including the majority of cells critical to the pathophysiology of HIV infection, for prolonged and potentially life-long periods (Visser and Van Bekkum, 1990). In addition to lymphoid cells, both monocyte/macrophages and marrow derived dendritic cells, two cell populations now felt to be critical to sustaining HIV infection *in vivo*, will be replaced following transplantation (*ibid*). Brain macrophages and at least a minority of the microglial cells in the brain appear to be bone marrow-derived and replacement following transplantation may impact on HIV associated neurologic disease (de Groot *et al.*, 1992).

To date, a small number of syngeneic and allogeneic bone marrow transplants have been performed in individuals infected with HIV with B-cell lymphoma or other diseases amenable to marrow transplantation (Vilmer *et al.*, 1987; Angelucci *et al.*, 1989; Holland *et al.*, 1989; Aboulafia *et al.*, 1990; Williams, 1991; Saral and Holland, 1993). The majority of patients tolerated the transplant-conditioning regimens; however, the long-term mortality has been virtually 100% due to recurrence of lymphoma, transplant-related complications, and recurrence of AIDS. In several patients, it was possible to eliminate HIV for periods of several months by treatment with high doses of AZT before the transplant followed by treatment with lower doses of AZT post-transplant (Holland *et al.*, 1989; Saral and Holland, 1993). These findings indicate that the HIV viral burden can be substantially diminished, at least transiently, by the combination of anti-retroviral agents and the transplant-conditioning regimen. A disadvantage to using allogeneic stem cells for transplantation of HIV individuals is the possibility of developing graft-versus-host disease and the requirement for immunosuppressive agents for prophylaxis against, and treating this problem. It may be possible to avoid this problem by using T cell-depleted allogeneic marrow or peripheral blood as a transplant source.

Since HIV does not appear to infect the most primitive totipotent human hematopoietic stem cells, it may be possible to perform genetic modification of autologous hematopoietic stem cells as well as allogeneic stem cells for treating HIV infection. Several recent large animal preclinical trials have served to establish the general safety and utility of autologous transplantation of hematopoietic stem cells genetically altered using retroviral-mediated gene transfer, albeit at efficiencies too low for clinical benefit in HIV infection (Einerhand and Valerio, 1992; van Beusechem *et al.*, 1992). Two clinical gene-marking studies indicate that it is possible to transduce autologous human hematopoietic stem cells capable of durably sustaining hematopoiesis at low efficiencies; however, it appears that prior treatment of donors with high-dose myeloablative chemotherapy may be required to achieve this effect (Brenner *et al.*, 1993).

An autologous stem cell transplantation gene therapy strategy would involve: (i) collection of hematopoietic cells and isolation of a population enriched for stem cells that are devoid of HIV infection; (ii) transduction of enriched stem cells with a vector active against HIV; (iii) treatment of the patient's endogenous HIV with chemotherapeutic or biotherapeutic agents; and (iv) transplant conditioning and transplantation of the transduced stem cells. Stem cell enrichment is important for technical and logistical reasons as well as for purging HIV-infected cells. A variety of methods to isolate enriched stem cells have been described based on positive selection for cells expressing the CD34 antigen (Berenson *et al.*, 1991). These cells are capable of durably reconstituting the hematopoietic system in both preclinical primate studies and human clinical trials and appear to not be infectable with HIV (Molina *et al.*, 1990). If further purification of hematopoietic stem cells is found to be necessary to completely eliminate residual HIV infected cells, a number of methods exist for eliminating cells committed to the lymphoid and monocyte lineages, including agglutination with soybean lectin and immunoadhesion to antibodies specific for appropriate surface markers such as CD3, CD4 for lymphocytes, and CD33 and CD15 for monocytes. The resulting lin-CD34+ population has been shown to be enriched for early hematopoietic cells in a variety of *in vitro* assays (Smith *et al.*, 1991). However, formal confirmation of the ability of these cells to initiate and durably sustain engraftment in humans has not yet been performed. Recently, a number of alternative sources of both autologous and allogeneic hematopoietic stem cells in addition to bone marrow have been identified which appear to be capable of reproducibly initiating engraftment in recipients. These include peripheral blood, umbilical cord blood and fetal tissue. These alternative stem cell sources may provide target cells with a number of advantages to bone marrow cells including ease of collection and more efficient gene transfer (Bregni *et al.*, 1992).

The issue of the appropriate conditioning regimen for stem cell transplantation involves a number of considerations, including the need to provide a physiologic niche for the reintroduced stem cells and the need to eliminate residual reservoirs of endogenous HIV infection. It may be possible to transplant autologous hematopoietic stem cells with little or no myeloablative conditioning, particularly if a competitive advantage is conferred to transduced cells *in vivo* as has been postulated in the initial gene therapy trials for ADA deficiency SCID (Bordignon *et al.*, 1993). In addition, a variety of novel methods for achieving myeloablation without significant extrahematopoietic toxicity such as treatment of the recipient with complement fixing or radiolabeled monoclonal antibodies directed at hematopoietic cells have been described (Caron *et al.*, 1992).

Enhancing HIV specific immunity

The two basic methods for enhancing immunity to HIV include adoptive transfer of HIV-specific CTL or immunization. As described above, Greenberg and colleagues have initiated a clinical trial involving infusing *ex vivo* expanded, HSV–TK-transduced, HIV-specific CD8+ CTL in patients undergoing bone marrow transplantation for HIV-associated B-cell lymphoma. A second clinical trial has been approved by the FDA to attempt to augment *in vivo* anti-HIV immunity by vaccination of HIV seropositive individuals with autologous fibroblasts transduced with retroviral vectors expressing HIV gene products including env (Jolly, 1991). Potential concerns with these vaccination studies is that induction of enhanced CTL responses could worsen the course of HIV infection if anti-CD4+ T-cell autoimmunity is induced. To date, however, this has not been observed in studies involving more traditional routes of vaccination.

11.7 CONCLUSIONS

A number of gene therapy strategies and gene delivery systems have been developed for inhibition of HIV. When tested in simple *in vitro* cell systems, several of the strategies have been shown to effectively inhibit HIV replication. However, preclinical and clinical evaluation of HIV gene therapy approaches continues to be complicated by the tremendous complexity of HIV infection, limitations in understanding critical aspects of the pathophysiology of HIV, the involvement of a wide array of immunologic and environmental factors in the development and course of HIV infection that cannot be replicated *in vitro*, and the current gaps in the characterization of the biology of peripheral blood lymphocytes and hematopoietic stem cells, particularly following transplantation. In addition, there is concern that much of the pathology of HIV infection may be due to phenomena which may not be effected by gene therapy strategies.

Significant progress is being made toward addressing these concerns, however. Clarification of the biology of HIV and the pathophysiology of infection is occurring at a rapid pace. A number of additional targets for gene therapy are being evaluated, including targeting the integration process and protease activity. Improved small animal models for evaluating potential gene therapy approaches to HIV infection are being created. Knowledge gained from adoptive immunotherapy trials for treatment of AIDS-related illnesses, from bone marrow stem cell transplant studies for AIDS lymphoma and other hematologic diseases, and from the initial studies involving infusions of genetically modified autologous T-cells will be extremely valuable in developing stem cell-based gene therapy strategies. In the future, it will be critical to develop methods to efficiently analyze the wide variety of different HIV treatment strategies and delivery systems in preclinical animal models and clinical trials, given the significant limitations of *in vitro* studies. In addition, these strategies must be evaluated in combination with other gene therapy or pharmacologic and immunotherapy strategies, since all the gene therapy strategies described to date have theoretical limitations. Finally, as one or more gene therapy approaches to treating HIV infection joins the rapidly expanding armamentarium of HIV treatment modalities, it will be critical to minimize the morbidity, complexity and expense of the procedure to widen its availability.

REFERENCES

Aboulafia, D., Mitsuyasu, R. and Miles, S. (1990) Bone marrow transplantation in an HIV positive thalassemic child following therapy with azidothymidine. *Haematologica*, 75(3), 285–7.

Angelucci, E., Lucarelli, G., Baronciani, D. *et al.* (1989) Bone marrow transplantation in an HIV positive thalassemic child following therapy with azidothymidine. *Haematologica*, 75:285–7.

Bahner, I., Zhou, C. *et al.* (1993) Comparison of transdominant HIV genes expressed by retroviral vectors in human T-lymphocytes. *J. Virol.*, **67**, 3199.

Baltimore, D. (1988) Intracellular immunization. *Nature*, **335**, 395–6.

Bednarik, D., Mosca, J., Raj, N. and Pitha, P. (1989) Inhibition of human immunodeficiency virus (HIV) replication by HIV-trans-activated alpha 2-interferon. *Proc. Natl Acad. Sci. USA*, **86**, 4958–62.

Benjouad, A., Mabrouk, K., Moulard, M. *et al.* (1993) Cytotoxic effect on lymphocytes of Tat from human immunodeficiency virus (HIV-1). *FEBS Lett.*, **319**, 119–24.

Berenson, R., Bensinger, W., Hill, *et al.* (1991) Engraftment after infusion of CD34+ marrow cells in patients with breast cancer or neuroblastoma. *Blood*, **77**, 1717–23.

Bevec, D., Dobrovnik, M., Hauber, J. and Bohnlein, E. (1992) Inhibition of human immunodeficiency virus type 1 replication in human T cells by retroviral-mediated gene transfer of a dominant-negative Rev trans-activator. *Proc. Natl Acad. Sci. USA*, **89**, 9870–4.

Bevec, D., Volc Platzer, B. *et al.* (1994) Constitutive expression of chimeric Neo-Rev response element transcripts suppresses HIV-1 replication in human CD4+ T-lymphocytes. *Human Gene Ther.*, **5** 193–00.

Blackburn, R., Clerici, M., Mann, D. *et al.* (1991) Common sequence in HIV 1 GP41 and HLA class II beta chains can generate crossreactive autoantibodies with immunosuppressive potential early in the course of HIV 1 infection. *Adv. Exp. Biol.*, **303**, 63–9.

Blaese, R.M., Culver, K.W., Anderson, W.F. *et al.* (1993) Treatment of severe combined immune deficiency disease due to adenosine deaminase deficiency with CD34+ selected autologous peripheral blood cells transduced with a human ADA gene. *Human Gene Ther.*, **4**, 521–7.

Bordignon, C., Mavillo, F., Ferrari, G. *et al.* (1993) Transfer of the ADA gene into bone marrow cells and peripheral blood lymphocytes for the treatment of patients affected by ADA deficient SCID. *Human Gene Ther.*, **4**, 513–20.

Bounocore, L. and Rose, J. (1993) Blockade of human immunodeficiency virus type 1 production in CD4+ T-cells by an intracellular CD4 expressed under control of the viral long terminal repeat. *Proc. Natl Acad. Sci. USA.*, **90**, 2695–9.

Branch, A. and Robertson, H. (1991) Efficient trans cleavage and a common structural motif for the ribozymes of the human hepatitis delta agent. *Proc. Natl Acad. Sci. USA.*, **86**, 10163–7.

Bregni, M., Magni, M., Siena, S., Di, N., Bonadonna, G. and Gianni, A. (1992) Human peripheral blood hematopoietic progenitors are optimal targets of retroviral-mediated gene transfer. *Blood*, **80**, 1418–22.

Brenner, M., Rill, D., Moen, R. *et al.* (1993) Gene-making to trace origin of relapse after autologous bone-marrow transplantation. *Lancet*, **341**, 85–6.

Buchschacher, G., Freed, E. and Panganiban, A., (1992) Cells induced to express a human immunodeficiency virus type 1 envelope gene mutant inhibit the spread of wild-type virus. *Human Gene Ther.*, **3**, 391–7.

Buchschacher, G. and Panganiban, A. (1992) Human immunodeficiency virus vectors for inducible expression of foreign genes. *J. Virol.*, **66**, 2731–9.

Bukrinsky, M., Haggarty, S. *et al.* (1992) A nuclear localization signal within HIV-1 matrix protein that governs non-dividing cells. *Nature*, **365**, 666–9.

Cameron, P.U., Forsum, U., Teppler, H., Granelli, P.A. and Steinman, R.M. (1992) During HIV-1 infection most blood dendritic cells are not productively infected and can induce allogenic CD4+ T cells clonal expansion. *Clin. Exp. Immunol.*, **88**, 226–36.

Caron, P.C., Co, M.S., Bull, M.K., Avdalovic, N.M., Queen, C. and Scheinberg, D.A. (1992) Biological and immunological features of humanized M195 (anti-CD33) monoclonal antibodies. *Cancer Res.*, **52** 6761–7.

Carroll, R., Peterlin, B.M. and Derse, D. (1992) Inhibition of human immunodeficiency virus type 1 Tat activity by coexpression of heterologous trans activators. *J. Virol.*, **66**, 2000–7.

Carroll, R., Lin, J.T., Kim, J.J.M., Vahey, M., Burke, D. and St. Louis, D. (1993) Construction and characterization of replication defective HIV-1 packaging cell lines. *J. Biol. Chem.*, **S17E**, 241–000.

Cech, T.R. (1988) Ribozymes and their medical implications. *JAMA*, **260**, 3030–4.

Cech, T.R., Herschlag, D., Piccirilli, J.A. and Pyle, A.M. (1992) RNA catalysis by a group I ribozyme. Developing a model for transition state stabilization. *J. Biol. Chem.*, **267**, 17479–82.

Chatterjee, S., Johnson, P. and Wong, K. (1992) Dual-target inhibition of HIV-1 *in vitro* by means of an adeno-associated virus antisense vector. *Science*, **258**, 1485–8.

Chen, C., Banerjea, A., Harmison, G., Haglund, K. and Schubert, M. (1992) Multitarget-ribozyme directed to cleave at up to nine highly conserved HIV-1 env RNA regions inhibits HIV-1 replication-potential effectiveness against most presently sequenced HIV-1 isolates. *Nucleic Acids Res.*, **20**, 4581–9.

Chin, J. (1991) Present and future dimensions of the HIV/AIDS pandemic. 7th International Conference on AIDS, Florence.

Constantoulakis, P., Campbell, M., Felber, B., Nasioulas, G., Afonina, E. and Pavlakis, G. (1993) Inhibition of Rev-mediated HIV-1 expression by an RNA binding protein encoded by the interferon inducible 9–27 gene. *Science*, **259**, 1314–17.

Cullen, B.R. (1992) Mechanism of action of regulatory proteins encoded by complex retroviruses. *Microbiol. Rev.*, **56**, 375–94.

Davis, K., Hayward, A., Ozturk, G. and Kohler, P. (1983) Lymphocyte Transfusion in case of Acquired Immunodeficiency Syndrome. *Lancet*, 599–603.

de Groot, C., Huppes, W., Sminia, T., Kraal, G. and Dijkstra, C. (1992) Determination of the origin and nature of brain macrophages and microglial cells in mouse central nervous system, using non-radioactive *in situ* hybridization and immunoperoxidase techniques. *Glia*, **6**, 301–9.

Donahue, R., Kessler, S., Bodine, D. *et al.* (1992) Helper virus induced T cell lymphoma in nonhuman primates after retroviral mediated gene transfer. *J. Exp. Med.*, **176**, 1125–35.

Droge, W. (1993) Cysteine and glutathione deficiency in AIDS patients: a rationale for the treatment with N-acetyl-cysteine. *Pharmacology*, **46**, 61–5.

Dropolic, B., Lin, N., Martin, M. and Jeang, K.T. (1992) Functional characterization of a U5 ribozyme: intracellular suppression of HIV expression. *J. Virol.*, **66**, 1432–41.

Dumenil, D., Jacquemin-Sablon, H., Neel, H., Frendal, E. and Dautry, F. (1989) Mock retroviral infection alters the developmental potential of murine bone marrow stem cells. *Mol. Cell. Biol.*, **9**, 4541–4.

Einerhand, M. and Valerio, D. (1992) Gene transfer into hematopoietic stem cells: prospects for human gene therapy. *Curr. Top. Microbiol. Immunol.*, **177**, 217–35.

Embretson, J., Zupanic, M., Ribas, J. *et al.* (1992) Massive covert infection of helper T-lymphocytes and macrophages during the incubation period of AIDS. *Nature*, **362**, 359–62.

Fauci, A.S. (1988) The human immunodeficiency virus: infectivity and mechanisms of pathogenesis. *Science*, **239**, 617–22.

Forster. A. and Symons, R. (1987) Self-cleavage of virusoid RNA is performed by the proposed 55-nucleotide active site. *Cell*, **50**, 9–16.

Freed, E., Delwart, E., Buschacher, G. and Panganaban, A. (1992) A mutation in the HIV transmembrane glycoprotein gp 41 dominantly interferes with fusion and infectivity. *Proc. Natl Acad. Sci. USA*, **89**, 70–4.

Graham, B., Matthews, T., Belshe, R. *et al.* (1993) Augmentation of human immunodeficiency virus type 1 neutralizing antibody by priming with gp160 recombinant vaccinia and boosting with rgp160 in vaccinia-naive adults. The NIAID AIDS Vaccine Clinical Trials Network. *J. Infect. Dis.*, **167**, 533–7.

Green, M., Ishino, M. and Loewenstein, P. (1989) Mutational analysis of HIV-1 tat minimal domain peptides: identification of trans-dominant mutants that suppress HIV-LTR driven gene expression. *J. Virol.*, **58**, 215–23.

Gruber, M., Hewlett, I., Simms, T., Vujcic, L., Manischewitz, J. and Golding, H. (1992) Study of viral replication in HIV-1-infected CEM T-cell subclones which are reduced in their ability to form syncytia. *Aids Res. Hum. Retroviruses*, **8**, 1139–46.

Hampel, A., Tritz, R., Hicks, M. and Cruz, P. (1990) Hairpin RNA model: evidence for helices and sequence requirement for substrate RNA. *Nucleic Acids Res.*, **18**, 299–304.

Hantzopoulos, P., Sullenger, B., Ungers, G. and Gilboa, E. (1989) Improved gene expression upon transfer of the adenosine deaminase minigene outside the transcriptional unit of retroviral vector. *Proc. Natl Acad. Sci. USA*, **86**, 3519–23.

Harrison, G., Maxwell, F., Long, C., Rosen, C., Glode, L. and Maxwell, I. (1991) Activation of a diphtheria toxin A gene by expression of human immunodeficiency virus-1 Tat ands Rev proteins in transfected cells. *Human Gene Ther.*, **2**, 53–60.

Harrison, G., Long, C., Maxwell, F., Glode, L. and Maxwell, I. (1992a) Inhibition of HIV production in cells containing an integrated HIV-regulated diphtheria toxin A chain gene. *Aids Res. Hum. Retroviruses*, **8**, 39–45.

Harrison, G., Long, C., Curiel, T., Maxwell, F. and Maxwell, I. (1992b) Inhibition of human immunodeficiency virus-1 production resulting from transduction with a retrovirus containing an HIV-regulated diphtheria toxin A chain gene. *Human Gene Ther.*, **3**, 461–9.

Haseloff, J. and Gerlach, W. (1988) Simple RNA enzymes with new and highly specific endoribonuclease activities. *Nature*, **334**, 585–91.

Haseltine, W. (1992) *The Molecular Biology of HIV-1. AIDS, Etiology, Diagnosis, Treatment, and Prevention*, 3rd edn, (eds V. DeVita, S. Hellman and S. Rosenberg), Lippincott, Philadelphia, pp. 39–59.

Heisig, V., Jahn, G., Ebeling, M. and Laufs, R. (1989) Use of a HIV-1 retroviral vector system for gene transfer into human cells. *Hamatol Bluttransfus.*, **32**, 406.

Helseth, E., Kowalski, M., Gabuzda, D., Oslshevsky, U., Haseltine, W. and Sodroski, J. (1990) Rapid complementation assays measuring replicative potential of HIV envelope glycoprotein mutants. *J. Virol.*, **64**, 2416–20.

Holland, H.K. *et al.* (1989) Allogeneic bone marrow transplantation, zidovudine and HIV infection. *Ann. Intern. Med.*, **111**, 973–81.

Hope, T., Klein, N., Elder, M. and Parslow, T. (1992) Trans-dominant inhibition of human immunodeficiency virus type 1 Rev occurs through formation of inactive protein complexes. *J. Virol.*, **66**, 1849–55.

Jolly, D. (1991) HIV infection and gene transfer therapy. *Human Gene Ther.*, **2**, 111–12.

Jolly, D. and Warner, J. (1990) Retroviral vectors as vaccines and immunotherapeutics. *Semin. Immunol.*, **2**, 329–39.

Joshi, S., Van Brunschot, A., Asad, S., van der Elst, I., Read, S. and Bernstein, A. (1991) Inhibition of human immunodeficiency virus type 1 multiplication by antisense and sense RNA expression. *J. Virol.*, **65**, 5524–30.

Kamel-Reid, S. and Dick, J. (1989) Engraftment of immunodeficient mice with human hematopoietic cells. *Science*, **242**, 1706–9.

Kestler, H., Ringler, D., Mori, K. *et al.* (1991) Importance of nef gene for the maintenance of high virus loads for the development of AIDS. *Cell*, **65**, 651–62.

Koch, J.A. and Ruprecht, R.M. (1992) Animal models for anti-AIDS therapy. *Antiviral Res.*, **19**, 81–109.

Koga, Y., Sasaki, M., Nakamura, K. *et al.* (1990) Intracellular distribution of the envelope glycoprotein of HIV and its role in the production of the cytopathic effects in CD4+ and CD4– cell lines. *J. Virol.*, **64**, 4661–71.

Krown, S., Niedzwiecki, D., Bhalla, R., Flomenberg, N., Bundow, D. and Chapman, D. (1991) Relationship and prognostic value of endogenous interferon-alpha, beta 2-microglobulin, and neopterin serum levels in patients with Kaposi sarcoma and AIDS. *J. Acquir. Immune Defic. Syndr.*, **4**, 871–80.

Kubota, S., Furuta, R., Maki, M. and Hatanaka, M. (1992) Inhibition of human immunodeficiency virus type 1 Rev mutant which interferes with nuclear/nucleolar localization of Rev. *J. Virol.*, **66**, 2510–13.

LaFace, D., Hermonat, P., Wakeland, E. and Peck, A. (1988) Gene transfer into hematopoietic progenitor cells mediated by an adeno-associated virus vector. *Virology*, **162**, 483–6.

Lane, H., Depper, J., Greene, W., Whalen, G., Waldmann, T. and Fauci, A. (1990) Preferential infection of CD4+ memory T cells by human immunodeficiency virus type 1: evidence for a role in the selective T-cell functional defects observed in infected individuals. *Proc. Natl Acad. Sci USA*, **87**, 6058–62.

Lapidot, T., Pflumio, F., Doedens, M., Murdoch, B., Williams, D. and Dick, J. (1992) Cytokine stimulation of multilineage hematopoiesis from immature human cells engrafted in SCID mice. *Science*, **255**, 1137–41.

Lebkowski, J., McNally, M., Okarma, T. and Lerch, L. (1988) Adeno-associated virus: a vector system for efficient introduction and integration of DNA into a variety of mammalian cell types. *Mol. Cell. Biol.*, **8**, 3988–96.

Lee, T., Sullenger, B., Gallardo, H., Ungers, G. and Gilboa, E. (1992) Overexpression of RRE-derived sequences inhibits HIV-1 replication in CEM cells. *New Biol.*, **4**(1), 66–74.

Letvin, N.L. (1992) Nonhuman primate models for HIV vaccine development. *Immunodefic. Rev.*, **3**(3), 247–60.

Lin, X., Dashti, A., Schinazi, R. and Tang, J. (1993) Intracellular diversion of glycoprotein gp160 of HIV to lysosomes as a strategy of AIDS gene therapy. *FASEB Journal*, **7**, 1070–80.

Lisziewicz, J., Rappaport, J. and Dhar, R. (1991) Tat-regulated production of multimerized TAR RNA inhibits HIV-1 gene expression. *New Biol.*, **3**(1), 82–9.

Lisziewicz, J., Sun, D., Smith, J. *et al.* (1993) Inhibition of HIV replication by regulated expression of polymeric TAR RNA decor as a strategy for gene therapy in AIDS. *Proc. Natl Acad. Sci. USA*, **90**, 8000–4.

Lo, K., Biasolo, M., Dehni, G., Palu, G. and Haseltine, W. (1992) Inhibition of replication of HIV-1 by retroviral vectors expressing tat-antisense and anti-tat ribozyme RNA. *Virology*, **190**, 176–83.

Lubin, I., Faktorowich, Y., Lapidot, T. *et al.* (1991) Engraftment and development of human T and B cells in mice after bone marrow transplantation. *Science*, **252**, 427–31.

Mace, K., Seif, I. *et al.* (1991) Enhanced resistance to HIV-1 replication in U937 cells stably transfected with the human IFN-β gene behind an MHC promoter fragment. *J. Immunol.*, **10**, 3553–9.

Malim, M., Bohnlein, E., Hauber, J. and Cullen, B. (1989) Functional dissection of the HIV Rev transactivator-derivation of a transdominant repressor of Rev function. *Cell*, **58**, 205–14.

Malim, M., McCarn, D., Tiley, L. and Cullen, B. (1991) Mutational analysis of the HIV rev activation domain. *J. Virol.*, **65**, 4248–54.

Malim, M., Freimuth, W., Liu, J. *et al.* (1992) Stable expression of transdominant Rev protein in human T cells inhibits human immunodeficiency virus replication. *J. Exp. Med.*, **176**, 1197–201.

Marasco, W., Haseltine, W.A. and Chen, S.Y. (1993) Design, intracellular expression, and activity of a human anti-immunodeficiency virus type 1 gp120 single chain antibody. *Proc. Natl Acad. Sci. USA*, **90**, 7889–93.

Markowitz, D., Goff, S. and Bank, A. (1989) Construction and use of a safe amphotropic packaging cell line. *Virology*, **167**, 400–6.

Matsuda, Z., Yu, X., Yu, Q.C., Lee, T.H. and Essex, M. (1993) A virion specific inhibitory molecule with therapeutic potential for human immunodeficiency virus type 1. *Proc. Natl Acad. Sci. USA*, **90**, 3544–8.

McCune, J., Kaneshima, H., Krowka, J. *et al.* (1991). The SCID-hu mouse: a small animal model for HIV infection and pathogenesis. *Annu. Rev. Immunol.*, **9**, 399–429.

Meyaard, L., Otto, S.A., Jonker, R.R., Mijnster, M.J., Keet, R.P. and Miedema, F. (1992) Programmed death of T cells in HIV-1 infection. *Science*, **257**, 217–19.

Meyer, J., Nick, S., Stamminger, T., Grummt, F., Jahn, G. and Lipps, J. (1993) Inhibition of HIV-1 replication by a high copy number vector expressing antisense RNA for reverse transcriptase. *Gene*, **129**, 263–8.

Michael, S., Huang, C., Romer, M., Wagner, E., Hu, P. and Curiel, D. (1993) Binding-incompetent adenovirus facilitates molecular conjugate-mediated gene transfer by the receptor-mediated endocytosis pathway. *J. Biol. Chem.*, **268**, 6866–9.

Miedema, F. (1992) Immunological abnormalities in the natural history of HIV infection: mechanisms and clinical relevance. *Immunodefic. Rev.*, 3(3), 173–93.

Miller, A.D. (1992) Retroviral vectors. *Curr. Top. Microbiol. Immunol.*, **158**, 1–24.

Miller, D., Adam, M. and Miller, D. (1990) Gene transfer by retrovirus vectors occurs only in cells that are actively replicating at the time of infection. *Mol. Cell. Biol.*, **10**, 4239–42.

Modesti, N., Garcia, J., Debouck, G., Peterlin, M. and Gaynor, R. (1991) Transdominant Tat mutants with alterations in the basic domain inhibit HIV-1 gene expression. *New Biol.*, 3, 759–68.

Molina, J., Scadden, D., Sakaguchi, M., Fuller, B., Woon, A. and Groopman, J.

(1990) Lack of evidence for infection of or effect on growth of hematopoietic progenitor cells after *in vivo* or *in vitro* exposure to human immunodeficiency virus. *Blood*, **76**, 2476–82.

Morgan, R., Looney, D., Muenchau, D., Wong, S., Gallo, R. and Anderson, W.F. (1990) Retroviral vectors expressing soluble CD4: a potential gene therapy for AIDS. *Aids Res. Hum. Retroviruses*, **6**(2), 183–91.

Mosier, D., Gulizia, R., Baird, S., Wilson, D., Spector, D. and Spector, S.A. (1991) Human immunodeficiency virus infection of human-PBL-SCID mice. *Science*, **251**, 791–4.

Muro, C., Samulski, R. and Kaplan, D. (1992) Gene transfer in human lymphocytes using a vector based on adeno-associated virus. *J. Immunother.*, **11**, 231–7.

Muzyczka, N. (1992) Use of adeno-associated virus as a general transduction vector for mammalian cells. *Curr. Top. Microbiol. Immunol.*, **158**, 97–129.

Nabel, E., Gordon, D., Yang, Z. *et al.* (1992) Gene transfer *in vivo* with DNA–liposome complexes: lack of autoimmunity and gonadal localization. *Human Gene Ther.*, **3**, 649–56.

Nable, G., Fox, B., Post, L., Thompson, C. and Woffendin, C. (1994) A molecular genetic intervention for AIDS-effects of a transdominant negative form of rev. *Human Gene Ther.*, **5**, 79–92.

Natsoulis, G. and Boeke, J. (1991) New antiviral strategy using capsid–nuclease fusion proteins. *Nature*, **352**, 632–5.

Natsoulis, G. and Boeke, J.D. (1993) Inhibition of MuLV by a gag-nuclease fusion protein. Conference on Gene Therapy Approaches to Treatment of HIV Infection. February 4–7: p.27.

Ojwang, J., Hampel, A., Looney, D., Wong, S. and Rappaport, J. (1992) Inhibition of human immunodeficiency virus type 1 expression by a hairpin ribozyme. *Proc. Natl Acad. Sci. USA*, **89**, 1802–6.

Orloff, S.L., Kennedy, M.S., Belperron, A.A., Maddon, P.J. and McDougal, J.S. (1993) Two mechanisms of soluble CD4 (sCD4)-mediated inhibition of human immunodeficiency virus type 1 (HIV-1) infectivity and their relation to primary HIV-1 isolates with reduced sensitivity to sCD4. *J. Virol.*, **67**, 1461–71.

Page, K., Landau, N. and Littman, D. (1990) Construction and use of a HIV vector for analysis of virus infectivity. *J. Virol.*, **64**, 5270–6.

Panganabian, A. and Fiore, D. (1988) Ordered interestrand and intrastrand DNA transfer during reverse transcription. *Science*, **241**, 1064–9.

Pantaleo, G., Graziosi, C. *et al.* (1993) HIV infection is active and progressive in lymphoid tissue during the clinically latent stage of disease. *Nature*, **362**, 355–8.

Pearson, L., Garcia, J., Wu, F., Modesti, N., Nelson, J. and Gaynor, R. (1990) A transdominant tat mutant that inhibits tat induced gene expression from the HIV long terminal repeat. *Proc. Natl Acad. Sci. USA*, **87**, 5079–83.

Plaeger, M., Hausner, M., Isacescu, V. and Giorgi, J. (1992) CD8 T-cell-mediated inhibition of HIV replication in HIV infected adults and children. *Aids Res. Hum. Retroviruses*, **8**, 1375–6.

Poli, G. and Fauci, A.S. (1992) The role of monocyte/macrophages and cytokines in the pathogenesis of HIV infection. *Pathobiology*, **60**, 246–51.

Pomerantz, R., Bagasra, O. and Baltimore, D. (1992) Cellular latency of human immunodeficiency virus type 1. *Curr. Opin. Immunol.*, **4**, 475–80.

Pozansky, M., Lever, A., Bergeron, L., Haseltine, W. and Sodroski, J. (1991) Gene transfer into human lymphocytes by a defective human immunodeficiency virus type 1 vector. *J. Virol.*, **65**, 532–6.

Purdy, B. and Plaisance, K.L. (1991) Anti-retroviral therapy of human immunodeficiency virus infection: current strategies and challenges for the future. *Blood*, **78**, 859–84.

Ratner, L., Polmar, S., Paul, N. and Ruddle, N. (1987) Cytotoxic factors secreted by cells infected by human immunodeficiency virus type 1. *Aids Res. Hum. Retroviruses*, **3**, 147–55.

Redfield, R., Birx, D. *et al.* (1991) A phase 1 evaluation of the safety and immunogenicity of vaccination with recombinant gp160 in patients with early human immunodeficiency virus infection. Military Medical Consortium for Applied Retroviral Research. *N. Engl. J. Med.*, **324**, 1677–84.

Rhodes, A. and James, W. (1990) Inhibition of HIV replication in cell culture by endogenously synthesized antisense RNA. *J. Gen. Virol.*, **71**, 1965–74.

Riddell, S.R., Rabin, M., Geballe, A. *et al.* (1991) Human cytomegalovirus specific cytotoxic T lymphocytes recognize infected cells in the absence of viral gene expression. *J. Immulol.*, **146**, 2795–804.

Riddel, S., Greenberg, P., Overell, *et al.* (1992) Phase 1 study of cellular adoptive immunotherapy using genetically modified CD8+ HIV specific T cells for HIV seropositive patients undergoing allogeneic bone marrow transplantation. *Human Gene Ther.*, **3**, 319–38.

Rosenberg, S. *et al.* (1990) Gene transfer into human-immunotherapy of patients with advanced melanoma, using tumor-infiltrating lymphocytes modified by retroviral gene transduction. *N. Engl. J. Med.*, **323**, 570–8.

Ruprecht, R., Koch, J., Sharma, P. and Armany, R. (1992) Development of antiviral treatment strategies in murine models. *Aids Res. Hum. Retroviruses*, **8**, 997–1011.

Saral, R. and Holland, K. (1993) Bone marrow transplantation for the acquired immune deficiency syndrome, in *Bone Marrow Transplantation*, (eds S. Foreman), Blackwell Scientific Publications, Boston, pp. 654–64.

Sarver, N., Cantin, E.M., Chang, P.S. *et al.* (1990) Ribozymes as potential anti-HIV-1 therapeutic agents. *Science*, **247**, 1222–5.

Schwartz, S., Felber, B.K., Fenyo, E.M. and Pavalkis, G.N. (1989) Rapidly and slowly replicating human immunodeficiency virus type 1 isolates can be distinguished according to target-cell tropism in T-cell and monocyte cell lines. *Proc. Natl Acad. Sci. USA*, **86**, 7200–3.

Sczakiel, G. and Pawlita, M. (1991) Inhibition of HIV replication in human T-cells stably expressing antisense RNA. *J. Virol.*, **65**, 468–72.

Sczakiel, G., Oppenlander, M., Rittner, K. and Pawlita, M. (1992) Tat- and Rev-directed antisense RNA expression inhibits and abolishes replication of human immunodeficiency virus type 1: a temporal analysis. *J. Virol.*, **66**, 5576–81.

Shimada, T., Fujii, H., Mitsuya, H. and Nienhuis, A.W. (1991) Targeted and highly efficient gene transfer into CD4+ cells by a recombinant human immunodeficiency virus retroviral vector. *J. Clin. Invest.*, **88**, 1043–7.

Shirazi, Y. and Pitha, P.M. (1992) Alpha interferon inhibits early stages of the human immunodeficiency virus type 1 replication cycle. *J. Virol.*, **66**, 1321–8.

Sioud, M. and Drlica, K. (1991) Prevention of human immunodeficiency virus type 1 integrase expression in *E. coli* by a ribozyme. *Proc. Natl Acad. Sci. USA*, **88**, 7303–7.

Smith, C., Gasparetto, C., Collins, N. *et al.* (1991) Purification and partial characterization of a human hematopoietic precursor population. *Blood*, 77, 2122–9.

Smith, C., Lee, S., Sullenger, B., Gallardo, M., Ungers, G. and Gilboa, E. (1992) Intracellular immunization against hiv using rna decoys. Third International Symposium on Targeted Gene Therapy for the Treatment of HIV Infection, p.6.

Spector, D.H., Wade, E., Wright, D.A. *et al.* (1990) Human immunodeficiency virus pseudotypes with expanded cellular and species tropism. *J. Virol.*, **64**, 2298–308.

Stanley, S.K. and Fauci, A.S. (1993) T cell homeostasis in HIV infection: part of the solution, or part of the problem? [Editorial comment]. *J. Acquir. Immune Defic. Syndr.*, **6**(2), 142–3.

Steffy, K.R. and Wong-Staal, F. (1993) Transdominant inhibition of wild-type human immunodeficiency virus type 2 replication by an envelope deletion mutant. *J. Virol.*, **67**, 1854–9.

Sullenger, B. and Cech, T. (1993) The importance of colocalization of ribozyme and substrate containing RNAs for trans cleavage *in vivo*. *Science*, **263**, 1566–9.

Sullenger, B., Gallardo, H., Ungers, G. and Gilboa, E. (1990a) Overexpression of TAR sequences renders cells resistant to human immunodeficiency virus replication. *Cell*, **63**, 601–8.

Sullenger, B., Lee, T., Smith, C., Ungers, G. and Gilboa, E. (1990b) Expression of chimeric tRNA-driven antisense transcripts renders NIH 3T3 cells highly resistant to Moloney murine leukemia virus replication. *Mol. Cell. Biol.*, **10**, 6512–23.

Sullenger, B., Gallardo, H., Ungers, G. and Gilboa, E. (1991) Analysis of trans-acting response decoy RNA-mediated inhibition of human immunodeficiency virus type 1 transactivation. *J. Virol.*, **65**, 6811–16.

Sun, X.-H. and Baltimore, D. (1989) Human immunodeficiency virus tat activated expression of poliovirus protein 2A inhibits mRNA translation. *Proc. Natl Acad. Sci. USA*, **86**, 2143–6.

Taylor, N.R. and Rossi, J.J. (1991) Ribozyme-mediated cleavage of an HIV-1 gag RNA: the effects of nontargeted sequences and secondary structure of ribozyme cleavage activity. *Antisense Res. Dev.*, **1**, 173–86.

Trono, D., Feinberg, M.B. and Baltimore, D. (1989) HIV-1 gag mutants can dominantly interfere with the replication of the wild type virus. *Cell*, **59**, 113–20.

Tuerk, C., MacDougal, S. and Gold, L. (1992) RNA pseudoknots that inhibit HIV reverse transcriptase. *Proc. Natl Acad. Sci. USA*, **89**, 6988–92.

Ulmer, J., Donnelly, J., Parker, S.E. *et al.* (1990) Heterologous protection against influenza by injection of DNA encoding a viral protein. *Science*, **247**, 1465.

van Beusechem, B., Kukler, A., Heidt, P. and Valerio, D. (1992) Long-term expression of human adenosine deaminase in rhesus monkeys transplanted with retrovirus-infected bone-marrow cells. *Proc. Natl Acad. Sci. USA*, **89**, 7640–4.

Venkatesh, A., Caruso, M. and Klatzman, D. (1992) Selective killing of CD4+ cells harboring a human immunodeficiency virus inducible suicide gene prevents viral spread in infected cell population. *Proc. Natl Acad. Sci. USA*, **182**,

Vilmer, E., Feuillette, A., Rabian, *et al.* (1987) Clinical and immunological restoration in patients with AIDS after marrow transplantation, using lymphocyte transfusions from the marrow donor. *Transplantation*, **44**, 25–9.

Visser, J. and Van Bekkum, D. (1990) Purification of pluripotent hematopoietic stem cells: past and present. *Exp. Hematol.*, **18**, 248–56.

Wang, A., Unge, D., Srikantan, *et al.* (1993) Gene inoculaion generates immune response against HIV-1. *Proc. Natl Acad. Sci. USA*, **90**, 4156–60.

Warner, J., Anderson, C., Laube, L. *et al.* (1991) Induction of HIV-specific CTL and antibody responses in mice using retroviral vector-transduced cells. *Aids Res. Hum. Retroviruses*, 7, 645–55.

Weerasinghe, M., Liem, S., Asad, S., Read, S. and Joshi, S. (1991) Resistance to human immunodeficiency virus type 1 (HIV-1) infection in human CD4+ lymphocyte-derived cell lines conferred by using retroviral vectors expressing an HIV-1 RNA-specific ribozyme. *J. Virol.*, **65**, 5531–4.

Williams, D. (1990) Expression of introduced genetic sequences in hematopoietic cells following retroviral-mediated gene transfer. *Human Gene Ther.*, **1**, 229–39.

Williams, D. (1991) Syngeneic bone-marrow transplantation and failure to eradicate HIV [letter]. *AIDS*, **5**, 344.

Yamada, O., Yu, M., Yee, J., Kraus, G., Looney, D. and Wong-Staal, F. (1994) Intracellular immunization of human T cells with a hairpin ribozyme against human immunodeficiency virus type 1. *Gene Therapy*, **1**, 38–45.

Yarchoan, R., Pluda, J.M., Porno, C.F., Mitsuya, H. and Broder, S. (1991) An idiotypic network model of AIDS immunopathogenesis. *Proc. Natl Acad. Sci. USA*, **88**, 3060–4.

Yu, M., Poeschla, E. and Wong-Staal, F. (1994) Progress towards gene therapy for HIV infection. *Gene Therapy*, **1**, 13–26.

12

Cystic fibrosis and lung diseases

SAMUEL C. WADSWORTH and ALAN E. SMITH

12.1 INTRODUCTION

Although gene therapy is still in its infancy, the lung has received a good deal of attention as a target organ. This reflects the fact that the lung appears to be a favorable site for gene therapy; it is a relatively simple organ with the target cells in most instances lining its lumen, vectors can be delivered to a portion of the lung or the whole lung without a surgical procedure, and the success of delivery and efficacy of a vector can be assessed non-invasively. Another measure of the perceived practicality of lung gene therapy is the relatively large number of applications dealing with this organ that have been submitted in the United States to the Recombinant Advisory Committee (RAC), a regulatory agency of the National Institutes of Health that publicly reviews all such protocols. In addition to having important applications for lung disease, there is the clear hope that important lessons can be learned from trials in the lung that will aid in the design of gene therapy treatments for other organs.

The respiratory epithelium that forms the mucosal lining of the lung, and also of the nasal and sinus cavities, is a pseudostratified epithelium composed of relatively few cell types, most of which are terminally differentiated and non-dividing. Ciliated epithelial cells are the predominant cell type in the upper airways of the bronchial tree; additional cell types include basal cells of uncertain function and secretory cells such as Clara cells. In the deeper airways the ciliated cells decrease progressively in proportion while Clara cells are present in greater numbers. The airways terminate in specialized alveolar cells that make intimate contact with the circulatory system

Molecular and Cell Biology of Human Gene Therapeutics
Edited by George Dickson
Published in 1995 by Chapman & Hall. ISBN 0 412 62550 4

to effect gas exchange. Other cell types in the basal layer of the airway epithelium, such as cells of the submucosal glands, appear to be less accessible to current gene therapy vectors. Here we will discuss gene therapy of the lung exclusively by way of the lumen of the lung, not by way of the circulatory system. Gene therapy of the endothelial cells lining the circulatory system in general is a separate topic and transcytosis of gene therapeutics from the circulation to the airway epithelium is not well documented, though there are reports of this process.

Because of the similarity between nasal and airway respiratory epithelial cells, the nose has proved a very useful site for preliminary administration of gene therapy vectors. Treatment and examination of target tissue is non-invasive and because the target tissue is limited, vector doses can be kept to a minimum. Administration of vectors to the airways of the lung can be accomplished by direct instillation using a bronchoscope or by aerosolization. Bronchoscopic administration of gene therapeutics is desirable for early human trials because administration can be restricted to a portion of the lung, thereby minimizing the potential for an adverse event affecting the entire lung. As vectors are shown to be safe in the human lung, aerosol administration to the entire lung can be attempted. Clearly in the long run, aerosol delivery for lung gene therapy would be desirable because it is non-invasive and amenable to lower cost drug delivery, perhaps even home self-administration.

Three broad classes of lung disorders are potentially amenable to treatment by the gene therapy approach: (i) genetic disorders requiring long-term, probably life-long, replacement of a protein that has been compromised by mutation; (ii) acute disorders where only transient activity of a therapeutic gene, perhaps delivered in a single dose, might be required; and (iii) cancers. This chapter will not focus on acute disorders or cancers. Treatment of acute lung disorders via the gene therapy route is at an early stage of development, few gene products have been identified for a potential gene therapy approach, and clear treatment strategies are presently lacking. Cancer gene therapy is considered separately in this volume.

12.2 GENE THERAPY FOR CYSTIC FIBROSIS

This chapter will focus on gene therapy approaches for cystic fibrosis (CF). CF is a lethal, autosomal recessive disease of Caucasians and affects approximately 60 000 CF patients in the world today. Median life expectancy of patients with optimal care is 29 years.

Although CF affects the epithelia of several organs, the greatest morbidity and mortality is associated with lung abnormalities. The lung will therefore be the main target for gene therapy. The hallmark abnormality of the disease is altered electrolyte transport, defective chloride secretion and increased Na^+ absorption, by epithelial cells including those that line the airways (Knowles *et al.*, 1983; Quinton, 1990). It is thought that these alterations in electrolyte transport in the airways lead to changes in the composition of respiratory tract fluid and mucus, and hence to defective mucociliary clearance and lung damage from repeated cycles of infection and inflammation.

The genetic defect resulting in CF has been identified, the gene sequences encompassing the gene have been cloned, and the protein affected by CF-associated mutations identified (Collins, 1992). The structure of this protein, referred to as the cystic fibrosis transmembrane conductance regulator (CFTR) has been reviewed extensively elsewhere (Welsh *et al.*, 1992). Briefly, CFTR is a membrane Cl^- channel regulated by cAMP-dependent phosphorylation and intracellular nucleotides. The properties of the CFTR chloride channel are the same as those of the chloride channels in the apical membrane of airway epithelial cells and transfer of the cDNA for wild-type CFTR to lung cells corrects the Cl^- channel defect in primary cultures of CF epithelia (Drumm *et al.*, 1990; Rich *et al.*, 1990).

12.3 TARGET CELLS IN THE LUNG FOR CYSTIC FIBROSIS GENE THERAPY

Within the lung, the airway epithelium is the target tissue for gene transfer that is most likely to be of benefit to a patient with CF. This means in practical terms that the target would appear to be the mature cells that line the pulmonary airways. These are not rapidly dividing cells; rather, most of them are non-proliferating and many may be terminally differentiated.

Because CFTR is a chloride channel and chloride can move through tight junctions between adjacent cells in the respiratory epithelial layer, it appears that not all cells in the epithelium necessarily need be corrected by delivery of CFTR to effect a physiological correction. Indeed, mixing experiments using CF epithelial cells, some of which had been stably transduced with a CFTR-encoding retrovirus, suggests that as little as 5% of monolayer cells need be corrected (Johnson *et al.*, 1992).

We are unaware of any method for removal of the target cell from the patient, treatment by gene therapy *ex vivo*, and then re-implantation into the patient. Gene therapy for CF appears to require *in vivo* treatment of the airway epithelium.

12.4 VECTORS FOR CYSTIC FIBROSIS GENE THERAPY

Two broad classes of gene therapy vectors for CF and other lung diseases appear promising, namely viruses and plasmid DNA–lipid complexes. Each of these vector classes has advantages and disadvantages for consideration in designing the optimal gene therapy preparation.

12.4.1 Virus vectors

That the target cell for CF gene therapy is non-dividing and requires treatment *in situ* argues against the use of current retrovirus vectors and suggests the use of adenovirus (Ad) and adeno-associated virus (AAV) vectors. Furthermore, because between 10^9 and 10^{10} cells constitute the likely target in a patient, a high-titer virus preparation of substantial purity is required. This requirement eliminates retrovirus vectors from consideration and favors the use of adenovirus vectors over AAV vectors as we currently know them.

Adenovirus vectors

All of the adenovirus vectors currently under consideration for use in CF gene therapy are based on either the Ad2 or Ad5 serotypes. Ad2 and Ad5 serotypes are involved in relatively mild human disease, predominantly in the respiratory tract. All current vector backbones have been modified by deletion of the E1 region, the region responsible for regulating virus replication, for high-level expression of the majority of other virus genes, and for the limited oncogenic potential of these viruses in rodent model systems. Thus the current E1-deleted, and therefore replication-defective, Ad2 and Ad5-based gene therapy vectors are unlikely to cause serious disease in humans based on viral growth.

The Ad2 and Ad5 genomes are well characterized at the molecular level with complete nucleotide sequences and clones of subgenomic regions being available. Moreover, molecular details of the replication cycle and gene regulation program of these viruses are well documented. The detailed molecular biology of adenoviruses also allows for ready incorporation of therapeutic genes and/or cDNA molecules. Deletion of the E1 region allows the incorporation of a relatively large cDNA such as the CFTR cDNA. Deletion in addition of non-essential gene sequences such as those within the E3 region theoretically allows the incorporation of approximately 7 kb of exogenous DNA sequences. Importantly, expression of therapeutic genes can occur in the infected cell in the absence of cell division or viral genome integration. Additional information on adenovirus vectors can be found elsewhere in this volume (Chapter 2).

Production of Ad2 and Ad5 vectors is feasible on a large scale. E1-deleted vectors are routinely grown in the human 293 cell line, a cell line derived from human embryonic kidney which contains integrated copies of the Ad5 E1 region and which constitutively expresses E1 viral proteins. Active

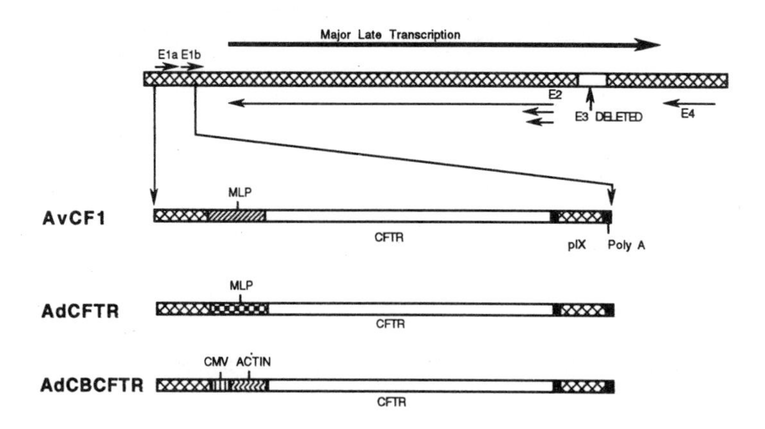

Figure 12.1 The Ad5-based vectors produced for use in human clinical cystic fibrosis (CF) gene therapy trials. Each of the Ad5-based vectors bears deletions of the E3 region as well as the E1 region. The major differences between these vectors relate to the regulatory sequences used to drive expression of the CFTR cDNA. In the AvCF (Yei *et al.*, 1994) and AdCFTR (Rosenfeld *et al.*, 1992) vectors, the adenovirus major late promoter (MLP; from Ad5 in AvCF and from Ad2 in AdCFTR) is used, while in AdCBCFTR (Engelhardt *et al.*, 1993) the β actin promoter is used in combination with enhancer sequences from cytomegalovirus (CMV). Each of these regulatory regions has the potential to promote very high-level expression CFTR mRNA.

preparations of high-titer adenovirus vectors can be purified readily from infected cell lysates.

Adenovirus vectors for CF

The Ad5-based vectors produced for use in human clinical CF gene therapy trials are illustrated in Figure 12.1. Each of the Ad5-based vectors bears deletions of the E3 region as well as the E1 region. The major differences between these vectors relate to the regulatory sequences used to drive expression of the CFTR cDNA. In the AvCF (Trapnell *et al.*, 1994) and AdCFTR (Rosenfeld *et al.*, 1992) vectors, the adenovirus major late promoter (MLP; from Ad5 in AvCF and from Ad2 in AdCFTR) is used, while in AdCBCFTR (Engelhardt *et al.*, 1993) the β actin promoter is used in combination with enhancer sequences from cytomegalovirus (CMV). Each of these regulatory regions has the potential to promote very high-level expression CFTR mRNA.

The Ad2-based CF vectors are shown diagrammatically in Figure 12.2. Aside from the difference in the virus serotype, these vectors differ from the Ad5-based CF vectors by the retention of the E3 region. E3 sequences were

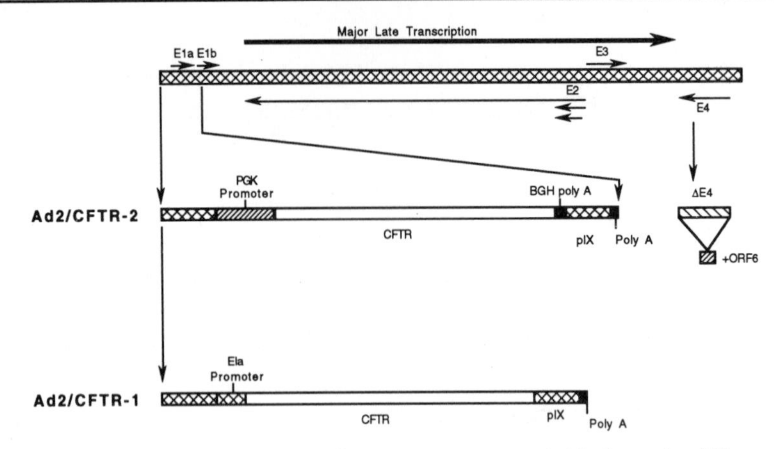

Figure 12.2 The Ad2-based cystic fibrosis (CF) vectors. Aside from the difference in the virus serotype, these vectors differ from the Ad5-based CF vectors by the retention of the E3 region. E3 sequences were retained in these vectors because E3-deleted, but otherwise wild-type adenoviruses have been shown to be more pathogenic for rodents. It should be noted however that expression of E3 in E1-deleted vectors has not been reported. Ad2/CFTR-1 and Ad2/CFTR-2 differ from each other in three respects. First Ad2/CFTR-1 uses the Ad2 E1a promoter (Rich *et al.*, 1993) while Ad/CFTR-2 uses the phosphoglycerate kinase (PGK) promoter. These are not strong promoters, but the endogenous levels of CFTR in human airway cells are not high, perhaps meaning that strong promoters are not required. Second, to render the vector more defective, the majority of the E4 region has been deleted from Ad2/CFTR-2, leaving only open reading frame 6 (ORF6) of E4, the minimal E4 function required for virus replication *in vitro*. Third, the bovine growth hormone poly-adenylation signal has been incorporated into Ad2/CFTR-2 downstream from the CFTR cDNA; the Ad2/CFTR-1 vector uses the endogenous E1b poly-adenylation signal.

retained in these vectors because E3-deleted, but otherwise wild-type adenoviruses have been shown to be more pathogenic for rodents. It should be noted however that expression of E3 in E1-deleted vectors has not been reported. Ad2/CFTR-1 and Ad2/CFTR-2 differ from each other in three respects. First Ad2/CFTR-1 uses the Ad2 E1a promoter (Rich *et al.*, 1993) while Ad/CFTR-2 uses the phosphoglycerate kinase (PGK) promoter. These are not strong promoters, but the endogenous levels of CFTR in human airway cells are not high, perhaps meaning that strong promoters are not required. Second, to render the vector more defective, the majority of the E4 region has been deleted from Ad2/CFTR-2, leaving only open reading frame 6 (ORF6) of E4, the minimal E4 function required for virus replication *in vitro*. Third, the bovine growth hormone poly-adenylation signal has been incorporated into Ad2/CFTR-2 downstream from the CFTR cDNA; the Ad2/CFTR-1 vector uses the endogenous E1b poly-adenylation signal.

Adeno-associated virus (AAV) vectors

As noted above, one advantage of AAV vectors is their ability to integrate into the cell genome, an event which may lead to prolonged expression of the therapeutic gene product which in turn would lengthen the interval between treatments. In contrast to adenovirus vectors, AAV vectors, because they have a single-stranded genome, must either replicate in the treated cell or integrate into the cell genome before expression can take place. The ability to integrate could be interpreted as a disadvantage of AAV since this event could theoretically be mutagenic. For wild-type AAV, this does not appear to be the case since AAV infection, which is widespread in the human population, is not associated with disease. (see Muzyczka, 1992 for a review of AAV.)

One clear limitation of AAV vectors is the genome size, approximately 4680 nucleotides. Because the terminally repeated sequences totaling 290 nucleotides must be retained for replication, the longest DNA sequence that can be inserted without exceeding 100% of the wild-type genome length is 4370 nucleotides. Recombinant genomes greater than 100% of the wild-type genome length have an increased probability of suffering deletions during virus propagation.

Propagation of recombinant AAV vectors requires the usual adenovirus helper functions and in addition the normal AAV gene products because all AAV genes are deleted from the recombinant vector genome. Thus production of AAV vectors is a more complex process than the production of adenovirus vectors and presently only limited amounts of virus can be made and tested.

Adeno-associated virus (AAV) vectors for CF

A minimal CFTR cDNA is approximately 4500 nucleotides in length, too large for optimal incorporation into an AAV vector. To date there is only one literature report of an AAV-based CF vector (Flotte *et al.*, 1993). This AAV–CFTR vector incorporates two unusual properties to accommodate the genome length limitation. First, expression of CFTR is driven by promoter activity reportedly present in the AAV inverted terminal repeat sequences at the right end of the genome. Second, the first 118 amino acids of CFTR are deleted with translational initiation taking place within a synthetic sequence fused onto the 5′ end of the truncated CFTR cDNA. This truncated CFTR protein reportedly functions as a chloride channel in human CF cells.

12.4.2 Preclinical studies of virus vector safety

Three known types of adverse reactions can occur following administration of wild-type virus, and presumably virus vectors: inflammatory responses upon virus administration; immunological responses to proteins expressed by the virus; and immunological responses to administered virus structural proteins. These responses are by no means mutually exclusive, nor are they necessarily dependent on extensive viral gene expression.

Inflammatory response to virus administration

Introduction of a virus preparation into the lung can trigger both direct and indirect reactions. Direct reactions may be provoked by binding of virus to receptors on the surface of cells lining the airways and/or the subsequent entry of virus into cells. For adenoviruses there is good evidence that binding of virus to the cell is a separate event from entry into the cell, and that the latter involves integrins on the cell surface (Wickham *et al.*, 1993). There are insufficient data available to determine whether the binding and/or entry events stimulate a pathologically significant response in airway epithelial cells.

Introduction of virus preparations into the lung almost certainly results in uptake of virus particles non-specifically by macrophages in the lung triggering the release of cytokines. Consistent with this expectation is the demonstration by Ginsberg *et al.* (1991) that in mice, which are non-permissive for the virus, the levels of the macrophage-derived cytokines IL-1, IL-6 and TNFα are elevated following the administration of wild-type adenovirus to the lung. At the cellular level, mice and cotton rats (which are semi-permissive for the virus) respond to high doses ($>10^8$ infectious units) of wild-type adenovirus with a robust lymphocytic lung infiltrate. The immediate cellular response (up to day 3 post-inoculation) is believed to be due to cytokine expression. It is unclear whether highly purified, disabled recombinant adenoviruses will provoke an analogous response in humans.

Immunological response to virus administration

The lymphocytic response in rodents to virus administration beyond 3 days is believed to be a cytotoxic T cell (CTL) response to limited viral protein synthesis within the cells in the lung. A substantial body of evidence indicates that the CTL response is specific for and restricted to the E1a and/or E1b proteins (Rawle *et al.*, 1991), the proteins that are deleted from current adenovirus vectors. If the response of the human immune system to the expression of adenovirus proteins mirrors the reported response of the rodent immune system, then expression in the human lung of virus genes from an E1-deleted vector should not provoke a CTL response, although expression of CFTR potentially could if it were polymorphic with respect to the patient's CFTR. A CTL response presumably would also be dependent on the degree to which viral protein synthesis is disabled in the recombinant vector.

A humoral antibody response to the virus inoculum is predicted and has been observed in animals after exposure to recombinant adenoviruses (Zabner *et al.*, 1994). What are the consequences of the presence of anti-adenovirus antibodies in the lung for repeat administration of virus? Will anti-adenovirus antibodies impede efficient gene transfer and will such antibodies lead to immunopathological effects? The answers to these questions are not known, but will clearly affect the prospects for CF gene therapy. It is known, however, that antibodies to a wide variety of human proteins, both natural and recombinant, are elicited in 10–15% of treated patients, usually with little or no effect.

Studies with adenovirus gene therapy vectors

Safety studies

Two major studies on the safety of adenovirus CF vectors have been published. Simon *et al.* (1993) employed baboons to assess the safety of lung administration of AdCBCFTR. All post-treatment parameters measured from non-respiratory organs were normal, although serum antibody titers were not reported. In general no replication or spread of the vector was found, the exception being the presence of virus in one bronchiolar–alveolar lavage (BAL) specimen at 4 days. In the lung there was a dose and time post-administration-dependent mononuclear cell infiltrate in the alveoli revealed on necropsy which was severe at the highest dose tested. Alveolar infiltrates were visible on chest X-ray at virus doses greater than 6×10^9 IU/kg bodyweight, but only at 2 to 3 weeks post-administration. BAL fluids were also analyzed and revealed a transient increase in neutrophils at 4 days post-administration and a steady increase in lymphocytes up to 21 days post-administration, the last point analyzed.

Zabner *et al.* (1994) carried out a series of safety studies with cotton rats and with rhesus monkeys. Ad2/CFTR-1 was administered to the lungs of cotton rats either as a single or repeat dose up to 4×10^{10} infectious units/kg bodyweight. Neither significant systemic nor intrapulmonary inflammation was observed as compared with untreated animals. Serum and BAL antibodies, including neutralizing antibodies, were detected in treated animals. The only spread of virus was detected to the bowel of several treated cotton rats, presumably due to swallowing of the virus following normal mucociliary clearance from the lung. Ad2/CFTR-1 was also administered repeatedly to the nasal respiratory epithelium of rhesus monkeys. Serum antibodies, including neutralizing antibodies following the second administration, were detected in all treated monkeys but neutralizing antibodies were not detected in nasal washings. No clinical signs or abnormalities were observed in the treated animals, nor was there local nasal inflammation above that seen in sham-treated animals.

Efficacy

In preclinical animal studies, the surrogate efficacy parameters of reporter gene expression and CFTR mRNA expression are commonly measured. Engelhardt *et al.* (1993) reported β galactosidase reporter activity and the presence of CFTR mRNA in baboons treated with AdCBCFTR but CFTR mRNA persisted for less than 21 days. Zabner *et al.* (1994), using Ad2/CFTR-1, reported expression of CFTR mRNA and protein in cotton rat lung and rhesus monkey nasal epithelium for 2 to 3 weeks post-treatment and CFTR protein in rhesus monkey nasal epithelium for up to 6 weeks post-treatment. Rosenfeld *et al.* (1992) reported expression of CFTR mRNA for up to 6 weeks in lung airway epithelium in cotton rats treated with AdCFTR, though protein expression was observed for only 2 weeks.

The results of these safety and efficacy studies highlight the great promise of adenovirus vectors for CF gene therapy and at the same time indicate areas where more information is required to complete development of a safe and effective CF therapeutic. Some of the information that is required can only be acquired through experimental trials in humans, and in particular in CF patients. Results of the preclinical studies summarized above have formed the basis of clinical trials that are reviewed below.

12.4.3 Plasmid DNA–cationic lipid complexes

Complexes of cationic lipids and DNA molecules are widely used for the efficient introduction of plasmid DNA molecules into cells in culture. Assuming that the high efficiency of gene transfer seen *in vitro* can be achieved *in vivo*, cationic lipid–DNA complexes have a number of potential advantages over viruses as gene therapy vectors including the ability to package longer gene sequences in plasmid vectors, the likely lack of antigenicity of the complex, and the presumed safety of cationic lipids by analogy with neutral liposomes.

Two routes of administration of plasmid DNA–cationic lipid complexes to the lung have been reported, namely direct delivery to the airways (see below) and injection into the circulatory system (Brigham *et al.*, 1989; Zhu *et al.*, 1993). For treatment of pulmonary epithelial cells, delivery into the airways would be the logical route of administration. Curiously, Zhu *et al.* (1993) report the efficient delivery and expression in the respiratory epithelium of intravenously-injected plasmid DNA–cationic lipid complexes.

Lipid–DNA efficacy

Several reports describe transfer of cationic lipid–DNA complexes containing CFTR cDNA and reporter DNA sequences to mouse lung by way of the airways (Yoshimura *et al.*, 1992; Alton *et al.*, 1993; Hyde *et al.*, 1993; Sorscher *et al.*, 1993). Significant differences between experiments exist with regard to recipient animals, lipids, lipid : DNA formulations, methods of delivery, DNA constructs, and methods for detection of gene expression. Nevertheless, it appears that cationic lipid–DNA complexes are capable of effecting gene transfer in lung. However, although expression of plasmid-borne DNA sequences has been reported to occur in a substantial proportion, 40% or more, of respiratory epithelial cells, and CFTR RNA sequences reported to persist for prolonged periods of time (up to 28 days), the efficacy and duration of expression *in vivo* requires further documentation.

Lipid–DNA safety

Lipid–DNA complexes, being composed of a synthetic lipid and a bacterially-produced DNA molecule, should have a different safety profile from that of animal viruses. The major issues remain the same however, namely acute toxicity to the lung at the time of administration and effects of long-term administration of the complex.

Relatively little is known about the acute toxicity to the lung of cationic lipids, although to date no published reports have highlighted specific problems. Alton *et al.* (1993) report no toxic effects or epithelial damage after administration of one dose of DOTAP, DOTMA : DOPE, or DC-Chol : DOPE variously complexed with plasmid DNA molecules encoding *E. coli* β gal or human CFTR. Sorscher *et al.* (1993) also report no toxic effects of administration of DIMRIE : DOPE–DNA complexes.

Additional toxicity issues that must be evaluated include, the potential for contamination of the plasmid DNA with bacterial endotoxin, the potential for toxicity due to repeat administration of lipid–DNA complexes and to lipid load within the lung, and the potential metabolic breakdown products of lipids.

12.5 CLINICAL TRIALS WITH CYSTIC FIBROSIS GENE THERAPY VECTORS

A total of eight CF gene therapy protocols have been reviewed and approved by public regulatory agencies: seven by the RAC in the United States, six employing adenovirus vectors and one employing cationic lipid–plasmid DNA vectors; and one employing cationic lipid–plasmid DNA vectors in England. Despite the differences in the vectors, there are certain features common to all of the trials. Each of the trials is designed to assess vector safety in a relatively small number of CF patients. The gene being transferred in each of the trials is the wild-type CFTR cDNA, with the eventual goal of restoring chloride channel activity lacking in CF patients. To this end measurements of CFTR channel function are also planned to assess biochemical efficacy of the gene transfer. To date only one of the trials has been completed and reported (Zabner *et al.*, 1994). Only a brief overview of each study will be discussed here, details of the trial plans can be found in the publicly-available RAC proposals and in an upcoming issues of *Gene Therapy* for the UK trial.

Two different strategies for Phase I clinical trials for CF gene therapy have been proposed: one more aggressive and proceeding directly to lung administration, and one more conservative assessing safety of vector administration in the upper airways before proceeding to lung administration.

To date, all of the lung administration protocols involve adenovirus vectors and are designed to assess the safety of a single dose of vector administered to the lung by way of a bronchoscope. Typically, virus will be administered directly to only a single lobe of the lung to minimize distribution of the vector to the whole lung. One or two patients will be treated with the lowest dose of vector before treatment of additional patients with higher doses. This strategy reduces risk to patients since no individual is given multiple treatment, but it still allows an assessment of increasing dose of vector in patients. Trials involving AdCBCFTR and AdCFTR are underway and an adverse reaction of a patient involved in the AdCFTR trials has been widely reported in the popular press; however, the details of the incident have not yet been published.

The more conservative protocols assessing vector safety in the upper airways involve adenovirus and lipid–DNA vectors. With one exception, these protocols involve administration of a low dose of vector to the nasal respiratory epithelium of one or two patients followed by subsequent administration of a higher dose of vector to the nasal respiratory epithelium of another group of patients. As with the lung administration protocols this strategy allows an assessment of vector safety at increasing vector doses but without exposing any patient to multiple vector doses. The additional margin of safety in this type of protocol over the lung administration protocols is that any potential adverse reaction to the procedure will be minimized due to the site of treatment. Results of one of the nasal administration trials with an adenovirus vector have been published (Zabner *et al.*, 1994) and are briefly summarized here.

Patients were treated with low doses of Ad2/CFTR-1 (2×10^6; 6×10^6; or 5×10^7 IU) over a limited area of the nasal respiratory epithelium and monitored for indications of local and/or systemic reactions to vector administration. All measured post-treatment parameters were normal. Other than minor inflammation at the site of treatment that was also seen in sham-treated volunteers, no treatment-associated abnormalities were observed. Biopsy of the treatment site revealed no histological abnormalities beyond those commonly observed in CF patients. Importantly, there was also no evidence of virus spread beyond the administration site.

In addition to the assessment of safety, indications of CFTR gene transfer were also sought by measuring chloride channel function in the nasal epithelium at the site of virus treatment. At each dose of virus, restoration of chloride channel activity was observed. Basal transepithelial voltage measurements in vector-treated CF patients as compared with non-CF patients were intermediate or approaching normal values and remained at the higher levels for at least 3 weeks. This clinical study established the important principle that an adenovirus vector is capable of transferring a CFTR cDNA to cells in a CF patient and that the resulting CFTR protein can transiently restore chloride channel activity. Furthermore, the multiplicity of infection at the lowest dose corresponded to only one infectious unit per cell. If this could be achieved in all cells of the airway, the calculated total dose of approximately 5×10^9 IU would seem not unreasonable.

Having established that CFTR gene transfer and restoration of channel activity is feasible *in vivo*, a follow-up trial has been planned with two major goals; first, to assess the impact of repeated dosing with increasing dose of vector and second, to gain evidence of clinical efficacy. This protocol also involves treatment of the upper airways, nasal and sinus epithelium, with Ad2/CFTR-2. Safety and biochemical efficacy parameters to be measured are as in the first nasal protocol. Additional efficacy parameters to be analyzed in the sinus as a function of treatment include changes in cytokine profile, bacterial colonization of the sinus cavity and CT scan of the sinus cavity to measure the extent of mucus occlusion.

12.6 CONCLUSIONS

It is perhaps extraordinary to reflect that the time that elapsed from the discovery of the CF gene to the treatment of the first patients with gene therapy, albeit in a trial with limited goals, was less than 4 years. This achievement is notable not only for its speed but also because it provides a model for the treatment of other genetic diseases and indeed for the use of other information generated for the analysis of the human genome.

The results with the first CF patients clearly show that gene therapy as an approach is more than just a theoretical possibility. Correction of the basic defect was demonstrated over a period of a few weeks following treatment with an adenovirus-based vector. That is not to say that major hurdles do not remain. Over-riding among these are: the issue of the safety of long-term administration of viral vector to an organ as potentially sensitive as the lung; the duration of the effect and the need for repeat dosing; and not least the issue of what parameters can reasonably be used as indicators of clinical benefit. Nevertheless, it is difficult at this stage not to be optimistic of the outcome of the next 4 years' research.

REFERENCES

Alton, E.W.F.W., Middleton, P.G., Smith, S.N. *et al.* (1993) Non-invasive liposome-mediated gene delivery can correct the ion transport defect in cystic fibrosis mutant mice. *Nature Genet.*, **5**, 135–42.

Brigham, K.L., Meyrick, B., Christman, B. *et al.* (1989) Rapid communication: *in vivo* transfection of murine lungs with a functioning prokaryotic gene using a liposome vehicle. *Am. J. Med. Sci.*, **4**, 278–81.

Collins, F.S. (1992) Cystic fibrosis: molecular biology and therapeutic implications. *Science*, **256**, 774–9.

Drumm, M.L., Pope, H.A., Cliff, W.H. *et al.* (1990) Correction of the cystic fibrosis defect *in vitro* by retrovirus-mediated gene transfer. *Cell*, **62**, 1227–33.

Engelhardt, J.F., Simon, R.H., Yang, Y. *et al.* (1993) Adenovirus-mediated transfer of the CFTR gene to lung of nonhuman primates: biological efficacy study. *Human Gene Ther.*, **4**, 759–69.

Flotte, T.R., Afione, S.A., Solow, R. *et al.* (1993) Expression of cystic fibrosis transmembrane conductance regulator from a novel adeno-associated virus promoter. *J. Biol. Chem.*, **268**, 3781–90.

Ginsberg, H.S., Moldawer, L.L., Sehgal, P.B. *et al.* (1991) A mouse model for investigating the molecular pathogenesis of adenovirus pneumonia. *Proc. Natl Acad. Sci. USA*, **88**, 1651–5.

Hyde, S.C., Gill, D.R., Higgins, C.F. *et al.* (1993) Correction of the ion transport defect in cystic fibrosis transgenic mice by gene therapy. *Nature*, **362**, 250–5.

Johnson, L.G., Olsen, J.C., Sarkadi, B. *et al.* (1992) Efficiency of gene transfer for restoration of normal airway epithelial function in cystic fibrosis. *Nature Genet.*, **2**, 21–5.

Knowles, M.R., Stutts, M.J., Spock, A. *et al.* (1983) Abnormal ion permeation through cystic fibrosis regulatory epithelium. *Science*, **221**, 1067–70.

Muzyczka, N. (1992) Use of adeno-associated virus as a general transduction vector for mammalian cells. *Curr. Top. Microbiol. Immunol.*, **158**, 97–129.

Quinton, P.M. (1990) Cystic fibrosis: a disease in electrolyte transport. *FASEB J.*, **4**, 2709–17.

Rawle, F.C., Knowles, B.B., Ricciardi, R.P. *et al.* (1991) Specificity of the mouse cytotoxic T lymphocyte response to adenovirus 5-E1A is immunodominant in H-2^b, but not H-2^d of H-2^k mice. *J. Immunol.*, **146**, 3977–84.

Rich, D.P., Anderson, M.P., Gregory, R.J. *et al.* (1990) Expression of cystic fibrosis transmembrane conductance regulator corrects defective chloride channel regulation in cystic fibrosis airway epithelial cells. *Nature*, **347**, 358–63.

Rich, D.P., Couture, L.A., Cardoza, L.M. *et al.* (1993) Development and analysis of recombinant adenoviruses for gene therapy of cystic fibrosis. *Human Gene Ther.*, **4**, 461–76.

Rosenfeld, M.A., Yoshimura, K., Trapnell, B.C. *et al.* (1992) *In vivo* transfer of the human transmembrane conductance regulator gene to the airway epithelium. *Cell*, **68**, 143–55.

Simon, R.H., Engelhardt, J.F., Yang, Y. *et al.* (1993) Adenovirus-mediated transfer of the CFTR gene to lung of nonhuman primates: toxicity study. *Human Gene Ther.*, **4**, 771–80.

Sorscher, E.J., Logan, J.J., Frizzell. R.A. *et al.* (1993) Gene therapy for cystic fibrosis using cationic liposome mediated gene transfer: a phase I trial of safety and efficacy in the nasal airway. RAC Protocol #9312–066; RAC approved December 3, 1993, pending final NIH approval.

Welsh, M.J., Anderson, M.P., Rich, D.P. *et al.* (1992) Cystic fibrosis transmembrane conductance regulator: a chloride channel with novel regulation. *Neuron*, **8**, 821–9.

Wickham, T.J., Mathias, P., Cheresh, D.A. *et al.* (1993) Integrins $\alpha_v\beta_5$ promote adenovirus internalization but not virus attachment. *Cell*, **73**, 309–19.

Yei, S., Mittereder, N., Wert, S., Whitsett, J.A., Wilmott, R.W. and Trapnell, B.C. (1994) *In vivo* evaluation of the safety of adenovirus-mediated transfer of the human cystic fibrosis transmembrane conductance regulator cDNA to the lung. *Human Gene Ther.*, **5**, 731–44.

Yoshimura, K., Rosenfeld, M.A., Nakamura, H. *et al.* (1992) Expression of the human transmembrane conductance regulator gene in the mouse lung after *in vivo* intratracheal plasmid-mediated gene transfer. *Nucleic Acids Res.*, **20**, 3233–40.

Zabner, J., Petersen, D.M., Puga, A.P. *et al.* (1994) Safety and efficacy of repetitive adenovirus-mediated transfer of CFTR cDNA to airway epithelia of primates and cotton rats. *Nature Genet.*, **6**, 75–83.

Zhu, N., Liggitt, D., Liu, Y. *et al.* (1993) Systemic gene expression after intravenous DNA delivery into adult mice. *Science*, **261**, 209–11.

13

Arthritis

CHRISTOPHER H. EVANS and PAUL D. ROBBINS

13.1 INTRODUCTION

Arthritis is common, crippling, chronic and incurable. Many patients respond poorly to conventional treatments and thus the pursuit of novel therapies is encouraged. For reasons defended elsewhere (Evans and Robbins, 1994a, b) we suggest the development of gene treatments for arthritis.

There are two contexts in which genes can be used as anti-arthritic agents. In the first of these, a gene or a combination of appropriate genes, is introduced locally into each diseased joint; at present, the synovial lining of the joint seems the best target tissue. Expression of transgenes within synovial cells leads to accumulation of the gene products either intracellularly or, if the product is a secreted protein, extracellularly within the joint space. With present technology, the transfer of genes encoding secreted, anti-arthritic proteins would seem to have a better chance of therapeutic success than those encoding intracellular proteins, antisense RNAs, decoy RNAs or ribozymes. Under these conditions, the synovium of the diseased joint then becomes the site of synthesis of the joint's anti-arthritic proteins. The advantages of this as a drug delivery system are many: it obviates the problems typically associated with delivering drugs, especially proteinaceous drugs, to joints, as well as limiting side effects and providing a chronicity of effect. Local gene delivery in this manner may be accomplished by *ex vivo* or *in vivo* techniques (Bandara *et al.*, 1992; Evans and Robbins 1994a, b).

In the second context of using genes as anti-arthritic agents, genes are not introduced locally into individual diseased joints, but instead are delivered

Molecular and Cell Biology of Human Gene Therapeutics
Edited by George Dickson
Published in 1995 by Chapman & Hall. ISBN 0 412 62550 4

Table 13.1 *In vitro* expression of *lacZ* gene in primary cultures of lapine synovial fibroblasts following transduction with various vectors

Vector	*Status of transgene*	*Promoter*	*LacZ⁺ cells (% of total)*
Retrovirus (MFG)	Integrated	LTR	20–30
Herpes simplex virus	Episomal	CMV	~1 (toxic)
Adenovirus	Episomal	CMV	100
Liposomes (DC-Chol)	Episomal	CMV	0–30
Plasmid DNA (control)	Episomal	CMV	0

LTR, Long terminal repeat; CMV, Cytomegalovirus

to extra-articular locations where the secreted gene products become systemically available. This is termed systemic delivery and may also be accomplished by *ex vivo* or *in vivo* techniques. Possible target cells include hematopoietic and mesenchymal stem cells, stromal cells, skeletal muscle cells, keratinocytes, hepatocytes and others (Evans and Robbins 1994a, b).

13.2 LOCAL GENE DELIVERY TO DISEASED JOINTS

13.2.1 Transducing synoviocytes *ex vivo* and *in vivo*

In the healthy joint, rates of synovial cell division are so low as to be difficult to measure (Mohr *et al.*, 1975; Coulton *et al.*, 1980; Lalor *et al.*, 1987). Although there is hyperplasia of the type B, fibroblastic synoviocytes in inflamed joints (Qu *et al.*, 1994), it remains likely that any vector for use in local, *in vivo* gene delivery will need to be able to transduce a population of cells where mitoses are infrequent. However, this low frequency of cell division may well permit the use of vectors which deliver DNA episomally.

As the *ex vivo* method involves cell division, especially if a selection step is included, vectors employed in this approach need to integrate their transgenes into the chromosomal DNA of the cell or provide episomal DNA that is capable of replication. Based upon these criteria, we have screened, or are in the process of screening, several different vectors for their ability to transduce rabbit synoviocytes *in vitro* (Table 13.1) and *in vivo* (Table 13.2) with the *lacZ* marker gene.

As shown in Table 13.1, adenovirus was clearly the most effective *in vitro* vector, able to convert 100% of the synovial fibroblasts to a lacZ$^+$ phenotype. The ICP4$^-$ herpes simplex virus (DeLuca *et al.*, 1985) proved toxic under our culture conditions, even at a multiplicity of infection of less than 1. Liposomes gave variable results. Cultures lipofected with the cationic liposome DC-Chol (Gao and Huang, 1991) contained anywhere from

Table 13.2 Expression of *lacZ* gene in rabbit synovium following intra-articular injection of vector

Vector	*LacZ⁺ staining*	
	Intensity[a]	*Duration*
Retrovirus (MFG)	0	0
Herpes simplex virus	+ + +	5–7 days
Adenovirus	+ + +	>1 month
Liposomes (DC-Chol)	+	1–2 days
Plasmid DNA	+/–	1–2 days

[a] Intensity of staining for β-galactosidase was assessed by subjective visual inspection on a scale of 0 to + + +

0–30% lacZ$^+$ cells in different experiments. The reasons for this variability are not known.

Of these vectors (Table 13.1), only the retrovirus efficiently integrates its genetic material into the genomic DNA of the host cell, and is thus the best suited for *ex vivo* gene transfer. Up to one-third of the synoviocytes were rendered lacZ$^+$ by a single infection with MFG-*lacZ*. As retroviruses require division of the host cell for infection and integration, the *in vitro* transduction efficiency was probably limited by the proportion of dividing cells in these unsynchronized cultures. We have also successfully transduced a rabbit synoviocyte cell line, HIG-82 (Georgescu *et al.*, 1988) with recombinant adeno-associated virus (AAV), suggesting a possible future use for AAV in the transfer of genes to synovium. Difficulties in generating large amounts of high-titer recombinant AAV presently limit progress in this direction.

Each of the vectors used in the *in vitro* experiments was then injected directly into the knee joints of healthy adult rabbits. At various times after injection, rabbits were killed, their knee joints carefully dissected and the internal surfaces of the knees stained for expression of β-galactosidase (Table 13.2). As expected the retrovirus was inactive in these normal knees, where cell turnover is low. However, both adenovirus and herpes simplex virus (HSV) gave very strong expression of β-galactosidase. This effect of HSV contrasts strongly with its toxic effects *in vitro* (Table 13.1). A similar dichotomy has been seen with neuronal cells (J. Glorioso, personal communication). However, expression following injection of HSV-*lacZ* was lost within a week, whereas that following injection of the adenoviral vector persisted for at least a month. Similar experiments by Roessler *et al.* (1993) have shown similar results for adenovirus, with expression of *lacZ* persisting for 8 weeks. Transient expression with HSV-*lacZ* probably reflects a combination of cytotoxicity and promoter shut-off.

Lipofection *in vivo* produced a low level of β-galactosidase activity which was transient. Further investigation revealed that the liposomes were preferentially taken up by the type A synoviocytes, which resemble macrophages. Lipofection of macrophages is known to be inefficient due to the rapidity with which these cells degrade exogenous DNA (Singhal and Huang, 1994). It was surprising to note a low, but real, level of *in vivo* transfection of synoviocytes with naked DNA. A similar effect has been noted in other laboratories (F. Ledley, personal communication).

Based upon these observations we decided to develop further the *ex vivo* delivery system using the retroviral vector, MFG, and the *in vivo* delivery system using HSV, adenovirus and liposomes. Although liposomes are presently inefficient gene delivery systems in joints (Table 13.2), their potential advantages as stable, non-viral, non-toxic, non-antigenic vectors are such as to encourage further development. Progress has been most rapid with the *ex vivo* system.

13.2.2 *Ex vivo* gene therapy in rabbit knee joints

As a potentially anti-arthritic gene product, we selected the interleukin-1 receptor antagonist protein (IL-1ra or IRAP; Arend, 1993; Evans and Robbins, 1994c) for our initial studies. A cDNA encoding the human form of this protein was inserted into the MFG retrovirus to produce the vector MFG-IRAP. Infection of HIG-82 cells or primary cultures of rabbit synoviocytes with MFG-IRAP led to the *in vitro* production of several hundred ng/10^6 cells of recombinant human IRAP during a 3-day incubation.

Transduced cells were introduced into the knee joints of recipient rabbits by intra-articular injection (Evans *et al.*, 1992). *In vivo* expression of human IRAP was then monitored by ELISA measurements of lavage fluid. An inoculum of 10^7 cells/knee led to the intra-articular accumulation of over 10 ng/knee of human IRAP 3 days later. With allografted cells, this value fell to zero within 2 weeks due to immune recognition of the allograft. However, with autografted cells, IRAP expression could be detected for as long as 5–6 weeks (Bandara *et al.*, 1993). Of the possible explanations for the temporal loss of IRAP expression, using autografted cells, promoter shut-off is the most likely. The time-course for the phenomenon resembles that seen by other investigators attempting to express transgenes driven by viral promoters in fibroblasts *in vivo* (Palmer *et al.*, 1991). Temporary gene expression is clearly a limitation that will need to be overcome before any clinical utilization of these methods in chronic arthritis.

Levels of IRAP produced intra-articularly by *ex vivo* gene delivery were sufficient to block the intra-articular pathologies that otherwise follow injection of interleukin-1 (IL-1) into the knee joints of rabbits. These pathologies included the influx of leukocytes into the joint space, the release of proteoglycan breakdown fragments from cartilage, and synovial hyperplasia and hypertrophy (Hung *et al.*, 1994). This result is quite surprising, because the degree of inhibition exceeded that predicted on theoretical grounds. It suggests that delivery of an anti-arthritic gene might indeed produce a better therapeutic effect than delivery of its cognate protein.

With this encouraging result, we are now in the process of evaluating IRAP gene therapy in antigen-induced arthritis in the rabbit. The literature does not encourage optimism on the outcome, as two groups have independently reported a negative result when using IRAP protein against antigen-induced arthritis in the mouse (Wooley *et al.*, 1993) and rabbit (Lewthwaite *et al.*,1993). Nevertheless, our preliminary data are most promising. In short-term studies, *ex vivo* transfer of the human IRAP gene has reproducibly inhibited the influx of inflammatory cells into the joint and, in particular, damage to the articular cartilage. These data again indicate the possible greater effectiveness of gene delivery over protein delivery. One unexpected, but welcome, finding was the higher level of human IRAP present in arthritic knees compared with control knees.

13.2.3 Human gene therapy for arthritis: *ex vivo* versus *in vivo*

At first sight, the *in vivo* method of gene delivery to diseased joints would seem to offer overwhelming advantages over *ex vivo* delivery in human joint diseases. The *ex vivo* method is invasive, costly, cumbersome, labor-intensive and requires sophisticated technologies which are probably only available at major tertiary care centers. For a common, polyarticular disease, such as rheumatoid arthritis, *in vivo* delivery using a vector which can be simply injected into joints in the surgery seems far preferable. However, closer inspection of these issues reveals underlying complexities which blur this simple comparison of *ex vivo* and *in vivo* methods.

The major concern is that of safety. Gene treatment of human arthritis means the introduction of novel genetic information into individuals suffering from a chronic condition which, although unpleasant, is not lethal. Present human gene therapy is limited to cancers, AIDS and fatal genetic diseases, where the patient is likely to die under standard medical treatments. Because patients with arthritis will not die from their disease, the safety issues surrounding gene therapy are of paramount importance.

Once a vector is injected directly into a joint, the investigator has little further control over the types of cells that will be transduced. If the population of cells within joints were static, this might not matter. However, there is a constant trafficking of lymphocytes through joints as part of surveillance by the immune system. This is likely to be greatly increased in inflammatory arthritides. Transduction of these cells as they traffic through the joint could lead to carriage of the transgene to untargeted organs, with consequences that are presently unpredictable.

A second consideration concerns the injection of viruses into joints.

Although viruses used for *in vivo* gene therapy have been genetically disabled, they are nonetheless antigenic. Their use thus runs the risk of sensitizing a naive individual to the viral epitopes, thus engendering an immune reaction each time subsequent re-injection of the vector is required. As many individuals show evidence of prior infection with wild-type adenovirus (which causes colds), AAV (which is asymptomatic), and HSV (which causes cold-sores), their first use as vectors in patients is likely to provoke an immune flare. In the *ex vivo* method using the present generation of retroviruses, no viral proteins are expressed in the transduced cells. Although infection of cells with retroviruses runs the theoretical risk of triggering insertional mutagenesis and related complications, these have never been observed to occur. Likewise, the production of replication-compentent virus through recombination seems extremely rare. In any case, all and any analyses that may be required can be performed on the transduced cells before their re-implantation. Obvious tests would include screening for recombinant, replication-proficient virus, the presence of adventitious agents and, perhaps, any growth abnormalities of the cells.

Harvesting donor cells for *ex vivo* delivery may also be less of a problem than it might seem. Synovial tissue could be harvested at the time of prior, scheduled arthroplasty or during surgical synovectomy. Cells could be banked and retrieved as and when needed. Alternatively it may be possible to recover tissue from an easily accessible site, such as the skin, and to transplant transduced cells to the synovium. This has never been tried, but there is no obvious *a priori* reason why it should not work.

In vivo delivery using liposomes or some other non-viral, non-antigenic, injectable system might provide the optimal method, once efficient transduction and sustained expression have been achieved.

13.3 SYSTEMIC GENE DELIVERY FOR ARTHRITIS

For polyarticular diseases, such as rheumatoid arthritis, treating each joint individually by local gene delivery would be tedious. Although this would be much less tedious than the present practice of replacing each joint surgically with a prosthesis, a more general method of delivering anti-arthritic genes would be helpful. As mentioned earlier, so-called systemic gene delivery would accomplish this. However, this increases safety concerns as the highest concentration of the gene produce is no longer within the diseased joint, as in local gene therapy, but elsewhere. Thus the selection of the therapeutic genes and the need to regulate their expression become key issues.

Our experience with systemic delivery of the IRAP gene in mice suggests that this approach can be safe. MFG-IRAP has been used to infect hematopoietic stem cells in mice (S. Boggs *et al.* submitted). Sera of mice treated in this way can contain human IRAP concentrations in excess of 200 ng/ml.

Yet the animals remain healthy, put on weight in the normal manner, do not fall victim to animal-room pathogens and live the normal lifespans of laboratory mice in Pittsburgh. Additional mice expressing a soluble tumor necrosis factor (TNF) receptor in the same manner have recently been generated. The susceptibility of these mice to arthritis is presently being evaluated.

Although transduction of hematopoietic stem cells permits proof of principle, the complexities of the technique limit its use both in laboratory animals and humans. Various simpler alternatives exist. Perhaps the simplest is the intramuscular injection of plasmid DNA (Wolff *et al.*, 1990). Raz *et al.* (1993) were able to regulate immune function in the MRL/1pr autoimmune mouse by this delivery method, but we have failed to obtain measurable levels of circulating human IRAP in mice using this approach.

Future use might be made of keratinocytes, skin fibroblasts, myoblasts, hepatocytes, stromal cells or cells trapped within organoids as means of systemic gene delivery for arthritis.

13.4 CONCLUSIONS AND FUTURE DIRECTIONS

Gene therapy holds considerable promise as a future treatment for human arthritis. Either local or systemic gene delivery may be used, employing *ex vivo* or *in vivo* techniques. Considerable success has already been achieved in treating arthritis in rabbits' knees by transfer of a gene encoding human IRAP. There are many other candidate genes for treating arthritis, including those encoding soluble IL-1 and TNF-α receptors, anti-adhesion molecules, and chondroprotective agents such as IGF-1, as well as the immunomodulatory and anti-inflammatory cytokines IL-4, IL-10 and IL-13 (Evans and Robbins 1994a, b). Future improvements will include prolonged, regulated expression of the transgenes.

Arthritis is not the only disorder of joints amenable to gene treatment. Additional applications include ligament repair (Evans *et al.*, 1993), cartilage repair and arthrofibrosis. Finally, the gene transfer concepts discussed in this chapter can be applied to generate novel animal models of joint disease.

ACKNOWLEDGEMENTS

The authors' work in this area has been supported by a UERP grant from Procter & Gamble, and grants RO1 DK46640 and PO1 DK44935 from NIDDK. We also thank Mrs Mary Lou Duerring for typing the manuscript.

REFERENCES

Arend, W.P. (1993) Interleukin-1 receptor antagonist. *Adv. Immunol.*, **54**, 167–227.

Bandara, G., Robbins, P.D., Georgescu, H.I. *et al.* (1992) Gene transfer to synoviocytes: prospects for gene treatment of arthritis. *DNA Cell Biol.*, **11**, 227–31.

Bandara, G., Mueller, G.M., Galea-Lauri, J. *et al.* (1993) Intraarticular expression of biologically active interleukin-1 receptor antagonist protein by *ex vivo* gene transfer. *Proc. Natl Acad. Sci. USA*, **90**, 10764–8.

Coulton, L.A., Henderson, B., Biterski, L. *et al.* (1980) DNA synthesis in human rheumatoid synovial lining. *Ann. Rheum. Dis.*, **39**, 241–7.

DeLuca, N.A., McCarthy, A. and Schaffer, P.A. (1985) Isolation and characterization of deletion mutants of herpes simplex virus type I in the gene encoding immediate–early regulatory protein ICP4. *J. Virol.*, **56**, 558–70.

Evans, C.H. and Robbins, P.D. (1994a) Gene therapy for arthritis, in *Gene Therapeutics: Methods and Applications of Direct Gene Transfer*, (ed. J.A. Wolff), Birkhauser, Boston, pp. 320–43.

Evans, C.H. and Robbins, P.D. (1994b) Prospects for treating arthritis by gene therapy. *J. Rheumatol.*, **21**, 779–82.

Evans, C.H. and Robbins, P.D. (1994c) The interleukin-1 receptor antagonist and its delivery by gene transfer. *Receptor*, **4**, 9–15.

Evans, C.H., Bandara, G., Mueller, G.M. *et al.* (1992) Synovial cell transplants for gene transfer to joints. *Transplantation Proc.*, **24**, 2966.

Evans, C.H., Bandara, G., Robbins, P.D. *et al.* (1993) Gene therapy for ligament healing, in *The Anterior Cruciate Ligament: Current and Future Concepts*, (eds S. Arnoczsky, S.L. Woo, C.B. Frank *et al.*), Raven Press, New York, pp. 419–22.

Gao, X. and Huang, L. (1991) A novel cationic liposome reagent for efficient transfection of mammalian cells. *Biochem. Biophys. Res. Commun.*, **179**, 280–5.

Georgescu, H.I., Mendelow, D. and Evans, C.H. (1988) HIG-82: an established line from rabbit periarticular soft tissue which retains the "activatable" phenotype. *In Vitro*, **24**, 1015–22.

Hung, G.L., Galea-Lauri, J., Mueller, G.M. *et al.* (1994) Suppression of intra-articular responses to interleukin-1 by transfer of the interleukin-1 receptor antagnoist gene to synovium. *Gene Therapy*, **1**, 64–9.

Lalor, P.A., Mapp, P.I., Hall, P.A. *et al.* (1987) Proliferative activity of cells in the synovium as demonstrated by monoclonal antibody, Ki-67. *Rheumatol. Int.*, **7**, 183–6.

Lewthwaite, J.C., Blake, S.M., Hardingham, T.E. *et al.* (1993) The effect of interleukin-1 receptor antagonist on the progression of antigen-induced arthritis in the rabbit. *Trans. Orthop. Res. Soc.*, **18**, 237.

Mohr, W., Beneke, G. and Mohing, W. (1975) Proliferation of synovial lining cells and fibroblasts. *Ann. Rheum. Dis.*, **34**, 219–24.

Palmer, T.D., Rosman, G.J., Osborne, W.B.A. *et al.* (1991) Genetically modified skin fibroblasts persist long after transplantation but gradually inactivate introduced genes. *Proc. Natl Acad. Sci. USA*, **88**, 1330–4.

Qu, Z., Garcia, C.H., O'Rourke, L.M. *et al.* (1994) Local proliferation of fibroblast-like synoviocytes contributes to synovial hyperplasia. *Arthritis Rheum.*, **37**, 212–20.

Raz, E., Watanabe, A., Baird, S.M. *et al.* (1993) Systemic immunological effects of cytokine genes injected into skeletal muscle. *Proc. Natl Acad. Sci. USA*, **90**,

4523–7.

Roessler, B.J., Allen, E.D., Wilson, J.M. *et al.* (1993) Adenoviral-mediated gene transfer to rabbit synovium *in vivo*. *J. Clin. Invest.*, **92**, 1085–92.

Singhal, A. and Huang, L. (1994) Gene transfer in mammalian cells using liposomes as carriers in *Gene Therapeutics: Methods and Applications of Direct Gene Transfer*, (ed. J.A. Wolff), Birkhauser, Boston, pp. 118–42.

Wolff, J.A., Malone, R.W., Williams, P. *et al.* (1990) Direct gene transfer into mouse muscle *in vivo*. *Science*, **247**, 1465–8.

Wooley, P.H., Whalen, J.D., Chapman, D.L. *et al.* (1993) The effect of an interleukin-1 receptor antagonist protein on type II collagen-induced arthritis and antigen-induced arthritis in mice. *Arthritis Rheum.*, **36**, 1305–14.

14

Duchenne muscular dystrophy

GEORGE DICKSON and SUSAN C. BROWN

14.1 DUCHENNE MUSCULAR DYSTROPHY

Duchenne and Becker muscular dystrophy are X-linked muscle-wasting disorders that arise from mutations in the gene coding for dystrophin. The incidence of Duchenne muscular dystrophy (DMD) is approximately 1 in 3500 live male births, one-third of which are sporadic with no previous family history. In the absence of dystrophin, patients with DMD exhibit progressive muscle fiber necrosis which, despite the regenerative capacity of the muscle during the early stages, leads to the gradual deposition of fibrotic and fat tissues and the loss of ambulation around 11 years of age. Death due to respiratory failure usually occurs by the third decade (Emery, 1993). Becker muscular dystrophy (BMD), the incidence of which is approximately 10-fold less than DMD, follows a generally much milder course due to the presence of a partially functional form of dystrophin. Generally the BMD phenotype tends to be much more varied, both with regard to the age of onset and clinical presentation, and some patients never lose the ability to walk.

Several animal models of DMD have been identified, all of which lack dystrophin and display repeated cycles of muscle fiber necrosis and regeneration (Partridge,1991). In the *cxmd* dog the myopathy follows a similar course to that in humans, with fibrous and fatty infiltration of the muscle structure leading to severe disability and premature death due to respiratory insufficiency (Valentine *et al.*, 1990). By contrast, effective regeneration and the absence of fibrosis in the *mdx* mouse (Bulfield *et al.*, 1984) results in clinical and pathological features which are remarkably mild compared with DMD. The only exceptions to this are the diaphragm and intercostal

Molecular and Cell Biology of Human Gene Therapeutics
Edited by George Dickson
Published in 1995 by Chapman & Hall. ISBN 0 412 62550 4

muscles which in older mice exhibit a myopathy comparable with the human disease (Stedman *et al.*, 1991). Dystrophin deficiency in the cat leads to marked diaphragmatic and glossal hypertrophy (Gaschen *et al.*, 1992), but unlike the human disease and the *cxmd* dog, there is no progressive loss of muscle fibers, fibrosis or weakness. These differences between species limit the usefulness of models such as the *mdx* mouse for therapeutic testing, but provide some insight into the immediate as opposed to the secondary consequences of dystrophin deficiency (Partridge, 1991).

14.2 THE DYSTROPHIN GENE

The dystrophin gene is the largest identified so far and extends over 2400 kb, of which 14 kb is transcribed into an mRNA comprising 79 exons (Roberts *et al.*, 1993) (Figure 14.1a). Transcription is controlled in a cell-specific and developmentally controlled manner by at least five independent promoters (Ahn and Kunkel, 1993). Three of these promoters, namely the cortical (C), muscle (M) and Purkinje cell (P) encode full-length forms of dystrophin, while the Schwann cell (S) and general or glial (G) encode truncated forms and are referred to as Dp116 and Dp71 respectively (Figure 14.1a, d). These truncated forms have a wide tissue distribution as outlined in Table 14.1. The alternative splicing of six exons of the C-terminus generates further diversity (Feener *et al.*, 1989; Barnea *et al.*, 1990; Blake *et al.*, 1993), and may indicate different roles for products of the dystrophin gene in various locations (Ahn and Kunkel, 1993). However, this remains speculative as at the present time the function of dystrophin is unknown.

A large proportion of DMD cases (60%) are due to partial gene deletions, 40% of which have breakpoints mapping to a single large intron between exons 44 and 45 (Ahn and Kunkel, 1993). The most important factor in determining the severity of the phenotype (i.e. whether the dystrophy is DMD or BMD) is not the size of the deletion, but whether it causes a shift in the open reading frame of the mRNA transcript (Monaco *et al.*, 1988; Koenig *et al.*, 1989). Deletions that cause a shift in the open reading frame, and the synthesis of a C-terminally truncated form of dystrophin, generally lead to DMD (Monaco *et al.*, 1988). By contrast, the milder BMD phenotype results from deletions which leave the open reading frame intact and which therefore permit the synthesis of an internally deleted form of dystrophin that is at least partially functional.

Exceptions to this general rule do occur and are attributed to variant alternative splicing patterns or the skipping of juxtaposed exons that are able to overcome the effect of the original deletion on the open reading frame. Other less common types of mutation include gene duplications which may account for 5–10% of DMD cases (Emery, 1993) and point mutations which, by either causing a stop codon or an amino acid substitution, can lead to severe disease.

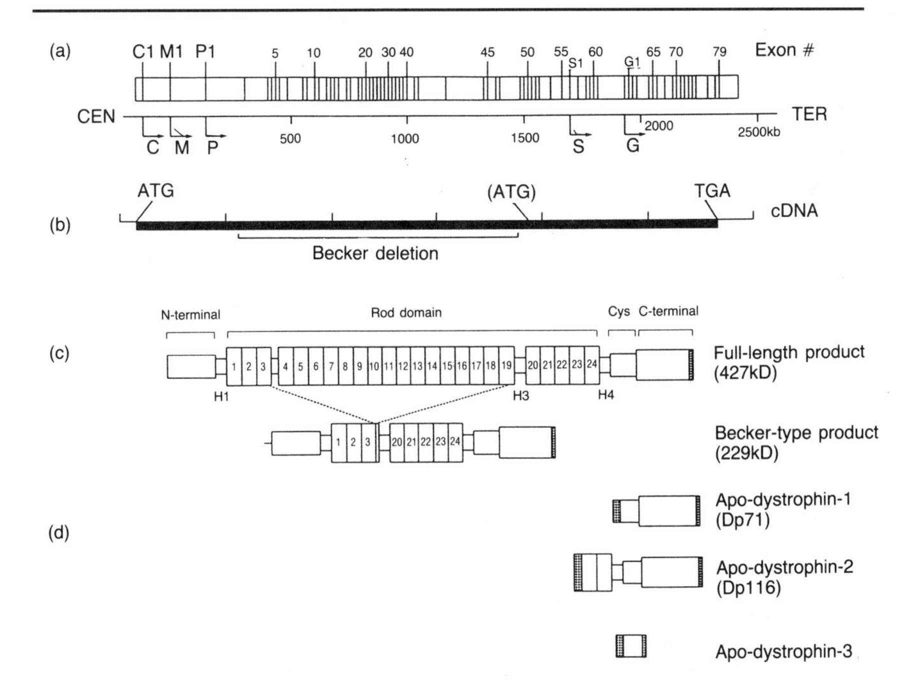

Figure 14.1 Diagrammatic representation of the DMD gene and its products. The 79 exons span 2.4 million bases. At least five distinct promoters drive the cell type-specific expression of different dystrophin isoforms (a). The C (cortical), M (muscle) and P (Purkinje cell) promoters all express full-length forms of dystrophin, while the S (Schwann cell) and G (general or glial) dystrophin promoters encode the C-terminal apo-dystrophins Dp116 and Dp71 respectively. (b) Diagrammatic representation of human dystrophin full-length cDNA indicating the extent of the central deletions in the BMD mini-dystrophin gene. The domain organization of the resulting protein is shown in (c). The diagram of the domain organization of full-length dystrophin shows: (i) the N-terminal region (N-ter) which binds to F-actin; (ii) the long internal region consisting of 24 homologous repeats of approximately 109 amino acids which form a long rod-like domain; (iii) the cysteine-rich region (Cys) which is homologous to the Ca^{2+}-binding EF hand region of α-actinin; and (iv) the C-terminal domain (C-ter) which is homologous only to the dystrophin-related protein, utrophin. The four putative hinge structures in the rod domain (H1–H4), and the occurrence of alternatively-spliced regions at the extreme N- and C-termini of dystrophin and apo-dystrophins are indicated as filled areas. (d) Diagrammatic representation of the C-terminal apo-dystrophins. Apo-dystrophin-1 (Dp 71) and apo-dystrophin-3 are expressed in non-muscle tissues, including brain, lung, liver and kidney. Apo-dystrophin-2 (Dp116) is expressed in peripheral nerve tissue.

Table 14.1 Expression of dystrophin isoforms and homologs

Tissue	*427 kDa*	*Dp71*	*Dp116*	*Utrophin*
Muscle				
Skeletal	+	–	–	+/–
Cardiac	+	+		+
Smooth	+	nd	–	+
Organs				
Kidney	–	+	–	+
Liver	–	+	–	+
Lung	–	+	–	+
Pancreas	–	+	–	+
Testis	–	+	–	+
Central nervous system				
Neurons	+	nd	–	+
Glia	+	+	–	+
Peripheral nervous system				
Retina	+	–	–	nd
Neurons	nd	nd	–	nd
Schwann cell	–	–	+	+
Cochlea	+	nd	+	nd

14.3 DYSTROPHIN

14.3.1 Structural domains

Dystrophin is a large cytoskeletal protein with a molecular weight of 427 kDa, found beneath the plasma membrane of all muscle types and some neurons (Ahn and Kunkel, 1993; Hoffman *et al.*, 1987). The amino acid sequence of dystrophin suggests that it is composed of four contiguous domains namely (Figure 14.1c): (i) a N-terminal region of 240 amino acids, which is similar to the conserved actin-binding domain of α-actinin, spectrin and *Dictyostelium* actin binding protein 120, and which may contain two putative actin-binding sites (Hemmings *et al.*, 1992; Levine *et al.*, 1992), although recent work suggests that the major site is in the N-terminal 90 amino acids (Corrado *et al.*, 1994); (ii) a large domain comprising 24 triple-helical sequences which resemble the repeats in spectrin, and are interrupted by four proline-rich and proteolytically-sensitive hinge regions which may confer flexibility (Koenig and Kunkel, 1990); (iii) a cysteine-rich region between residues 3080 and 3360 which shows significant homologies to the C-terminal domain of α-actinin in *Dictyostelium discoideum* and may contain two calcium-binding sites although this is by no means proven; and (iv) a sequence of 420 amino acids at the C-terminal end which appears to be unique apart from a similarity to the carboxyl terminus of the dystrophin-related protein (DRP) or utrophin encoded on chromosome 6

(Love *et al.*, 1989). Based upon the amino acid sequence similarity to spectrin, it was predicted that dystrophin would form a flexible rod-shaped molecule with a length of 125 nm (Koenig *et al.*, 1988). This has since been confirmed by rotary-shadowed electron microscopy of isolated dystrophin (Sato *et al.*, 1992).

14.3.2 Distribution

Dystrophin locates to the cytoplasmic face of the normal adult muscle fiber membrane (Byers *et al.*, 1991), where it comprises approximately 5% of the membrane cytoskeleton (Ohlendieck and Campbell, 1991). As well as being present at both the neuromuscular and myotendinous junctions (Byers *et al.*, 1991; Ohlendieck *et al.*, 1991), dystrophin is specifically enriched in three distinct domains along the muscle fiber plasma membrane, namely in transverse rib-like structures overlying both the I bands and M lines of peripheral sarcomeres, and in occasional strands running along the longitudinal axis of the fiber (Porter *et al.*, 1992). This distribution is similar to the costameric lattice of skeletal muscle fibers which is thought to mediate the trans-membrane interaction of contractile elements with the extracellular basal lamina (Pardo *et al.*, 1983; Minetti *et al.*, 1992; Porter *et al.*, 1992). In cardiac muscle dystrophin localizes to the plasma membrane but is notably absent from the areas which overlay the adherens junctions of the intercalated disks (Byers *et al.*, 1991). The distribution is similar, though less abundant, in smooth muscle where dystrophin is excluded from the membrane above densities that mark adherens junctions (Byers *et al.*, 1991).

14.3.3 Association with other proteins

Correlations between domain-specific 'in frame' deletions with phenotype have provided important information regarding the functional importance of different domains of the dystrophin molecule. For example, whereas mutations in the putative actin binding domain tend to be associated with a severe or intermediate BMD phenotype, the absence of the cysteine-rich and proximal half of the carboxyl terminal domain invariably leads to a severe DMD phenotype (Koenig *et al.*, 1989). Therefore, although both the N- and C-terminal regions are known to independently localize to the membrane (Dunckley *et al.*, 1994), it is the interaction between the cysteine-rich and C-terminal domains of dystrophin and the membrane that are of the greatest functional importance (Suzuki *et al.*, 1992a; Dunckley *et al.*, 1994).

The cysteine-rich and first half of the C-terminal domains of dystrophin associate with the membrane via a large oligomeric complex of glycoproteins and proteins (Ervasti and Campbell, 1991; Suzuki *et al.*, 1992a, b; Ahn and Kunkel, 1993). These dystrophin-associated glycoproteins (DAGs) and proteins (DAPs) are referred to according to their molecular weights of

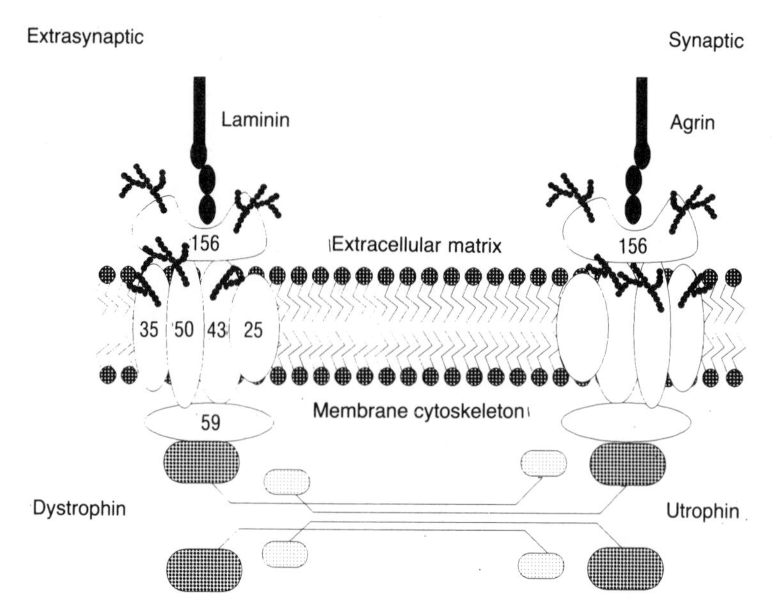

Figure 14.2 A model for the interaction of dystrophin and utrophin with the extrasynaptic and synaptic regions of the plasma membrane in skeletal muscle. The N-terminal region of dystrophin/utrophin interacts with cytoskeletal actin filaments while the cysteine-rich and C-terminal domains bind to a complex of membrane-spanning glycoproteins (DAGC). α-Dystroglycan (156 kDa component) binds to both laminin and agrin, indicating that a common linkage may act to stabilize the membrane cytoskeleton along the length of the muscle fiber.

156 kDa-DAG, 59 kDa-DAP, 50 kDa-DAG, 43 kDa-DAG, 35 kDa-DAG and 25 kDa-DAP (Ervasti and Campbell, 1991) (Figure 14.2). The integral membrane glycoproteins, namely the 50 kDa-DAG (adhalin), 43 kDa-DAG (β-dystroglycan) and the 35 kDa-DAG bind to the cysteine-rich and C-terminal regions of dystrophin (Suzuki *et al.*, 1992b). The 59 kDa-DAP is a peripheral membrane protein triplet associated with the cytoplasmic face of the complex, and may also directly associate with dystrophin (Ahn and Kunkel, 1993; Yang *et al.*, 1994a). Recent work shows this DAP to be a heterogeneous group of phosphorylated proteins consisting of an acidic (α-A1) and distinct basic component (β-A1) (Ahn *et al.*, 1994).

The 156-kDa component (α-dystroglycan) is a peripheral surface proteoglycan (Ervasti and Campbell, 1991) that is able to bind to both laminin (Dickson *et al.*, 1992; Ibraghimov-Beskrovnaya *et al.*, 1992), and agrin (Campanelli *et al.*, 1994; Gee *et al.*, 1994) in the extracellular matrix. This has led to the hypothetical model shown in Figure 14.2 where α-dystroglycan binds to laminin along the length of the fiber, except at the neuromuscular junction where utrophin interacts with agrin via a complex, identical or similar to that which associates with dystrophin. The hypothesis is that this arrangement may function to allow extracellular ligands such as laminin and agrin to stabilize the extrasynaptic and synaptic regional specializations of the membrane cytoskeleton (Campanelli *et al.*, 1994). While these studies are based upon data obtained from skeletal muscle, an identical or antigenically similar DAG complex also exists in cardiac muscle, where it locates to the transverse tubules as well as to the plasma membrane (Ervasti and Campbell, 1993; Klietsch *et al.*, 1993).

Provided the C-terminal domains are synthesized correctly, deletions in the rod domain of dystrophin give rise to BMD phenotype (Koenig *et al.*, 1989). This may indicate that the rod domain merely functions as a 'spacer' to separate the functionally important N- and C-terminal domains. However, there are indications that phenotype may be influenced by the position of the deletion in the rod domain, i.e. proximal, central or distal. This could be due to a number of factors such as deletions in this region altering either the elasticity of dystrophin (Cross *et al.*, 1990), or its ability to mediate interactions with itself or other proteins (Sato *et al.*, 1992). However, this is as yet unproven and in the absence of more detailed information regarding the function of this region, and indeed the whole molecule, patients with mild phenotypes effectively define the minimal requirements for a recombinant dystrophin protein.

14.4 FUNCTION OF DYSTROPHIN IN SKELETAL MUSCLE

The absence of dystrophin in DMD leads to an approximate 80–90% reduction in all components of the dystrophin-associated complex in the extrasynaptic regions of the muscle fiber membrane (Matsumura and Campbell, 1993). An absence or reduction of this complex is also a feature of another form of muscular dystrophy, namely severe childhood autosomal recessive muscular dystrophy (SCARMD). SCARMD is characterized by a DMD-like phenotype despite the normal expression of dystrophin, and is associated with a specific absence of the 50 kDa (adhalin) component of the complex (Matsumura *et al.*, 1992b). This leads to a moderate reduction in the 35 kDa DAG during the early stages of the disease, but later results in a reduction in all components of the complex, including dystrophin (Matsumura and Campbell, 1993). The fundamental cause of muscle fiber necrosis in both DMD and SCARMD appears to be the disruption of the dystrophin-associated complex (Ahn and Kunkel, 1993). That other forms of muscular dystrophy will be found to be due to defects in various components of this complex seems likely in view of recent findings of a specific deficiency of merosin, the laminin M chain in a classical non-Japanese form of congenital muscular dystrophy (Tome *et al.*, 1994) and dy (dystrophia muscularis) mice (Sunada *et al.*, 1994; Xu *et al.*, 1994). Thus each component of the dystrophin-associated complex becomes a candidate for identifying the source of neuromuscular disease (Ahn *et al.*, 1994).

In spite of the aforementioned observations functional evidence regarding the precise cellular role of dystrophin and its associated complex is currently lacking. One long-standing hypothesis is that in the absence of dystrophin sarcolemmal integrity is compromised, so as to render the fiber more susceptible to mechanical damage during normal contractile activity (Weller *et al.*, 1990; Stedman *et al.*, 1991). However, the data are so far contradictory, possibly due to the fact that different exercise regimes and various parameters have been used to assess fiber damage (Hutter, 1992). Direct measurements of the tensile strength of the membrane of either cultured myotubes, isolated muscle fibers, or sarcolemmal vesicles have failed to reveal any significant differences between dystrophin-deficient membrane and controls (Hutter, 1992). The only significant finding has been that *mdx* myotubes and muscle fibers *in vitro* are more susceptible to lysis when exposed to hypotonic solutions than are normal controls (Menke and Jockusch, 1991). However, these observations have not as yet been confirmed by others.

An alternative hypothesis is that dystrophin and its associated complex may act to organize membrane specializations, such as those involved in ion channel regulation (Turner *et al.*, 1991). This is supported by observations of abnormally elevated levels of total calcium in DMD muscle as determined by histochemical staining and X-ray microanalysis (Emery, 1993). In addition a number of studies using fluorescent Ca^{2+} probes have reported elevated levels of intracellular free calcium in myotubes derived from the muscle of *mdx* mice and DMD patients compared with normal myotubes (Franco and Lansman, 1990; Turner *et al.*, 1991; Bakker *et al.*, 1993). This influx might occur either through lesions in the membrane caused by the absence of dystrophin or alternatively through a channel whose function has been compromised by the absence of dystrophin (Franco and Lansman, 1990). However, a direct link between the dystrophin and its associated complex and the activity of calcium channels in the muscle fiber membrane remains to be shown.

14.5 THERAPEUTIC STRATEGIES FOR DUCHENNE MUSCULAR DYSTROPHY

The frequency and severity of DMD, together with the fact that about one-third of the cases are sporadic, has necessitated the need for efficient therapeutic approaches that will compensate for the deleterious effect of the absence of dystrophin and prevent the recurrent phases of degeneration that are so characteristic of DMD. The approaches that have been adopted so far fall into two broad categories, namely myoblast transfer and gene transfer using direct DNA injection or viral vectors.

14.5.1 Myoblast transfer

Mononucleated satellite cells fuse with muscle fibers that are either actively growing or undergoing repair. Consequently the intramuscular injection of genetically engineered or donor satellite cells should be a highly efficient way of introducing the dystrophin gene into dystrophic muscle. Experiments in *mdx* mice have lent support to this idea by showing that the fusion of normal satellite cells with existing *mdx* fibers results in the expression of dystrophin and increased fiber survival (Morgan, 1994).

However, to date the efficiency of myoblast transfer in DMD patients as judged by maximal voluntary force generation, dystrophin content of the muscle and the absence of donor-derived DNA and dystrophin mRNA in the injected muscle, has been poor (Engel, 1993; Karpati *et al.*, 1993a). A number of contributory factors have been cited, such as the poor mobility of the injected cells through connective tissue barriers within the muscle, a failure of myoblasts to fuse with non-regenerating muscle fibers, the poor proliferative capacity of the donor myoblasts, and immune rejection of the donor cells (Engel, 1993; Karpati *et al.*, 1993a). These problems, together with the general inaccessibility of some muscles, such as the diaphragm and heart, suggest that the clinical usefulness of myoblast transplantation may be limited.

14.5.2 Recombinant dystrophin genes

The construction of cDNA-based genes encoding recombinant human and mouse dystrophins provided the first steps towards evaluating direct DNA-mediated gene transfer as a potential treatment for DMD (Dickson *et al.*, 1991; Lee *et al.*, 1991). Full-length recombinant cDNAs encoding human (Dickson *et al.*, 1991) and mouse (Lee *et al.*, 1991) skeletal muscle dystrophin were initially cloned into constitutive high level eukaryotic expression plasmids, which following transfection into 3T3 fibroblasts and COS cells, were shown to encode recombinant dystrophin polypeptides of appropriate size (427 kDa) corresponding to the major form of dystrophin observed in adult muscle. Full-length mouse (Cox *et al.*, 1993) and human cDNAs (G. Dickson *et al.*, unpublished data) have now been expressed in transgenic *mdx* mice using muscle-specific M-creatine kinase or skeletal actin promoters. Both the human and mouse recombinant dystrophins correct the early necrotic and later fibrotic pathophysiology in *mdx* skeletal muscle.

However, the size of full-length dystrophin cDNAs (>12 kb) present some problems for somatic gene transfer studies in that it is above the packaging limits for adenovirus-based vectors, and is widely believed but not yet proven, to be too large for efficient retroviral vector encapsidation. One possible solution to this is to use recombinant cDNA constructions based upon the dystrophin gene of BMD patients that have been shown to produce an internally deleted but partially functional form of dystrophin.

One such patient has a deletion of 5106 nucleotides of coding sequence, that effectively removes 63% of the rod domain, but which still results in a relatively mild phenotype (England *et al.*, 1990) (Figure 14.1b, c).

The cDNA corresponding to this BMD minidystrophin has been reconstructed (Acsadi *et al.*, 1991), and examined by plasmid, retroviral and adenoviral-mediated somatic gene transfer into *mdx* muscle (Acsadi *et al.*, 1991; Dunckley *et al.*, 1992, 1993; Ragot *et al.*, 1993; Vincent *et al.*, 1993). Transgenic *mdx* mice were also made by microinjection of the construct into fertilized mouse ova (Wells *et al.*, 1992). In each case the internally deleted form of dystrophin localized to the plasma membrane, and in the transgenic mice its expression was correlated with a significant reduction in muscle fiber necrosis and regeneration. Moreover, the severe fibrotic changes that normally occur in the diaphragm of older *mdx* mice were prevented (Wells *et al.*, 1992; G. Dickson *et al.*, unpublished data), thus demonstrating that the minigene directed the expression of a functional dystrophin molecule.

14.5.3 Physical gene transfer

The direct intramuscular injection of purified plasmid DNA is perhaps one of the simplest techniques for mediating gene transfer into mammalian muscle fibers. Direct intramuscular injection of the luciferase gene has been shown to result in gene expression for at least 12 months in individual muscle fibers (Wolff *et al.*, 1990, 1991). Similarly, the injection of dystrophin gene constructs into *mdx* skeletal muscle results in the expression of recombinant dystrophin in clusters of muscle fibers around the injection site for at least 3 months (Acsadi *et al.*, 1991; Danko *et al.*, 1993; Dickson and Dunckley, 1993).

However, success with DMD will ultimately be measured in terms of the ability to transduce large numbers of muscle fibers and only between 1–2% of myofibers display a dystrophin-positive phenotype following a single injection of DNA into the *rectus femoris* of *mdx* mice (Acsadi *et al.*, 1991; Dickson and Dunckley, 1993). While there is evidence that this efficiency may be influenced by the developmental and physiological status of the targeted muscle (Wells and Goldspink, 1992; Davis *et al.*, 1993; Wells, 1993), it is of too low a level to be of potential clinical use at the present time. Nonetheless, the unusual stability of direct DNA-mediated gene transfer into skeletal muscle combined with its unique advantages in terms of simplicity and safety, makes it likely that increasingly efficient procedures will be developed in the future.

14.5.4 Virus-mediated gene transfer

A large number of viruses have been considered with a view to their use as potential high-efficiency somatic gene transfer vectors. However, the most studied systems and those upon which almost all human gene therapy protocols are currently based, are those of the disabled murine retrovirus (Morgan and Anderson, 1993) or of replication-defective adenovirus (Kozarsky and Wilson, 1993).

Recombinant retroviruses are able to stably introduce genes into the chromosomal DNA of many cell types, and are also able to simultaneously infect large numbers of cells (Eglitis *et al.*, 1985; Miller, 1992). Efficient retroviral-mediated gene transfer is however, dependent on host cell division which, in view of the postmitotic nature of muscle fiber nuclei, means that muscle transduction must be effected via infection of activated satellite cells. This has nonetheless been shown to be a feasible route of transfer *in vitro* and *in vivo* (Dunckley *et al.*, 1992, 1993; Dunckley and Dickson, 1994).

One further limitation of retroviral vectors is their limited ability to accept foreign DNA. However, as discussed above recombinant cDNA constructions based upon those which give rise to a mild BMD phenotype can be much smaller yet lead to the expression of a form of dystrophin that is at least partially functional. A single intramuscular injection of retrovirus containing the 6.3-kb human BMD dystrophin minigene has been shown to be sufficient to transduce up to 6% of *mdx* muscle fibers (Dunckley *et al.*, 1993). Expression of the mini-dystrophin was observed for over 9 months and was found to be associated with the reappearance of a least one component (43 kDa) of the dystrophin-associated complex (Dunckley *et al.*, 1993). Experimentally induced regeneration enhanced the efficiency of the transduction up to 12% of fibers, as might have been expected if activated satellite cells were the means by which the dystrophin was introduced into the muscle. This being the case then one long-term therapeutic approach for DMD might be to establish a permanent endogenous retrovirally-transduced stem cell population, the maintenance of which could be enhanced by previously described implant systems (Gilbert *et al.*, 1993).

Adenovirus infects most cell types with no requirement for cell division. This, together with the fact that it can be prepared in high titers and shows a high efficiency of gene transfer, make it a promising system for gene therapy. Adenovirus type 2 and type 5 are those currently being used for *in vivo* gene transfer into lung, liver, heart and skeletal muscle (Quantin *et al.*, 1992; Rosenfeld *et al.*, 1992; Stratford-Perricaudet *et al.*, 1992; Zabner *et al.*, 1993; Barr *et al.*, 1994). The transferred adenoviral genome appears to remain as an episomal element in infected cells until it is lost either through normal cell turnover or as the result of host immune responses. However, if problems associated with host immune surveillance can be circumvented, very high titer recombinant adenovirus preparations can be used to transduce cells without the requirement for division.

Adenoviral vectors have a limited ability to accept foreign DNA. At the present time the amount of foreign DNA that can stably be incorporated in an adenoviral vector is less than 7 kb. This therefore excludes the possibility of transferring full-length dystrophin cDNAs using these vectors. The BMD mini-dystrophin gene has, however, been cloned into a replication-defective adenoviral vector and injected into *mdx* skeletal muscle (Ragot *et al.*, 1993). Between 5–50% dystrophin-positive fibers have been reported following injections of very large numbers of adenoviral particles ($\sim 10^9$) in neonatal *mdx* mice. Furthermore, genetically-corrected fibers appeared to be stable (Vincent *et al.*, 1993) whereas control *mdx* myofibers infected only with a β-galactosidase expressing virus were lost, suggesting that satellite cells were not being transduced. While the initial results using adenoviral-mediated mini-dystrophin gene transfer in neonatal *mdx* appear to be most encouraging, long-term stability and high efficiency of gene transfer has not been achieved in older animals where inflammatory responses to adenoviral vectors have been observed (H. Davis, personal communication).

14.5.5 Utrophin as a substitute for dystrophin

The clinical pathophysiology in DMD arises largely from the progressive degeneration and loss of dystrophin-deficient skeletal myofibers (Wessel, 1990; Emery, 1993). Fiber necrosis is also a feature of *mdx* muscle but leads to only mild muscle dysfunction due to efficient muscle regeneration, lack of fibrosis and the distinct attenuation of myonecrosis in older mice (Partridge, 1991). The molecular basis for these differences in phenotype between DMD and the *mdx* mouse are unclear but may be due to the up-regulation of one or more proteins that can function similarly to dystrophin. In support of this, recent work has shown that vinculin, talin and utrophin are all up-regulated at the myotendinous junction of adult *mdx* mice relative to controls (Law *et al.*, 1994). It is, however, utrophin which has generated the most interest due to its extensive homology with dystrophin.

Utrophin is the 397-kDa product of a large gene spanning almost 1 Mb on the long arm of the human chromosome 6 and mouse chromosome 10. The actin-binding domains of dystrophin and utrophin have 85% similarity and the putative DAG binding regions have 88% similarity (Tinsley and Davies, 1993). In human fetal muscle utrophin attains a maximum level of expression around 17–18 weeks' gestation, after which it is down-regulated to negligible levels by 26 weeks, seemingly replaced by dystrophin (Clerk *et al.*, 1993). Many adult tissues express utrophin, but in normal skeletal muscle its distribution is restricted to the neuromuscular and myotendinous junctions, peripheral nerves and the vasculature. However, regenerating muscle fibers of DMD/BMD patients and *mdx* mice display a general staining of the muscle fiber plasma membrane (Khurana, 1991; Karpati *et al.*, 1993b; Love *et al.*, 1993).

It does nonetheless remain to be shown whether the re-expression of utrophin can lead to the restoration of the linkage between the muscle fiber cytoskeleton and the extracellular matrix. Comparisons between *mdx* muscles showing variable degrees of fiber necrosis, initially seemed to suggest that this might be the case. Elevated levels of utrophin were found to be correlated with normal levels of at least some components of the dystrophin-associated complex in cardiac and small-caliber skeletal muscles (Matsumura *et al.*, 1992a). However, more recent work based on a comparison of the extensor *digitorum longus*, soleus and diaphragm in the *mdx* mouse failed to find any correlation between the relative amount of utrophin and the severity of the pathology (Pons *et al.*, 1994). Similar comparisons made by others seem to confirm these results (Takemitsu *et al.*, 1994).

Heterogeneity in the dystrophin-associated complex in regard to both composition and function may be one reason for the inability of utrophin to substitute for dystrophin. For example, the up-regulation of utrophin in *mdx* cardiac muscle, preserves all dystrophin-associated proteins except the 59-kDa DAP, suggesting that this component may associate more specifically with dystrophin than utrophin (Yang *et al.*, 1994a). Furthermore the expression of utrophin in DMD muscle seems to be able to preserve the 43-kDa DAG to a greater extent than the 50 kDa component (Mizuno *et al.*, 1994).

14.6 FUTURE PERSPECTIVES

The identification and cloning of the DMD gene has led to major advances in diagnostic and genetic counseling procedures. However, due to the high spontaneous mutation rate of the gene new cases will continually arise making the search for some form of therapy a high priority. In spite of quite detailed knowledge regarding the distribution of dystrophin and its related gene products, the functional role of this protein and associated complex is still unknown. This information may prove to be of fundamental importance in the future, both in regard to the design of dystrophin gene constructs for incorporation into viral vectors, and also with respect to the possibility of up-regulating other cytoskeletal proteins to compensate for the absence of dystrophin.

Since it remains unclear as to why dystrophin deficiency leads to muscle fiber necrosis effective pharmacological intervention in this disease has not been possible and most work in recent years has focused on the development of methods for introducing functional dystrophin into dystrophic muscle fibers. Three main approaches have been adopted, namely: (i) the transplantation of myoblasts (Partridge *et al.*, 1989); (ii) the direct injection of expression plasmids bearing dystrophin cDNAs (Acsadi *et al.*, 1991); and (iii) the transfer of functional recombinant genes by viral vector systems (Dunckley *et al.*, 1992, 1993; Kozarsky and Wilson, 1993; Acsadi *et al.*, 1994). The overall efficiency of these various procedures in mature muscle is at the present time too low to offer any clinical potential. This work has nonetheless afforded some insight into the problems associated with each procedure and in doing so has provided some indication as to the direction of further work.

Myoblast transfer in DMD patients has so far produced disappointing results (Karpati *et al.*, 1993a), and direct injection of expression plasmids into human subjects in the near future is unlikely due to the overall low efficiency and the multiple injection sites that would be required. The trend has therefore been towards the development of more efficient viral vectors. This approach has the advantage of offering the opportunity to use a systemic mode of delivery (Zhu *et al.*, 1993), which while requiring high titers would nonetheless be more appropriate for a tissue as widely distributed as skeletal muscle. However, there are disadvantages such as the induction of an immune response that would make repeated administration impossible. Improvements in vector design may partially overcome this problem, since it has recently been shown that E1 deleted recombinant adenoviruses that also contain a temperature-sensitive mutation in the E2a gene are able to give rise to substantially longer recombinant gene expression and less inflammation when introduced into the lungs of mice (Yang *et al.*, 1994b). A further problem associated with systemic delivery is the possibility of retroviral vectors altering the germ-line genome which would be unacceptable on ethical grounds. One way of circumventing this would be to target skeletal muscle specifically. The envelope proteins of standard retroviral vectors confer widespread cell type infectivity in most mammalian species, but it may be possible to modify the envelope protein specificities, by fusion to specific cell-binding ligands such as immunoglobulins or cell adhesion molecules. In the case of the adenoviral vectors, while it remains theoretically possible that the tropism of this viral system could be specified by genetically engineering the viral coat proteins, this represents an immense technical challenge at present.

Transgene activity could alternatively be controlled by the use of muscle-specific promoter/enhancer elements. For example, it has been shown that large 2–6-kb fragments of the muscle creatine kinase and skeletal actin genes are able to direct muscle-specific dystrophin expression in transgenic *mdx* mice (Cox *et al.*, 1993; G. Dickson *et al.*, unpublished data). Smaller regions from these types of regulatory elements, and indeed the promoter/enhancer structures of the DMD gene itself (Barnea *et al.*, 1990) may retain appropriate tissue specificity and be sufficiently small to allow inclusion in retroviral and adenoviral dystrophin cDNA constructions. The design and production of viral gene transfer vectors which allow widespread transgene dissemination but which possess appropriate cell-type specific tropism and/or gene transcriptional activity are one of the most important areas of gene therapy research, and will be a vital component in the future development of gene therapeutic strategies for the treatment of DMD.

REFERENCES

Acsadi, G., Dickson, G., Love D.R. *et al.* (1991) Human dystrophin expression in *mdx* mice after intramuscular injection of DNA constructs. *Nature*, **352**, 815–18.

Acsadi, G., Jani, A., Massie, B. *et al.* (1994) A differential efficiency of adenovirus-mediated *in vivo* gene transfer into skeletal muscle cells of different maturity. *Hum. Mol. Genet.*, **3**, 579–84.

Ahn, A.H. and Kunkel, L.M. (1993) The structural and functional diversity of dystrophin. *Nature Genet.*, **3**, 283–91.

Ahn, A.H., Yoshida, M., Anderson, M.S. *et al.* (1994) Cloning of human basic A1, a distinct 59-kDa dystrophin-associated protein encoded on chromosom 8q23–24. *Proc. Natl Acad. Sci. USA*, **91**, 4446–50.

Bakker, A.J., Head, S.I., Williams, D.A. and Stephenson, D.G. (1993) Ca^{2+} levels in myotubes grown from the skeletal muscle of dystrophic (*mdx*) and normal mice. *J. Physiol. (Lond)*, **460**, 1–13.

Barnea, E., Zuk, D., Simantov, R., Nudel, U. and Yaffe, D. (1990) Specificity of expression of the muscle and brain dystrophin gene promoters in muscle and brain cells. *Neuron*, **5**, 881–8.

Barr, E., Carroll, J., Kalyncych, A.M. *et al.* (1994) Efficent catheter-mediated gene transfer into the heart using replication-defective adenovirus. *Gene Ther.*, **1**, 51–8.

Blake, D.J., Tinsley, J.M. and Davies, K.E. (1993) The emerging family of dystrophin-related proteins. *Trends Cell Biol.*, **4**, 19–23.

Bulfield, G., Siller, W.G., Wight, P.A.L. and Moor, K.J. (1984) X chromosome-linked muscular dystrophy (*mdx*) in the mouse. *Proc. Natl Acad. Sci. USA*, **81**, 1189–92.

Byers, T.J., Kunkel, L.M. and Watkins, S.C. (1991) The subcellular distribution of dystrophin in mouse skeletal cardiac and smooth muscle. *J. Cell Biol.*, **115**, 411–21.

Campanelli, J.T., Roberds, S.L., Campbell, K.P. and Scheller, R.H. (1994) A Role for dystrophin-associated glycoproteins and utrophin in agrin-Induced AChR clustering. *Cell*, **77**, 663–74.

Clerk, A., Morris, G.E., Dubowitz, V., Davies, K. and Sewry, C.A. (1993) Dystrophin-related protein, utrophin, in normal and dystrophic human fetal skeletal muscle. *Histochem. J.*, **25**, 554–61.

Corrado, K., Mills, P.L. and Chamberlain, J.S. (1994) Deletion analysis of the dystrophin-actin binding domain. *FEBS Lett.*, **344**, 255–60.

Cox, GA., Cole, N.M., Matsumura, K. *et al.* (1993) Over expression of dystrophin in transgenic *mdx* mice eliminates dystrophic symptoms without toxicity. *Nature*, **364**, 725–9.

Cross, R.A., Stewart, M. and Kendrick-Jones, J. (1990) Structural predictions for the central domain of dystrophin. *FEBS Lett.*, **262**, 87–92.

Danko, I., Fritz, J.D., Latendresse, J.S., Herweijer, H., Schultz, E. and Wolff, J.A. (1993) Dystrophin expression improves myofibre survival in *mdx* muscle following intramuscular plasmid DNA injection. *Hum. Mol. Genet.*, **2**, 2055–61.

Davis, H.L., Demeneix, B.A., Quantin, B., Coulombe, J. and Whalen, R.G. (1993) Plasmid DNA is superior to viral vectors for direct gene transfer into adult mouse skeletal muscle. *Human Gene Ther.*, **4**, 733–40.

Dickson, G. and Dunckley, M.G. (1993) Human dystrophin gene transfer:genetic correction of dystrophin deficiency, in *Molecular and Cell Biology of Muscular Dystrophy*, (ed. T. Partridge), Chapman & Hall, London, pp. 283–302.

Dickson, G., Love, D.R., Davies, K.E., Wells, K.E., Piper, T.A. and Walsh, F.S. (1991) Human dystrophin gene transfer: production and expression of a functional recombinant DNA-based gene. *Human Genet.*, **88**, 53–8.

Dickson, G., Azad, A., Morris, G.E., Simon, H., Noursadeghi, M. and Walsh, F.S. (1992) Co-localization and molecular association of dystrophin with laminin at the surface of mouse and human myotubes. *J. Cell Sci.*, **103**, 1223–33.

Dunckley, M.G and Dickson, G. (1994) Retroviral-mediated gene transfer and Duchenne muscular dystrophy, in *Gene Therapeutics*, (ed. J.A. Wolff), Birkhauser Boston, pp. 391–410.

Dunckley, M.G., Love, D.R., Davies, K.E., Walsh, F.S., Morris, G.E. and Dickson, G. (1992) Retroviral-mediated transfer of a dystrophin minigene into *mdx* myoblasts *in vitro*. *FEBS Lett.*, **296**, 128–34.

Dunckley, M.G.,Wells, D.J., Walsh, F.S. and Dickson, G. (1993) Direct retroviral-mediated transfer of a dystrophin minigene into *mdx* mouse muscle *in vivo*. *Hum. Mol. Genet.*, **2**, 717–23.

Dunckley, M.G., Wells, K.E., Piper, T.A., Wells, D.J. and Dickson, G. (1994) Independent localization of dystrophin N- and C-terminal regions to the sarcolemma of *mdx* mouse myofibres *in vivo*. *J. Cell Sci.*, **107**, 1469–75.

Eglitis, M.A., Kantoff, P., Gilboa, E. and Anderson, W.F. (1985) Gene expression in mice after high efficiency retroviral-mediated gene transfer. *Science*, **230**, 1395–8.

Emery, A.E.H. (1993) *Duchenne Muscular Dystrophy*, 2nd edn, Oxford University Press, Oxford.

Engel, A.G. (1993) Gene therapy for Duchenne dystrophy. *Ann. Neurol.*, **34**, 3–4.

England, S.B., Nicholson, L.V.B., Johnson, M.A. *et al.* (1990) Very mild muscular dystrophy associated with deletion of 46% of dystrophin. *Nature*, **343**, 180–2.

Ervasti, J.M. and Campbell, K.P. (1991) Membrane organization of the dystrophin–glycoprotein complex. *Cell*, **66**, 1121–31.

Ervasti, J.M. and Campbell, K.P. (1993) Dystrophin and the membrane skeleton. *Curr. Opin. Cell Biol.*, **5**, 82–7.

Feener, C.A., Koenig, M. and Kunkel, L.M. (1989) Alternative splicing of human dystrophin mRNA generates isoforms at the carboxyterminus. *Nature*, **338**, 509–11.

Franco, A. and Lansman, J.B. (1990) Calcium entry through stretch-inactivated ion channels in *mdx* myotubes. *Nature*, **344**, 670–3.

Gaschen, F.P., Hoffman, E.P., Gorospe, J.R. *et al.* (1992) Dystrophin deficiency causes lethal muscle hypertrophy in cats. *J. Neurol. Sci.*, **110**, 149–59.

Gee, S.H., Montanaro, F., Lindebaum, M.H. and Carbonetto, S. (1994) Dystroglycan-α, a dystrophin-associated glycoprotein, is a functional agrin receptor. *Cell*, **77**, 675–86.

Gilbert, J.C., Takada, T., Stein, J.E., Langer, R. and Vacanti, J.P. (1993) Cell transplantation of genetically altered cells on biodegradable polymer scaffolds in syngeneic rats. *Transplantation*, **56**, 423–7.

Hemmings, L., Kuhlman, P.A. and Critchley, D.R. (1992) Analysis of the actin-binding domain of α-actinin by mutagenesis and demonstration that dystrophin contains a functionally homologous domain. *J. Cell Biol.*, **116**, 1369–80.

Hoffman, E.P., Brown, R.H. and Kunkel, L.M. (1987) Dystrophin: the protein product of the Duchenne muscular dystrophy locus. *Cell*, **51**, 919–28.

Hutter, O.F. (1992) The membrane hypothesis of Duchenne muscular dystrophy: quest for functional evidence. *J. Inherit. Met. Dis.*, **15**, 565–77.

Ibraghimov-Beskrovnaya, O., Ervasti, J.M., Leveille, C.J., Slaughter, C.A., Sernett, S.W. and Campbell, K.P. (1992) Primary structure of dystrophin-associated glycoproteins linking dystrophin to the extracellular matrix. *Nature*, **355**, 696–702.

Karpati, G., Adjukovic, D., Arnold, D. *et al.* (1993a) Myoblast transfer in Duchenne muscular dystrophy. *Ann. Neurol.*, **34**, 8–17.

Karpati, G., Carpenter, S., Morris, G.E, Davies, K.E., Guerin, C. and Holland, P. (1993b) Localization and quantitation of the chromosome 6-encoded dystrophin-related protein in normal and pathological human muscle. *J. Neuropathol. Exp. Neurol.*, **52**, 119–28.

Khurana, T.S. (1991) Immunolocalization and developmental expression of dystrophin related protein in skeletal muscle. *Neuromusc. Disord.*, **1**, 185–94.

Klietsch, R., Ervasti, J.M., Arnold, W., Campbell, K.P. and Jorgensen, A.O. (1993) Dystrophin-glycoprotein complex and laminin colocalise to the sarcolemma and transverse tubules of cardiac muscle. *Circ. Res.*, **72**, 349–60.

Koenig, M., Monaco, A.P. and Kunkel, L.M. (1988) The complete sequence of dystrophin predicts a rod-shaped cytoskeletal protein. *Cell*, **53**, 219–28.

Koenig, M. and Kunkel, L.M. (1990) Detailed analysis of the repeat domain of dystrophin reveals four potential hinge segments that may confer flexibility. *J. Biol. Chem.*, **265**, 4560–6.

Koenig, M., Beggs, A.H., Moyer, M. *et al.* (1989) The molecular basis for Duchenne versus Becker muscular dystrophy: correlation of severity with type of deletion. *Am. J. Hum. Genet.*, **45**, 498–506.

Kozarsky, K.F. and Wilson, J.M. (1993) Gene therapy: adenovirus vectors (review). *Curr. Opin. Genet. Dev.*, **3**, 499–503.

Law, D.J., Allen, D.L. and Tidball, J.G. (1994) Talin vinculin and DRP (utrophin) concentrations are increased at *mdx* myotendinous junctions following onset of necrosis. *J. Cell Sci.*, **107**, 1477–83.

Lee, C.C., Pearlman, J.A., Chamberlain, J.S. and Caskey, C.T. (1991) Expression of recombinant dystrophin and its localization to the cell membrane. *Nature*, **349**, 334–6.

Levine, B.A., Moir, A.J.G., Patchell, V.B. and Perry, S.V. (1992) Binding sites involved in the interaction of actin with the N-terminal region of dystrophin. *FEBS Lett.*, **298**, 44–8.

Love, D.R., Hill, D.F., Dickson, G. *et al.*(1989) An autosomal transcript in skeletal muscle with homology to dystrophin. *Nature*, **339**, 55–8.

Love, D.R., Blyth, B.C., Tinsley, J.M., Blake, D.J. and Davies, K.E. (1993) Dystrophin and dystrophin-related proteins: a review of protein and RNA studies. *Neuromusc. Disord.*, **3**, 5–21.

Matsumura, K. and Campbell, K.P. (1993) Deficiency of dystrophin-associated proteins: a common mechanism leading to muscle cell necrosis in servere childhood muscular dystrophies. *Neuromusc. Disord.*, **3**, 109–18.

Matsumura, K., Ervasti, J.M., Ohlendieck, K., Kahl, S.D. and Campbell, K.P. (1992a) Association of dystrophin-related protein with dystrophin-associated proteins in *mdx* mouse muscle. *Nature*, **360**, 588–91.

Matsumura, K., Tome, F.M.S., Collin, H. *et al.* (1992b) Deficiency of the 50K dystrophin-associated glycoprotein in severe childhood autosomal recesssive muscular dystrophy. *Nature*, **359**, 320–2.

Menke, A. and Jockusch, H. (1991) Decreased osmotic stability of dystrophin-less muscle cells from the *mdx* mouse. *Nature*, **349**, 69–71.

Miller, A.D. (1992) Human gene therapy comes of age. *Nature*, **357**, 455–60.

Minetti, C., Beltrame, F., Marcenaro, G. and Bonilla, E. (1992) Dystrophin at the plasma membrane of human muscle fibers shows a costameric localization. *Neuromusc. Disord.*, **2**, 99–109.

Mizuno, Y., Yoshida, M., Nonaka, I., Hirai, S. and Ozawa, E. (1994) Expression of utrophin (dystrophin-related protein) and dystrophin-associated glycoproteins in muscles from patients with Duchenne muscular dystrophy. *Muscle Nerve*, **17**, 206–16.

Monaco, A.P., Bertelson, C.J., Liechti-Gallati, S., Moser, H. and Kunkel, L.M. (1988) An explanation for the phenotypic differences between patients bearing partial deletions of the DMD locus. *Genomics*, **2**, 90–5.

Morgan, J.E. (1994) Cell and gene therapy in Duchenne muscular dystrophy. *Hum. Gene Ther.*, **5**, 165–73.

Morgan, R.A and Anderson, W.F. (1993) Human gene therapy (review). *Annu. Rev. Biochem.*, **62**, 191–217.

Ohlendieck, K. and Campbell, K.P. (1991) Dystrophin constitutes 5% of membrane cytoskeleton in skeletal muscle. *FEBS Lett.*, **283**, 230–4.

Ohlendieck, K., Ervasti, J.M., Matsumura, K. *et al.* (1991) Dystrophin related protein is localised to neuromuscular junctions of adult skeletal muscle. *Neuron*, **7**, 499–508.

Pardo, J.V., Siliciano, J.D. and Craig, S.W. (1983) A vinculin-containing cortical lattice in skeletal muscle: transverse lattice elements ('costameres') mark sites of attachment between myofibrils and sarcolemma. *Proc. Natl Acad. Sci. USA*, **80**, 1008–12.

Partridge, T. (1991) Animal models of muscular dystrophy – what can they teach us? *Neuropathol. Appl. Neurobiol.*, **17**, 353–63.

Partridge, T.A., Morgan, J.E., Coulton, G.R., Hoffman, E.P. and Kunkel, L.M. (1989) Conversion of *mdx* myofibers from dystrophin-negative to -positive by injection of normal myoblasts. *Nature*, **337**, 176–9.

Pons, F., Robert, A., Marini, J.F. and Leger,.J.J. (1994) Does utrophin expression in muscles of *mdx* mice during postnatal development functionaly compensate for dystrophin deficiency? *J. Neurol. Sci.*, **122**, 162–70.

Porter, G.A., Dmytrenko, G.M., Winkelmann, J.C. and Block, R.J. (1992) Dystrophin colocalises with β-spectrin in distinct subsarcolemmal domains in mammalian skeletal muscle. *J. Cell Biol.*, **117**, 997–1005.

Quantin, B., Perricaudet, L.D., Tajbakhsh, S. and Mandel, J.L. (1992) Adenovirus as an expression vector in muscle cells *in vivo*. *Proc. Natl Acad. Sci. USA*, **89**, 2581–4.

Ragot, T., Vincent, N., Chafey, P. *et al.* (1993) Efficient adenovirus mediated transfer of a human minidystrophin gene to skeletal muscle of *mdx* mice.

Nature, **361**, 647–50.

Roberts, R.G., Coffey, A.J., Bobrow, M. and Bentley, D.R. (1993) Exon structure of the human dystrophin gene. *Genomics*, **16**, 536–8.

Rosenfeld, M.A., Yoshimura, K., Trapnell, B.C. *et al.* (1992) *In vivo* transfer of human cystic fibrosis transmembrane conductance regulator gene to the airway epithelium. *Cell*, **68**, 143–55.

Sato, O., Nonomura, Y., Kimura, S. and Maruyama, K. (1992) Molecular shape of dystrophin. *J. Biochem. (Tokyo)*, **112**, 631–6.

Stedman, H.H., Sweeney, H.L., Shrager, J.B. *et al.* (1991) The *mdx* mouse diaphragm reproduces the degenerative changes of Duchenne muscular dystrophy. *Nature*, **352**, 536–8.

Stratford-Perricaudet, L.D., Makeh, I., Perricaudet, M. and Briand, P. (1992) Widespread long-term gene transfer to mouse skeletal muscles and heart. *J. Clin. Invest.*, **90**, 626–30.

Sunada, Y., Bernier, S.M., Kozak, C.A., Yamada, Y. and Campbell, K.P. (1994) Deficiency of merosin in dystrophic dy mice and genetic linkage of laminin M chain gene to dy locus. *J. Biol. Chem.*, **269**, 13729–32.

Suzuki, A., Yoshida, M.,Yamamoto, H. and Ozawa, E. (1992a) Glycoprotein-binding site of dystrophin is confined to the cysteine-rich domain and the first half of the carboxy-terminal domain. *FEBS Lett.*, **308**, 154–60.

Suzuki, A., Yoshida, M., Hayashi, K., Mizuno, Y., Hagiwara, Y. and Ozawa, E. (1992b) Molecular organization at the glycoprotein-complex-binding site of dystrophin. Three dystrophin-associated proteins bind directly to the carboxy-terminal portion of dystrophin. *Eur. J. Biochem.*, **220**, 283–92.

Takemitsu, M., Koga, R., Ishiura, S., Nonaka, I., Arahata, K. and Sugita, H. (1994) Dystrophin-related protein in diaphragm, limb and myoblast transferred muscles of *mdx* mouse. *Clin. Neurol.*, **34**, 141–6.

Tinsley, J.M. and Davies, K.E. (1993) Utrophin: a potential replacement for dystrophin? *Neuromusc. Disord.*, **3**, 537–9.

Tome, F.M.S., Evangelista, T., Leclerc, A. *et al.* (1994) Congenital muscular dystrophy with merosin deficiency. *C. R. Acad. Sci. [111]*, **317**, 351–7.

Turner, P.R., Fong, P.Y., Denetclaw, W.F. and Steinhardt, R.A. (1991) Increased calcium influx in dystrophic muscle. *J. Cell Biol.*, **115**, 1701–12.

Valentine, B.A., Cooper, B.J., Cummings, J.F. and DeLahunta, A. (1990) Canine X-linked muscular dystrophy: morphologic lesions. *J. Neurol. Sci.*, **97**, 1–23.

Vincent, N., Ragot, T., Gilgencrantz, H. *et al.* (1993) Long-term correction of mouse dystrophic degeneration by adenovirus-mediated transfer of a minidystrophin gene. *Nature Genet.*, **5**, 130–4.

Weller, B., Karpati, G. and Carpenter, S. (1990) Dystrophin-deficient *mdx* muscle fibers are preferentially vulnerable to necrosis induced by experimental lengthening contractions. *J. Neurol Sci.*, **100**, 9–15.

Wells, D.J. (1993) Improved gene transfer by direct plasmid injection associated with regeneration in mouse skeletal muscle. *FEBS Lett.*, **332**, 179–82.

Wells, D.J. and Goldspink, G. (1992) Age and sex influence expression of plasmid DNA directly injected into mouse skeletal muscle. *FEBS Lett.*, **306**, 203–5.

Wells, D.J., Wells, K.E., Walsh, F.S. *et al.* (1992) Human dystrophin expression corrects the myopathic phenotype in transgenic *mdx* mice. *Hum. Mol. Genet.*, **1**(1), 35–40.

Wessel, H.B. (1990) Dystrophin: a clinical perspective. *Ped. Neurol.*, **6**, 3–12.
Wolff, J.A., Malone, R.W., Williams, P. *et al.* (1990) Direct gene transfer into mouse muscle *in vivo*. *Science*, **247**, 1465–8.
Wolff, J.A., Williams, P., Ascadi, G., Jiao, S., Jani, A. and Chong, W. (1991) Conditions affecting direct gene transfer into rodent muscle *in vivo*. *Biotechniques*, **11**, 474–85.
Xu, H., Christmas, P., Wu, X.R., Wewer, U.M. and Engvall, E. (1994) Defective muscle basement membrane and lack of M-laminin in the dystrophic *dy/dy* mouse. *Proc. Natl Acad. Sci USA*, **91**, 5572–6.
Yang, B., Ibraghivmov-Beskrovnaya, O., Moomaw, C.R., Slaugher, C.A. and Campbell, K.P. (1994a) Heterogeneity of the 59-kDa dystrophin-associated protein revealed by cDNA cloning and expression. *J. Biol. Chem.*, *269*, 6040–4.
Yang, Y., Nunes, F.A., Berencsi, K., Gonczol, E., Engelhardt, J.F. and Wilson, J.M. (1994b) Inactivation of *E2a* in recombinant adenoviruses improves the prospect for gene therapy in cystic fibrosis. *Nature Genet.*, 7, 362–9.
Zabner, J., Couture, L.A., Gregory, R.A. *et al.* (1993) Adenovirus-mediated gene transfer transiently corrects the chloride transport defect in nasal epithelia of patients with cystic fibrosis. *Cell*, 75, 207–16.
Zhu, N., Liggit, D., Liu, Y. and Debs, R. (1993) Systemic gene expression after intravenous DNA delivery into adult mice. *Science*, **261**, 209–11.

15

Cardiovascular disease

HOWARD PRENTICE and KEITH A. WEBSTER

15.1 INTRODUCTION

Cardiovascular disease is the greatest contributor to illness and death in the populations of the industrialized nations, exceeding in annual mortality that of all other diseases combined (Mangano, 1990). Although hypertension is the most prevalent form of cardiovascular disease, coronary artery disease (CAD) causes the highest morbidity and mortality. While progress has been made in understanding the events that lead to cardiovascular disease, the prognosis following the onset of congestive heart failure remains poor, despite the intervention of modern poly-pharmacology (Dzau *et al.*, 1991; Packer, 1993). As an alternative to conventional treatments for heart disease, gene therapies that target the vasculature and the myocardium may lead to novel and effective treatments that are long-lasting and require minimal surgical intervention. The implementation of gene therapy for cardiovascular disease will depend on the development of efficient gene transfer vehicles, appropriate cellular targeting technology, and the identification of genes that will express the requisite therapeutic proteins in a regulated manner. In recent years, significant advances have been made in the development of gene transfer techniques for the cardiovascular system, including optimization of the DNA carrier methods (viral vectors, liposomes, DNA–protein complexes, and injected naked DNA), development of macrodelivery systems [catheters, microprojectiles (Williams *et al.*, 1991), and stents], and cellular implantation. The field of cardiovascular gene transfer is moving rapidly, and while most research has been at the level of

Molecular and Cell Biology of Human Gene Therapeutics
Edited by George Dickson
Published in 1995 by Chapman & Hall. ISBN 0 412 62550 4

cell culture or animal models, there are now instances of on-going clinical trials in this area for example, in the treatment of familial hypercholesterolemia (Dzau *et al.*, 1993b). This chapter will focus on CAD-related cardiovascular diseases that may be amenable to gene therapy, gene transfer strategies for the vasculature and the myocardium, and future directions for research into cardiovascular gene transfer.

15.2 THERAPEUTIC POTENTIAL OF GENE TRANSFER FOR CARDIOVASCULAR DISEASE

Arterial gene transfer has potential for the treatment of a variety of vascular-related disorders, including atherosclerosis, thrombosis, restenosis after balloon angioplasty and hypertension. Direct transfer of expression vectors into the vessel wall by retroviral and liposome-mediated techniques has led to low-efficiency transfection but prolonged expression of recombinant proteins that may have therapeutic value (Nabel and Nabel, 1994). Vascular grafts of endothelial or smooth muscle cells expressing foreign gene products have resulted in foreign gene expression for longer than 6 months (Lynch *et al.*, 1992). In ischemic heart, the introduction of foreign genes in appropriately regulated expression vectors could be employed for modulating contractility, vascular tone and inotropic responsiveness (Cody *et al.*, 1992; Metzger *et al.*, 1993; Milano *et al.*, 1994). Following repeated ischemic episodes, the enhanced expression of recombinant proteins such as stress proteins (HSP70), anti-oxidants, or anti-inflammatory agents could limit reperfusion damage and infarct expansion. For heavily infarcted or aged myocardia, where scarred and fibrotic tissue has replaced cardiac myocytes and contractile capacity is compromised, a number of gene- or cell-based therapeutic interventions can be foreseen to stimulate angiogenesis and replace lost muscle. Cellular transplantation, including the grafting of neonatal, transformed, or stem cell generated cardiomyocytes or of skeletal muscle cells onto the heart wall, may not only enable strengthening of the ischemic myocardium but also provide a population of cells capable of expressing recombinant gene products (Soonpaa *et al.*, 1994). Finally, it should also be possible to use gene transfer to address cardiomyopathy such as that associated with Duchenne muscular dystrophy (Barr *et al.*, 1994).

Some prominent examples of metabolic-related diseases of the cardiovascular system that may be amenable to gene therapy include hypertension, familial hypercholesterolemia and diabetes mellitus. Gene therapy for hypertension may ultimately be possible through manipulating the components of the renin–angiotensin–aldosterone system, and by regulating vascular tone (Jeunemaitre *et al.*, 1992; Morishita *et al.*, 1993b; Paul *et al.*, 1994). Liver gene transfer has been employed in the Watanabe heritable hyperlipidemic (WHHL) rabbit model of familial hypercholesterolemia, an inherited disease that is marked by a deficiency of low-density lipoprotein (LDL) receptors (Wilson *et al.*, 1992). Transfer of the LDL-receptor gene as a DNA–polylysine complex targeted to the liver resulted in a transient reversal of the hypercholesterolemia. Human gene therapy trials for patients with familial hypercholesterolemia were among the first to be undertaken (Miller, 1992). A number of gene transfer and cellular strategies are under way for combating diabetes mellitus (reviewed in Newgard, 1992).

15.3 METHODS OF ARTERIAL GENE TRANSFER

The main methods that are being used to transfer DNA into cardiovascular tissues, and a summary of progress made in the expression of foreign genes in the vasculature are presented in Tables 15.1 and 15.2. Initial experiments employed DNA–liposome complexes or replication-defective retroviruses introduced into peripheral arteries by way of a double balloon catheter. It was originally reported by Nabel *et al.* (1990) that the cloned β-galactosidase gene could be expressed *in vivo* following infection of arterial walls with an amphotropic retrovirus or by liposome-mediated transfection. After retroviral infection, both endothelial and vascular smooth muscle cells expressed the foreign gene, with optimal expression after 2 to 3 months. Liposome-mediated transfer of a luciferase gene construct into peripheral arteries and into coronary arteries by a surgical technique also resulted in uptake and expression of the foreign gene *in vivo* (Lim *et al.*, 1991). Measurements of luciferase expression levels indicated that transfection efficiencies were low but suggested that local levels of protein production may be sufficient to produce therapeutically effective quantities of growth factors, growth factor inhibitors or thrombolytic agents (Lim *et al.*, 1991). Low transfection efficiencies have indeed been reported in a number of investigations using liposome-mediated approaches, with gene transfer into less than 1% of vascular smooth muscle cells (Losordo *et al.*, 1994). However, serial measurements of human growth hormone from transfected peripheral arteries suggested that secreted factors can have a paracrine effect that may induce the production of physiologically active levels. Vascular gene transfer may also be achieved in the absence of a transfection vehicle. Naked DNA delivered into femoral arteries through the use of a perfusion catheter resulted in levels of uptake and foreign gene expression similar to those observed with DNA–liposome complexes (Chapman *et al.*, 1992).

The use of percutaneous transluminal coronary angioplasty to treat coronary artery disease is increasing, even though its success rate is marred by a high incidence of restenosis (Chapman *et al.*, 1992; Dzau *et al.*, 1993a, b). No effective pharmacological treatment currently exists for restenosis (Brady and Warren, 1991; Lindner and Reidy, 1991). Although restenosis is

Table 15.1 Gene transfer vehicles for the cardiovascular system

Gene transfer vehicle	*Status of foreign gene*	*Efficiency of transfection*	*Cellular target in heart or vasculature*	*Reference*
Naked DNA	Episomal	Low	Cardiocytes	Acsadi *et al.* (1991)
			Vascular smooth muscle cells in balloon-injured artery	Riessen *et al.* (1993)
Liposomes	Episomal	Low	Endothelial cells	Nabel *et al.* (1990)
			Vascular smooth muscle cells	
Adenovirus	Episomal	High	Endothelial cells	Lemarchand *et al.* (1993)
			Vascular smooth muscle cells	Lee *et al.* (1993)
			Cardiomyocytes	Barr *et al.* (1994)
Retrovirus	Integrated	Low	Endothelial cells	Nabel *et al.* (1990)
			Smooth muscle cells	
			Cardiac fibroblasts and infiltrating monocytes in ischemic heart	Prentice *et al.* (1993)
Hemagglutinating virus of Japan (HVJ or Sendai virus)	Episomal	High	Vascular smooth muscle cells	Morishita *et al.* (1993a)

Table 15.2 Major innovations in arterial gene transfer

Method	*Result*	*Reference*
Arterial transfection with liposomes or retroviruses	Expression of foreign genes for 2–3 months	Nabel *et al.* (1990)
Implantation of endothelial cells or vascular smooth muscle cells expressing recombinant proteins	Successful implantation of endothelial cells giving foreign gene expression for several weeks. Smooth muscle cells grafted onto denuded arteries express an introduced gene for 6 months	Nabel *et al.* (1989); Wilson *et al.* (1989); Lynch *et al.* (1992)
Transfection of coronary arteries using a perfusion catheter.	Low but significant levels of gene expression obtained	Chapman *et al.* (1992); Flugelman *et al.* (1992)
Naked DNA delivered on hydrogel-coated angioplasty balloon	Successful DNA transfer following short balloon inflations	Riessen *et al.* (1993)
Arterial gene transfer using recombinant adenovirus	High-efficiency infection with foreign gene expression for up to 14 days	Lemarchand *et al.* (1993); Lee *et al.* (1993)
Phenotypic change caused by expression of recombinant FGF-1	Induction of hyperplasia and angiogenesis	Nabel *et al.* (1993)
Antisense oligonucleotides directed at preventing neointimal formation after angioplasty	Antisense oligonucleotides to c-myc or c-myb or to cdc-kinase and PCNA in combination inhibit neointimal formation	Simons *et al.* (1992); Bennett *et al.* (1994); Morishita *et al.* (1993a)

multifactorial and complex (Edgington, 1993; Morishita *et al.*, 1993a), numerous animal studies now suggest that therapeutic agents can reduce it if they are delivered locally, and at high concentrations that would not be tolerated systemically (Swain, 1989; Califf *et al.*, 1991; Chapman *et al.*, 1992; Dzau *et al.*, 1993b; Ohno *et al.*, 1994). While there are still questions concerning the relative contributions of hyperplasia and arterial wall remodeling to the development of restenosis, it seems likely that an appropriate combination of factors (or anti-factors) can be delivered to the coronary artery to prevent, or at least significantly postpone, restenosis following angioplasty.

Earlier methods for gene transfer into arteries required total artery occlusion for 30 minutes; however, much more rapid gene transfer can be achieved by using perfusion catheters in the coronary artery (Wolinski and Thung, 1991). Chapman *et al.* (1992) reported using a perfusion balloon catheter to introduce plasmid DNA–liposome complexes into canine coronary arteries *in vivo* that resulted in significant reporter gene expression. A similar catheter technique was used for introducing retroviral vectors into rabbit aortas (Flugelman *et al.*, 1992). Following a 1-minute retroviral infusion, foreign gene expression was observed in the rabbit aorta at 5 to 14 days, but transfection efficiencies were again low, with fewer than 100 transduced cells in a 2-cm region of the aorta.

During restenosis, a variety of chemokines, cytokines and growth factors, including thrombin, platelet-derived growth factor (PDGF), basic fibroblast growth factor (FGF), and epidermal growth factor (EGF) interact in a complex manner to promote smooth muscle cell proliferation and migration (Edgington, 1993; Morishita *et al.*, 1993a). These, along with other cell cycle regulatory factors, are currently the principal targets for gene-based approaches to inhibit restenosis. In considering possible gene transfer strategies, an important factor is the efficiency with which balloon-injured, denuded arteries can take up foreign genes. There is evidence that cells in the damaged arteries may in fact be more susceptible to gene transfer than those in undamaged normal or atherosclerotic arteries. The transfection efficiency of human smooth muscle cells by lipofection in culture is very low but improves in proliferating cultures (Pickering *et al.*, 1994). Higher transfection efficiencies are obtained *in vitro* for cells derived from restenotic atherosclerotic lesions than for those from primary atherosclerotic plaque. In the vasculature, liposome-mediated transfection may also be dependent on the cell cycle status of the target cells. In tissue explants from rabbit, a transfected gene was expressed in denuded arteries at 7- to 13-fold higher levels than in non-denuded arteries after up to 21 days in culture (Takeshita *et al.*, 1994). Iliac arteries of live rabbits subjected to angioplasty showed increased foreign gene expression compared with untreated arteries and demonstrated changes in neointimal proliferation that paralleled the observed alterations in gene expression (Takeshita *et al.*, 1994).

In an extension of previous studies employing naked DNA for arterial gene transfer, a relatively simple gene therapy strategy for restenosis has been developed that makes use of the angioplasty catheter as a delivery system. Naked DNA on a hydrogel-coated angioplasty balloon was successfully transfected percutaneously into external iliac arteries and into surgically exposed carotid arteries following inflations of 30 minutes or as short as 1 minute (Riessen *et al.*, 1993). The efficiency of this procedure was similar to that obtained with organ cultures. Thus, transfection during angioplasty in a single step may eliminate the need for introducing a second catheter.

15.3.1 Recombinant adenovirus in arterial gene transfer

Unlike other methods for gene transfer, recombinant adenovirus can infect normal arteries *in vivo* at high efficiency. Effective gene transfer occurred following exposure of the endothelium to a recombinant adenoviral vector for 15 minutes, with maximal expression at 7 days after injection (Lemarchand *et al.*, 1993). In balloon-injured rat carotid arteries, high-level infection of recombinant adenovirus was also observed, with transduction of approximately 30% of cells and foreign gene expression lasting up to 14 days (Lee *et al.*, 1993). Transduced cells from the injured arteries were identified immunohistochemically as smooth muscle cells. Over the relatively short periods that have been studied so far, it is probable that adenoviral genomes do not integrate significantly into the host genome. The high transfection/infection efficiency of adenoviral vectors has placed them at the forefront of the currently available gene transfer vehicles. Ohno *et al.* (1994) introduced an adenoviral vector encoding the herpes virus thymidine kinase (*tk*) gene into porcine arteries that had been injured by a balloon catheter and demonstrated infection of the smooth muscle cells and expression of the *tk* gene. As expected, the *tk*-expressing smooth muscle cells were shown to be selectively sensitized to the cytotoxic nucleoside analog ganciclovir, and intimal hyperplasia was correspondingly repressed in the tk-transfected and ganciclovir-treated arteries compared with that in untreated (tk^-) controls. No major local or systemic toxicity of the treatments was reported. Although these studies are extremely promising, further studies are required to determine the effects of long-term or repeated exposure of cells and tissues to recombinant adenoviruses.

15.4 ANALYSIS AND CONTROL OF VASCULAR REMODELING BY GENE TRANSFER

Gene transfer techniques have been used to assess the contributions of different growth factors and the renin–angiotensin system to hyperplasia and remodeling of the arterial vessel wall. By using a lipofectin technique, Nabel *et al.*, (1993b) reported that the over-expression of recombinant FGF-1 *in vivo* can promote intimal hyperplasia and angiogenesis in iliofemoral arteries. On the other hand, PDGF, and transforming growth factor-β1, (TGF-β1) were able to stimulate hyperplasia, but not angiogenesis (Nabel *et al.*, 1993a). Epidermal growth factor (EGF) is one of the key factors involved in smooth muscle cell (SMC) outgrowth. Pickering *et al.* (1993, 1994) demonstrated that proliferating SMCs from human athersclerotic plaque could be specifically neutralized by transfecting a gene containing a recombinant cytotoxin fused to a cDNA for EGF. Recombinant molecules were targeted specifically to the EGF receptor contained on the proliferating SMCs. Using gene transfer by the hemagglutinating virus of Japan (HVJ) method, it has been possible to study the effects of the vascular renin–angiotensin system (RAS) on the growth of vascular smooth muscle cells *in vitro* and *in vivo* (Morishita *et al.*, 1993b, 1994). Transfection of angiotensin-converting enzyme (ACE) or renin cDNA increased arterial DNA and RNA synthesis that was inhibited by a specific angiotensin II receptor antagonist (DuP 753). These experiments demonstrated that the increased local expression of ACE can cause vascular remodeling through endogenous production of angiotensin II.

15.5 APPLICATIONS OF ANTISENSE OLIGONUCLEOTIDES TO MODELS OF VASCULAR DISEASE

Arterial transfer of antisense oligonucleotides has been effective in decreasing or abolishing the expression of key proteins in the vasculature *in vivo*. This technology has been applied to a variety of gene products in models of restenotic lesions after angioplasty. Typically, 200 μl of pluronic solution is applied to the outside of the artery, from which the adventitia has been stripped (Simons *et al.*, 1992; Bennett *et al.*, 1994). In cell culture, antisense oligonucleotides complementary to the genes for c-myb, c-myc, non-muscle myosin and proliferating cell nuclear antigen (PCNA) are capable of reducing vascular smooth muscle cell proliferation (Bennett *et al.*, 1994). With the rat carotid artery model, *in vivo* application of a pluronic gel containing antisense oligonucleotides to c-myc or c-myb after angioplasty dramatically reduced neointimal formation (Simons *et al.*, 1992; Bennett *et al.*, 1994). Using a delivery vehicle consisting of the protein coat of HVJ in a liposome complex, it has also been shown that a combination of antisense oligonucleotides to the cell cycle proteins cdc2 kinase and PCNA will completely inhibit neointima formation (Morishita *et al.*, 1993a).

15.6 GRAFTING OF RETROVIRALLY TRANSDUCED ENDOTHELIAL OR SMOOTH MUSCLE CELLS ONTO INJURED ARTERIES

The pioneering studies by Nabel *et al.* (1989) and Wilson *et al.* (1989) demonstrating that vascular grafts with genetically modified endothelial cells could be successfully implanted into the arterial wall have opened the way for gene transfer strategies involving a variety of potentially therapeutic gene products. In a minipig model, clonally selected vascular endothelial cells transduced with a β-galactosidase-expressing retrovirus were introduced into denuded iliofemoral arteries by catheter (Nabel *et al.*, 1989). Some 2 to 4 weeks after removal of the catheter, arterial segments contained endothelial cells that had been successfully implanted into the vessel wall and that continued to express the foreign gene product. Using a dog model, it was shown that expanded populations of endothelial cells transduced with a β-galactosidase-expressing retrovirus could be seeded onto prosthetic Dacron vascular grafts, which were then implanted into carotid arteries

(Wilson *et al.*, 1989); 5 weeks after implantation, analysis of the graft revealed genetically modified endothelial cells lining the luminal surface. The results of transplantation studies on endothelial cells indicate that foreign genes can be expressed for several weeks after implantation. Such approaches have potential for the expression of thrombolytic, angiogenic or growth factor genes *in vivo*.

The use of intravascular stents has been effective in the treatment of iliac and coronary stenosis but is associated with early thrombotic stent closure. The incidence of stent closure can be decreased if the stent surface is seeded with endothelial cells before placement in patients (Dichek *et al.*, 1989; Swain, 1989). Endothelial cells transduced with a retrovirus encoding either tissue-type plasminogen activator (t-PA) or β-galactosidase continued to express the introduced gene after seeding on metal stents *in vitro*. When stents were expanded by balloon dilatation, the cells remained in place. Seeded cells transduced with a t-PA-encoding retrovirus secreted high levels of t-PA, possibly sufficient to produce thrombolysis at the stent surface. The expression of foreign genes in seeded endothelial cells has been reported to persist for at least 5 weeks (Nabel and Nabel, 1994). These results indicate that the introduction of transduced endothelial cells seeded onto intravascular stents may represent a viable means of obtaining recombinant gene expression in the vasculature, at least for a relatively short term.

SMCs are a target cell type for cell-based gene therapy; like endothelial cells they are in close proximity to the circulation, but have the advantage of being greater in number. In minipigs, vascular SMCs transduced with a β-galactosidase-encoding amphotropic retrovirus were implanted into iliofemoral arteries that had been denuded of endothelial cells. These arteries displayed β-galactosidase staining in cells of the intima and media at 11 days after implantation (Plautz *et al.*, 1991). SMCs transduced with the human adenosine deaminase (ADA) gene have been successfully seeded onto denuded carotid arteries (Lynch *et al.*, 1992). The expression of the introduced ADA gene for up to 6 months supported a potential role for SMCs as targets for gene therapy. Furthermore, retroviral-ADA-expressing SMCs that had been selected *ex vivo* by using the neomycin resistance marker and were seeded onto balloon-injured carotid arteries were found to cease proliferation after 28 days (Clowes *et al.*, 1994). Thus, retrovirally transduced SMCs do not become transformed. It is not known whether transduced endothelial or smooth muscle cells retain their correct phenotypes long term.

15.7 STRATEGIES FOR GENE AND CELL THERAPY IN THE MYOCARDIUM

The relative unresponsiveness of ischemia-related heart disease to classical pharmacology, particularly after the onset of congestive heart failure, has resulted in an increasing switch in the focus of research and development programs toward molecular, cellular and gene-based approaches. Progress in gene therapy strategies to target the myocardium has lagged behind that for other tissues, including the vasculature, primarily because of inefficient gene transfer techniques and also because the diseased heart poses a complex target. Can genes be introduced into the diseased heart that will improve overall function and preserve the viability of the cardiac myocytes? What genes should be inserted? How and when should the genes be delivered, and how can they be regulated?

There are two fundamental approaches to gene therapy for heart disease: (i) pre-emptive or cardioprotective strategies; and (ii) repair strategies through muscle restoration and replacement. Again, the technical difficulties and pathological complexities associated with the former approach have made progress slow, and more progress has been made on the latter approach. Replacement and restorative approaches include stimulating angiogenesis to initiate collateral vessel development in ischemic myocardia (Guzman *et al.*, 1993), directly replacing lost cardiac myocytes through the grafting of skeletal myocytes (Koh *et al.*, 1993), regeneration of cardiac myocytes (Soonpaa *et al.*, 1994), and converting other cell types within scarred, fibrotic myocardium to muscle by transfecting *in vivo* with myogenic transforming genes (Prentice *et al.*, 1993). Although direct gene-based cardioprotective strategies have not been developed so far, considerable progress has been made toward demonstrating that foreign genes can be targeted and appropriately expressed in cardiac myocytes *in vivo* (see below).

15.7.1 Cell techniques: intracardiac grafting

Because CAD-related heart disease is predominantly a disease of cardiac cell loss, and adult cardiac myocytes have negligible proliferative capacity, there is considerable interest in the development of intracardiac grafting techniques to replace lost cardiac muscle. Direct implantation of C2C12 skeletal myoblasts into syngeneic mouse hearts resulted in grafts that were viable for 3 months (Koh *et al.*, 1993). Grafted cells had withdrawn from the cell cycle by 14 days and appeared morphologically as differentiated myotubes. There was no evidence of cardiac arrhythmia resulting from graft formation or subsequent cardiac remodeling. Electrical coupling, force generation, and cardiac output were not measured in these preliminary studies. Although the long-term fate of the grafted skeletal myocytes is not clear, these studies may at least support a possible application for intracardiac myoblast grafting in the local delivery of recombinant molecules.

A series of reports from the laboratories of W.C. Calycomb and L.J. Field (see references in Soonpaa *et al.*, 1994) have documented the feasibility of transforming cardiac myocytes with the SV40 large T antigen oncogene.

Transformed myocytes maintain a highly differentiated state, including the ability to contract while actively dividing, and can be passaged *in vitro*. These myocytes also appeared to form long-term intracardiac grafts after introduction into syngeneic hosts, but as yet there are no reports on their anticipated ability to augment force generation or improve cardiac functions. Transformed myocytes have the severe limitation of uncontrolled proliferation *in vivo*. They will be of limited value for gene therapy purposes unless this property can be controlled. Other possible restorative approaches based on cell therapy include the generation of cardiac myocytes *in vitro* from pluripotential stem cells (Doetschman *et al.*, 1993) or the use of embryonic myocytes in grafts. It was shown recently that intracardiac grafts of embryonic day 15 cardiomyocytes resulted in the formation of stable grafts for at least 2 months (Soonpaa *et al.*, 1994). Interestingly, these studies presented evidence for intercalated discs between host and graft cells, suggesting that the transplanted cardiomyocytes were electrically coupled. Assuming that the intracardiac grafting techniques will be reproduced in the human heart, and that grafted muscle will stably improve cardiac functions, the principal limitation of the repair/replacement approach is the source of cardiac cells.

15.7.2 Gene techniques: direct DNA injection and regulation of foreign gene expression

Initial experiments on *in vivo* gene transfer into myocardial cells involved direct injection of DNA into the left ventricles of open-chest rats through a 26- to 30-gauge syringe needle (Acsadi *et al.*, 1991; Buttrick *et al.*, 1993). Remarkably, DNA was taken up and expressed locally by cardiac myocytes. The direct injection technique was originally applied to skeletal muscle, where long-term gene expression was obtained, although from relatively few transfected cells (Wolff *et al.*, 1990). In the heart, the efficiency of transfection is also low, perhaps 60 to 100 cells per injection within 1 to 2 mm of the needle track (Acsadi *et al.*, 1991; Buttrick *et al.*, 1993), but reporter expression has been shown to continue for more than 6 months (Gal *et al.*, 1993).

A number of investigators have examined the parameters that affect gene expression following direct DNA injection. In heart, Gal *et al.* (1993) observed that varying the quantity of DNA used did not significantly affect levels of gene expression, whereas increasing the volume of injectate significantly augmented expression of a luciferase reporter gene. This was not the case in the dog myocardium, where there is a linear dose–response relation between levels of gene expression and quantity of DNA injected, in the range of 10 to 200 μg per injection site (von Harsdorf *et al.*, 1993). In skeletal muscle, the mechanism of uptake of injected plasmid DNA has been investigated by electron microscopy studies with colloidal gold conjugated to plasmid DNA. The DNA does not appear to enter the cell through transient membrane disruptions or through endocytosis but may be taken up by some type of membrane transporter (Wolff *et al.*, 1992).

Although the direct injection method is unlikely to be widely used as a gene therapy tool, it has provided some valuable information on the expression and regulation of genes and promoters *in vivo* that may be important in the design of cardiac-specific vectors. The usefulness of direct DNA injection into myocardium as a means for determining activities of tissue-specific promoters *in vivo* has been established for a variety of genes, including muscle creatine kinase (Vincent *et al.*, 1993), cardiac troponin C (cTnC) (Parmacek *et al.*, 1992), and β-myosin heavy chain (β-MHC) (von Harsdorf *et al.*, 1993). By comparison of patterns of expression from the TnC fast promoter and the cTnC promoter, it has been further demonstrated that these muscle-specific genes retain their respective tissue-specification following direct DNA injection *in vivo* (Prentice *et al.*, 1994). Inconsistencies have also been revealed; for example, there appear to be some discrepancies in the responses of regulatory regions in the α-MHC promoter, depending on whether the analyses were carried out *in vitro* or by direct injection *in vivo*. In primary fetal cardiac myocytes, a region containing 161 base pairs (bp) of upstream sequence, including the thyroid hormone response element (TRE) from bp −156 to −136, was sufficient to confer thyroid hormone responsiveness to a reporter gene (Kitsis *et al.*, 1991; Buttrick *et al.*, 1993). *In vivo*, the TRE was again necessary for thyroid hormone induction, but additional sequences present in the first 388 bp were also required. Promoter regions responsible for negative regulation also differed between neonatal cardiocytes and the intact heart. Whereas the region from bp −540 to −613 appeared to contain a negative control regulatory element in neonatal cardiac myocytes, this region had no such role *in vivo*. In contrast, a region upstream of bp −613 was found to act as a negative control region *in vivo*.

15.8 MYOCARDIAL GENE TRANSFER USING REPLICATION-DEFECTIVE ADENOVIRUS

Adenovirus has the capacity to infect both replicating cells and non-replicating cells, including cardiac myocytes, with high efficiency (Kass-Eisler *et al.*, 1993). Direct injection of an adenoviral vector into the heart results in infection of all regions of the myocardium. Gene transfer with adenoviral vectors *in vivo*, through the wall of the heart from the coronary vasculature, can lead to foreign gene expression in >30% of cardiomyocytes (Barr *et al.*, 1994). Intravenous injection of β-galactosidase-expressing adenovirus into neonatal mice led to significant levels of expression, particularly in skeletal muscle and the heart (Stratford-Perricaudet *et al.*, 1992). Expression was sustained over a 12-month period, although a decrease in tissue staining was observed after 10 months. At 3 months after injection of neonatal mice, the viral DNA detected in the heart remained mostly extra-chromosomal. Adenovirus-mediated gene transfer by systemic delivery into adult mice was found to give a short-term pattern of expression similar to that in newborns, but with less efficient uptake and poorer β-galactosidase staining (Stratford-Perricaudet *et al.*, 1992). In contrast, when Guzman *et al.* (1993) injected recombinant adenovirus into the cardiac apex of the adult rat, maximal expression of β-galactosidase occurred after 3 days. After 1 week, β-galactosidase histological staining was positive but there was no detectable enzymatic activity. These studies suggested that expression of the injected β-galactosidase-encoding adenovirus in the heart was transient, persisting for no more than a few days. The transient nature of gene expression from adenovirus injected into adult myocardium may involve an immunological or inflammatory response to the injection procedure (Guzman *et al.*, 1993). Barr *et al.* (1994) found no evidence of an inflammatory response or of tissue remodeling as a result of high-titer adenovirus infusion into the coronary artery circulation. In these studies, foreign gene expression in the myocardial tissues persisted for at least 2 weeks.

15.9 TRANSGENIC STUDIES

Cardioprotective gene therapy will require functional studies on animals that are expressing the candidate therapeutic gene(s). Until better DNA delivery systems are developed, transgenic animals may be the only way to determine the effects of global, stable, and long-term expression of a gene on cardiovascular functions. Foreign genes can be targeted to the heart by standard transgenic techniques, using a cardiac-specific promoter to direct the expression of the gene (reviewed in Hunter *et al.*, 1993). For example, transgenic mice containing part of the ventricle-specific myosin light chain 2 (MLC_{2v}) promoter fused to the luciferase gene expressed luciferase enzyme exclusively in the myocardium (Franz *et al.*, 1993). Luciferase expression was also developmentally regulated in parallel with the predicted pattern of endogenous MLC_2. Several studies have demonstrated similar restricted and developmentally regulated expression from other muscle- and cardiac-specific promoters, including those of the α and β-myosin heavy chains, α-skeletal and cardiac actins, muscle creatine kinase, and atrial natriuretic peptide (Hunter *et al.*, 1993; Dunwoodie *et al.*, 1994). In addition to analyses of promoter functions *in vivo*, a limited number of transgenic studies have described phenotypic changes in the myocardium in response to the expression of the transgene.

Liu *et al.* (1993) described transgenic mice that over-expressed the human glucose transporter GLUT4. The endogenous GLUT4 promoter was used in the transgene, and mice expressed 10-fold higher GLUT4 protein than controls, with elevated levels in the plasma membranes of adipocytes and cardiac myocytes, consistent with the correct pattern of tissue specificity of the endogenous protein. Compared with the wild-type mice, the transgenic mice had significantly improved glycemic control compatible with increased basal glucose transport. This model may be useful in examining the role of glucose transporters during ischemic stress. Metzger *et al.* (1993) used a 650-bp fragment of the α-myosin heavy chain (α-MHC) promoter to target skeletal troponin C (sTnC) to the hearts of transgenic mice. They demonstrated that cardiac myocytes from the transgenic animals, in which sTnC replaced cTnC in the troponin complex, had reduced sensitivity to acidosis, as shown by a shift in the pH response of the tension–pCa curve, toward that of skeletal muscle. This study not only confirmed the role of troponin C in determining muscle sensitivity to acidosis but also demonstrated how a transfected foreign gene can radically modify a cardiac physiological function. From the gene therapy perspective, it is not clear whether a reduced pH sensitivity of contraction would be an advantage to the heart, since enhanced or prolonged contractility during severe ischemia may compromise the energy status of the myocytes and worsen the impact of the stress. In a comparably designed study, Milano *et al.* (1994) used a 5.5-kb fragment of the α-MHC promoter to over-express the β_2-adrenergic receptor in transgenic mice. Expression of the transgene receptor was restricted to the atria and ventricles and resulted in increased adenylate cyclase activity, enhanced atrial contractility, and increased left ventricular function *in vivo*. The authors argued that over-expressing β-receptors in this way, or by other means of gene transfer, may provide a gene-based approach to improve the inotropic responsiveness of the failing heart. Clearly it would be of interest to determine the function and fate of over-expressed β-receptors in a model of heart disease. Experiments such as these will be essential to identify and characterize the activities of candidate proteins for use in cardiac gene therapy.

15.10 FUTURE DIRECTIONS FOR CARDIOVASCULAR GENE TRANSFER

Major advances have been made over a relatively short period in the application of gene therapy to cardiovascular disease. The three basic approaches that are being tested include direct transfer of genes into cardiovascular tissues, grafting of cells that have been transfected *ex vivo* with a foreign gene, and grafting of new muscle to restore damaged myocardial tissue. The most progress has been made in techniques designed to prevent or postpone restenosis of arteries following angioplasty. Results from animal models have been very encouraging in this area and it seems probable that some of the protocols to supress restenosis will undergo clinical testing in the near future.

Each of the main-line approaches to cardiovascular gene therapy has limitations. Progress on direct gene transfer, which is perhaps the most promising approach, suffers principally on two accounts: (i) inefficient or unsatisfactory delivery; and (ii) undefined methods for regulating the expression of a potentially therapeutic protein, once it is delivered to diseased tissue. While low-efficiency techniques may be sufficient for DNA delivery to vascular endothelium and smooth muscle tissues, high-efficiency vehicles are required to target DNA to heart muscle. Replication-deficient, recombinant adenoviral vectors suffer inherently from being derived from an active virus: they have large genomes, and are potentially immunogenic (Mulligan, 1993); they may integrate at low levels into the host genome; and their longevity *in vivo*, in either dividing or non-dividing cells, is not known. The long-term effects of an apparently symbiotic relationship between adenoviral DNA and cardiac myocytes may be unpredictably adverse. Experiments are under way to improve adenoviral vectors by eliminating non-essential regions of the genome. Alternatively, high-efficiency non-viral methods may be preferable. In experiments in our laboratories at SRI International and Stanford University, we have obtained up to 60% transfection efficiency of cardiac myocytes in culture by using the HVJ/liposome-mediated DNA transfer method (Ellison *et al.*, 1994). It is probably only a matter of time before adequate delivery vehicles are available.

The second problem is to regulate a gene once it is delivered to the diseased tissue. In the ideal case, the therapeutic gene should be silent under normal conditions and become induced in response to a stimulus that results from the disease. As a means to regulate the expression of genes directed toward the ischemic myocardium, we have created plasmid constructs in which hypoxia response enhancer elements, derived from the erythropoietin gene, are inserted upstream of heart-specific promoters. In transient transfection and expression studies, these vectors were expressed only in heart cells and the expression was strongly inducible by hypoxia (Webster *et al.*, 1994). The ability to turn on the expression of an otherwise low-level or silent therapeutic protein in response to an ischemic episode could be a key factor for gene therapy protocols to target ischemic tissue.

All of the cell grafting procedures are quite demanding in terms of time, techniques and cost. Questions still remain concerning the stability of cell grafts, the long-term phenotype of the grafted cells, the regulation, and longevity of transfected therapeutic genes, and long-term effects on the host tissue. Cardiac muscle replacement approaches to treat ischemic and congestive heart diseases will require a source of histocompatible cardiac myocytes, in addition to demonstrations that the grafts are stable and functionally beneficial. Pluripotential stem cells may eventually become a suitable source of cardiac myocytes; alternatively, studies on the regulation of cardiac differentiation, and the cell cycle may lead to new methods that will allow cardiac cell division and regeneration under controlled conditions.

Despite some major concerns and limitations, the gene therapy approach to heart disease is a relatively new discipline that is likely to have an impact on treatment strategies very soon, possibly within the present decade (perhaps prematurely). An ever-expanding role for cardiovascular gene therapy can be predicted for the years to come.

REFERENCES

Acsadi, G., Jiao, S.S., Duke, A., *et al.*, (1991) Direct gene transfer and expression into rat heart *in vivo*. *New Biol.*, 3, 71–81.

Barr, E., Carroll, J., Kalynych, A.M. *et al.* (1994) Efficient catheter-mediated gene transfer into the heart using replication-defective adenovirus. *Gene Therapy*, **1**, 31–58.

Bennett, M.R., Anglin, S., McEwan, J. R. *et al.* (1994) Inhibition of vascular smooth muscle cell proliferation *in vitro* and *in vivo* by c-myc antisense oligodeoxynucleotides. *J. Clin. Invest.*, **93**, 820–8.

Brady, A. and Warren, J. (1991) Angioplasty and restenosis. *Br. Med. J.*, **303**, 729–30.

Buttrick, P.M., Kaplan, M.L., Kitsis, R.N. and Leinwand, L.A. (1993) Distinct behavior of cardiac myosin heavy chain constructs *in vivo*; discordance with *in vitro* results. *Circ. Res.*, **72**, 1211–17.

Califf, R.M., Fortin, D.F., Frid, D.J. *et al.* (1991) Restenosis after coronary angioplasty: an overview. *J. Am. Coll. Cardiol.*, **17**(6 suppl B), 2B–13B.

Chapman, G.D., Lim, C.S., Gammon, R.S. *et al.* (1992) Gene transfer into coronary arteries of intact animals with a percutaneous balloon catheter. *Circ. Res.*, **71**, 27–33.

Clowes, M.M., Lynch, C.M., Miller, A.D. *et al.* (1994) Long-term biological response of injured rat carotid artery seeded with smooth muscle cells expressing retrovirally introduced human genes. *J. Clin. Invest.*, **93**, 644–51.

Cody, R.J., Haas, G.H. and Binkley, P.F. (1992) Endothelin as a vasoconstrictor substance in congestive heart failure. *Heart Failure*, **8**, 135–41.

Dichek, D.A., Neville, R.F., Zwiebel, J.A. *et al.* (1989) Seeding of intravascular stents with genetically engineered endothelial cells. *Circulation*, **80**, 1347–53.

Doetschman, T., Shull, M., Kier, A. *et al.* (1993) Embryonic stem cell models systems for vascular morphogenesis and cardiac disorders. *Hypertension*, **22**, 618–29.

Dunwoodie, S.L., Joya, J.E., Arkell, R.M. *et al.* (1994) Multiple regions of the human cardiac actin gene are necessary for maturation-based expression in striated muscle. *J. Biol. Chem.*, **269**, 12212–19.

Dzau, V.J., Braunwald, E. and participants. (1991) Resolved and unresolved issues in the prevention and treatment of coronary artery disease: a workshop consensus statement. *Am. Heart J.*, **121**, 1244–63.

Dzau, V.J., Gibbons, G.H., Cooke, J.P. and Omoigui, N. (1993a) Vascular biology and medicine in the 1990s: scope, concepts, potentials, and perspectives. *Circulation*, **87**, 705–19.

Dzau, V.J., Morishita, R. and Gibbons, G.H. (1993b) Gene therapy for cardiovascular disease. *Tibtech*, **11**, 205–10.

Edgington, S.M. (1993) Chemokines in cardiovascular disease. *Bio/Technology*, **11**, 676–81.

Ellison, K., Bishopric, N.H., Webster, K.A. *et al.* (1994) Fusigenic liposome-mediated gene transfer into cardiac myocytes. *J. Mol. Cell. Cardiol.*, Submitted.

Flugelman, M.Y., Jaklitsch, M.T., Newman, K.D. *et al.* (1992) Low level *in vivo* gene transfer into the arterial wall through a perforated balloon catheter. *Circulation*, **85**, 1109–17.

Franz, W.-M., Breves, D., Klingel, K. *et al.*, (1993) Heart-specific targeting of firefly luciferase by the myosin light chain-2 promoter and developmental regulation in transgenic mice. *Circ. Res.*, **73**, 629–38.

Gal, D., Weir, L., Leclerc, G. *et al.* (1993) Direct myocardial transfection in two animal models. *Lab. Invest.*, **68**, 16–25.

Guzman, R.J., Lemarchand, P., Crystal, R.G. *et al.* (1993) Efficient gene transfer into myocardium by direct injection of adenovirus vectors. *Circ. Res.*, **73**, 1202–7.

Hunter, J.J., Zhu, H., Lee, K.J. *et al.* (1993) Targeting gene expression to specific cardiovascular cell types in transgenic mice. *Hypertension*, **22**, 608–17.

Jeunemaitre, X., Soubriere, F., Kotelevtsev, Y.V. *et al.* (1992) Molecular basis of human hypertension: role of angiotensinogen. *Cell*, **71**, 169–80.

Kass-Eisler, A., Falck-Pedersen, E., Alvira, M. *et al.* (1993) Quantitative determination of adenovirus-mediated gene delivery to rat cardiac myocytes *in vitro* and *in vivo*. *Proc. Natl Acad. Sci. USA*, **90**, 11498–502.

Kitsis, R.N., Buttrick, P.M., McNally, E.M. *et al.* (1991) Hormonal modulation of a gene injected into rat heart *in vivo*. *Proc. Natl Acad. Sci. USA*, **88**, 4138–42.

Koh, G.Y., Klug, M.G., Soonpaa, M.H. and Field, L.J. (1993) Differentiation and long-term survival of C2C12 myoblast grafts in heart. *J. Clin. Invest.*, **92**, 1548–54.

Lee, S.W., Trapneli, B.C., Rade, J.J. *et al.* (1993) *In vivo* adenoviral vector-mediated gene transfer into balloon-injured rat carotid arteries. *Circ. Res.*, **73**, 797–807.

Lemarchand, P., Jones, M., Yamada, I. and Crystal, R.G. (1993) *In vivo* gene transfer and expression in normal uninjured blood vessels using replication-deficient recombinant adenovirus vectors. *Circ. Res.*, **72**, 1132–8.

Lim, C.S., Chapman, G.D., Gammon, R.S. *et al.* (1991) Direct *in vivo* gene transfer into the coronary and peripheral vasculatures of the intact dog. *Circulation*, **83**, 2007–11.

Lindner, V. and Reidy, M.A. (1991) Proliferation of smooth muscle cells after vascular injury is inhibited by an antibody against basic fibroblast growth factor. *Proc. Natl Acad. Sci. USA*, **88**, 3739–43.

Liu, M.-L., Gibbs, E.M., McCoid, S.C. *et al.* (1993) Transgenic mice expressing the human GLUT4/muscle-fat facilitative glucose transporter exhibit efficient glycemic control. *Proc. Natl Acad. Sci. USA*, **90**, 11346–50.

Losordo, D.W., Pickering, J.G., Takeshita, S. *et al.* (1994) Use of the rabbit ear to serially assess foreign protein secretion after site-specific arterial gene transfer *in vivo*. *Circulation*, **89**, 785–92.

Lynch, C.M., Clowes, M.M., Osborne, W.R.A. *et al.* (1992) Long-term expression of human adenosine deaminase in vascular smooth muscle cells of rats: a model for gene therapy. *Proc. Natl Acad. Sci. USA*, **89**, 1138–42.

Mangano, D.T. (1990) Perioperative cardiac morbidity. *Anesthesiology*, **72**, 153–84.

Metzger, J.M., Parmacek, M.S., Barr, E. *et al.* (1993) Skeletal troponin C reduces contractile sensitivity to acidosis in cardiac myocytes from transgenic mice. *Proc. Natl Acad. Sci. USA*, **90**, 9036–40.

Milano, C.A., Allen, L.F., Rockman, H.A. *et al.*, (1994) Enhanced myocardial function in transgenic mice overexpressing the β_2-adrenergic receptor. *Science*, **264**, 582–6.

Miller, D.A. (1992) Human gene therapy comes of age. *Nature*, **357**, 455–60.

Morishita, R., Gibbons, G.H., Ellison, K.E. *et al.* (1993a) Single intraluminal delivery of antisense cdc2 kinase and proliferating-cell nuclear antigen oligonucleotides results in chronic inhibition of neointimal hyperplasia. *Proc. Natl Acad. Sci. USA*, **90**, 8474–8.

Morishita, R., Gibbons, G.H., Kaneda, Y. *et al.* (1993b) Novel and effective gene transfer technique for study of vascular renin angiotensin system. *J. Clin. Invest.*, **91**, 2580–5.

Morishita, R., Gibbons, G.H., Ellison, K.E. *et al.* (1994) Autocrine/paracrine angiotensin as a determinant of vascular structure: a gene transfer approach experiment. Abstracts of the International Society of Hypertension *S63*, 350.

Mulligan, R.C. (1993) The basic science of gene therapy. *Science*, **260**, 926–32.

Nabel, E.G. and Nabel, G.J. (1994) Complex models for the study of gene function in cardiovascular biology. *Annu. Rev. Physiol.*, **56**, 741–61.

Nabel, E.G., Plautz, G., Boyce, F.M. *et al.* (1989) Recombinant gene expression *in vivo* within endothelial cells of the arterial wall. *Science*, **244**, 1342–4.

Nabel, E.G., Plautz, G. and Nabel, G.J. (1990) Site-specific gene expression *in vivo* by direct gene transfer into the arterial wall. *Science*, **249**, 1285–8.

Nabel, E.G., Shum, L., Pompili, V. *et al.* (1993a) Direct transfer of transforming growth factor β1 gene into arteries stimulates fibrocellular hyperplasia. *Proc. Natl Acad. Sci. USA*, **90**, 10759–63.

Nabel, E.G., Yang, Z.-Y., Plautz, G. *et al.* (1993b) Recombinant fibroblast growth factor-1 promotes intimal hyperplasia and angiogenesis in arteries *in vivo*. *Nature*, **362**, 844–6.

Newgard, C.B. (1992) Cellular engineering for the treatment of metabolic disorders: prospects for therapy in diabetes. *Bio/Technology*, **10**, 1112–20.

Ohno, T., Gordon, D., San, H. *et al.* (1994) Gene therapy for vascular smooth muscle cell proliferation after arterial injury. *Science*, **265**, 781–4.

Packer, M. (1993) Positive inotropic agents for chronic heart failure: a new perspective. *Heart Failure*, **9**, 136–47.

Parmacek, M.S., Vora, A.J., Shen, T. *et al.* (1992) Identification and characterization of a cardiac specific transcriptional regulatory element in the slow/cardiac troponin C gene. *Mol. Cell. Biol.*, **12**, 1967–76.

Paul, M., Wagner, J., Hoffmann, S. *et al.* (1994) Transgenic rats: new experimental models for the study of candidate genes in hypertension research. *Annu. Rev. Physiol.*, **56**, 811–29.

Pickering, J.G., Bacha, P.A., Weir, L. *et al.* (1993) Prevention of smooth muscle cell outgrowth from human atherosclerotic plaque by a recombinant cytotoxin specific for the epidermal growth factor receptor. *J. Clin. Invest.*, **91**, 724–9.

Pickering J.G., Jekanowski, J., Weir, L. *et al.* (1994) Liposome-mediated gene transfer into human vascular smooth muscle cells. *Circulation*, **89**, 13–21.

Plautz, G., Nabel, E.G. and Nabel, G.J. (1991) Introduction of vascular smooth muscle cells expressing recombinant genes *in vivo*. *Circulation*, **83**, 578–83.

Prentice, H., Kloner, R.A., Sartorelli, V. *et al.* (1993) Transformation of cardiac fibroblasts into the skeletal muscle phenotype by injection of a MyoD-expressing retrovirus into ischemic heart. *Circulation*, **88**, I475.

Prentice, H., Kloner, R.A., Prigozy, T. *et al.* (1994) Tissue restricted gene expression assayed by direct DNA injection into cardiac and skeletal muscle. *J. Mol. Cell. Cardiol.*, (in press).

Riessen, R., Rahimizadeh, H., Blessing, E. *et al.* (1993) Arterial gene transfer using pure DNA applied to a hydrogel-coated angioplasty balloon. *Human Gene Ther.*, **4**, 749–58.

Simons, M., Edelman, E.R., DeKeyser, J.-L. *et al.* (1992) Antisense c-myb oligonucleotides inhibit intimal arterial smooth muscle cell accumulation *in vivo*. *Nature*, **359**, 67–70.

Soonpaa, M.H., Koh, G.Y., Klug, M.G. and Field, L.J. (1994) Formation of nascent intercalated disks between grafted cardiomyocytes and host myocardium. *Science*, **264**, 98–101.

Stratford-Perricaudet, L.D., Makeh, I., Perricaudet, M. and Briand, P. (1992) Widespread long term gene transfer to mouse skeletal muscles and heart. *J. Clin. Invest.*, **90**, 626–30.

Swain, J.L. (1989) Gene therapy. A new approach to the treatment of cardiovascular disease. *Circulation*, **80**, 1495–6.

Takeshita, S., Ga, D., Leclerc, G. *et al.* (1994) Increased gene expression after liposome-mediated arterial gene transfer associated with intimal smooth muscle cell proliferation. *J. Clin. Invest.*, **93**, 652–61.

Vincent, C.K., Gualberto, A., Patel, C.V. and Walsh, K. (1993) Different regulatory sequences control creatine kinase-M gene expression in directly injected skeletal and cardiac muscle. *Mol. Cell. Biol.*, **13**, 1264–72.

von Harsdorf, R., Schott, R.J., Shen, Y.-T. *et al.* (1993) Gene injection into canine myocardium as a useful model for studying gene expression in the heart of large mammals. *Circ. Res.*, **72**, 688–95.

Webster, K.A., Bishopric, N.H., Wu, X. *et al.* (1994) Targeting the expression of genes to hypoxic cardiac myocytes with hypoxia response enhancer elements. *Circ. Res.*, (submitted).

Williams, R.S., Johnston, S.A., Riedy, M. *et al* (1991) Introduction of foreign gene into tissues of living mice by DNA-coated microprojectiles. *Proc. Natl Acad. Sci. USA*, **88**, 2726–30.

Wilson, J.M., Birinyi, L.K., Salomon, R.N. *et al.* (1989) Implantation of vascular grafts lined with genetically modified endothelial cells. *Science*, **244**,1344–6.

Wilson, J.M., Grossman, M., Wu, C.H. *et al.* (1992) Hepatocyte-directed gene transfer *in vivo* leads to transient improvement of hypercholesterolemia in low density lipoprotein receptor-deficient rabbits. *J. Biol. Chem.*, **267**, 963–7.

Wolff, J.A., Malone, R.W., Williams, P. *et al.* (1990) Direct gene transfer into mouse muscle *in vivo*. *Science*, **247**, 1465–8.

Wolff, J.A., Dowty, M.E., Jiao, S. *et al.* (1992) Expression of naked plasmids by cultured myotubes and entry of plasmids into T tubules and caveolae of mammalian skeletal muscle. *J. Cell. Sci.*, **103**, 1247–59.

Wolinski, H. and Thung, S.N. (1991) Use of a perforated balloon catheter to deliver concentrated heparin into the wall of the normal canine artery. *J. Am. Coll. Cardiol.*, **15**, 475–81.

16

Degenerative and inherited neurological disorders

PEDRO R. LOWENSTEIN

The philosophers have only interpreted the world in various ways;
the point, however, is to change it
Karl Marx

16.1 INTRODUCTION

As of January 1995, a total of 95 human gene therapy/gene marking clinical protocols had been approved and/or presented worldwide; in addition many of these have already been started. Some 19% (18 protocols) are to treat inherited disorders, while 12% (eleven protocols) target CNS tumors (e.g. neuroblastomas and astrocytomas) (Summers, 1993; Anderson, 1994). None of the 95 protocols is to treat either inherited or sporadic neurological or neuromuscular disease proper. Also, of many new commercial gene therapy companies, only one is targeting gene therapy for neurological disease (Dodet, 1993; Summers, 1993). The current under-representation of human neurological gene therapy contrasts with its pre-eminent historical role, since human gene therapy was already proposed in 1972 for a metabolic disorder, the Lesch–Nyhan syndrome, which has devastating effects on brain function (Friedmann and Roblin, 1972). Nevertheless, conventional neurological therapies for both inherited and sporadic disorders are limited by factors such as drug side effects and the development of tolerance during long-term administration (Kopin, 1993), or technical shortcomings of neurosurgical procedures, for example due to tumor inaccessibility (Takamiya *et al.*, 1993).

Molecular and Cell Biology of Human Gene Therapeutics
Edited by George Dickson
Published in 1995 by Chapman & Hall. ISBN 0 412 62550 4

In this chapter a brief overview of current neurological therapies and some of their shortcomings (section 16.2) will be considered. Different disease mechanisms will require different therapeutic approaches. Thus, those specific neurological diseases which could be treated by gene therapy in the medium term will be discussed (section 16.3). Thirdly, neurological gene therapy strategies and gene transfer vectors will be reviewed as they apply specifically to gene transfer into the brain in general, and into neurons in particular (section 16.4; for further detailed reviews on individual vectors see Glorioso *et al.*, 1991; Breakfield *et al.*, 1992; Boris-Lawrie and Temin, 1993; Curiel, 1993; Leib and Olivo, 1993; Samulski, 1993; Wilkinson *et al.*, 1994; also see Chapters 2, 3 and 14). Finally, germ-line gene therapy and the future of neurological gene therapy (sections 16.5–16.6) will be considered.

The central issues concerning the future applications of gene therapy to the treatment of human neurological disorders, and in particular, inherited neurological disorders, are: (i) an understanding of the causes of disease; (ii) the elucidation of the molecular lesions and actions of gene mutations (in the case of inherited disease); (iii) the choice of vectors for gene transfer into the brain, with particular emphasis on the targeting of neurons; (iv) the existence of animal models which can be used to test novel therapeutic strategies; and (v) the development of strategies that can be applied to achieve the goal of human gene therapy for neurological disease, and to the selection of candidate disorders (Lowenstein, 1994).

The molecular lesions responsible for many inherited disorders, as well as rare familial forms of Alzheimer's disease, have been uncovered during the last 5 years. However, the causes underlying frequent sporadic degenerative disorders such as Parkinson's and Alzheimer's disease remain unknown, although certain alleles of the apolipoprotein E predispose to or protect against the development of Alzheimer's disease (Saunders *et al.*, 1993) (Appendix 16.A). Most recessive disorders (e.g. Tay–Sachs disease) are caused by the presence of defective gene products. Dominant disorders may arise through a variety of mechanisms: (i) by a reduction in the level of a protein product (e.g. hereditary neuropathy with liability to pressure palsies); (ii) the presence of an abnormal protein (e.g. Huntington's disease); (iii) an increase in the amount of gene product (e.g. Charcot–Marie–Tooth 1A); or (iv) the loss of heterozygosity, effectively a double mutation, where one is inherited and a second spontaneous one appears in the second allele (e.g. retinoblastoma, a disease which is dominant clinically, but is effectively recessive at the cellular level) (for a review, see Wilkie, 1994).

As with many new therapies, animal models which mimic important aspects of human disease, are very useful in demonstrating the potential feasibility of such new interventions. Many animal models for both inherited and sporadic brain disorders exist or are being developed. Even if some animal models of neurological disease do not exactly mimic the phenotype expressed in human patients, they will be of utmost importance in addressing questions of levels of expression, length of expression, and number of neurons or target cells that need to be transduced or transplanted into the brain in order to revert a pathological phenotype to normal.

Within this chapter, different gene therapy strategies applicable to *groups* of inherited, as well as non-inherited neurological disorders will be examined. To this end, criteria should be identified to define those disorders in which gene therapy is most likely to benefit patients, their corresponding therapeutic strategies, and to tailor available strategies to the particular requirements of individual disorders or groups of disorders. The aim of this process is to conceive therapeutic strategies for groups of disorders from the initial stages of the development of therapeutic gene transfer vectors. Gene therapy strategies (discussed in more detail in sections 16.3–16.4 and Table 16.2) would exploit the advantages provided by strategically designed vector backbones or gene transfer methods. Neuronal gene transfer methods should be effective in transferring genes to several target brain areas and neurons affected by various, possibly unrelated, disorders. DNAs or RNAs (e.g. entire genes, cDNAs, antisense DNA, etc.) of therapeutic interest could be cloned into such 'group-specific vector backbones', e.g. the use of a non-toxic HSV-1 vector for disorders affecting dorsal root ganglion neuron function. Given the diversity of diseases affecting brain function (Appendix 16.A), such strategies will allow to centralize available resources for neurological gene therapy and thus achieve the development of treatments for a large number of possible target disorders in the most efficient way.

16.1.1 What is gene therapy? Practical and ethical implications

Gene therapy has been classified by Anderson (1985) into four possible groupings: (1) somatic gene therapy; (2) germ-line gene therapy; (3) somatic gene enhancement; and (4) germ-line gene enhancement. In the first two groups, 'therapy' refers to treatments designed to relieve and/or prevent suffering in a patient, while the goal of groups 3 and 4 is 'enhancement' of characteristics designed to improve a particular attribute in an otherwise healthy individual. Current consensus of scientific and public opinion draws the line on the development and applicability of gene therapy to those applications described by group 1 (Anderson, 1985; Clothier *et al.*, 1992; Appendix 16.B).

The Clothier Committee (UK) (Clothier *et al.*, 1992; Appendix B) has defined somatic gene therapy as

> making good a defective gene in the body cells where it is needed. The aim is to provide the right genetic information, under proper control, in precisely those cells which need it for their normal function. Ideally the effect should be permanent so that no further therapy is required. It should also be permanent in the sense that the inserted gene is securely lodged. If successful, therapy would make good the genetic defect in the

treated cells and tissues of the individual. However, the germ cells, which give rise to sperm or ova, would retain the defective form of the gene, with the possibility of its transmission to future generations.

The purpose of somatic gene therapy in an individual patient is to alleviate disease in that individual, and that individual alone. There would be no intention to modify the gene in the germ-line. We consider that the development of safe and effective means of gene modification for this purpose is a proper goal for medical science.

As early as 1972, Friedmann and Roblin proposed plausible gene therapy strategies for inherited human disease, based on the then newly emerging knowledge of the molecular basis of both human disease, and the functioning of oncogenic viruses, including their ability to integrate their genomes into human DNA (see Wolff and Lederberg, 1994, for a recent review of the early history of gene therapy). More recently, Roemer and Friedmann (1992) described the different theoretical possibilities encompassed by gene therapy. Ideally what should be achieved for the treatment of inherited disorders is the *replacement* of the faulty copy of a disease-causing gene, by its normal counterpart, inserted into its own genomic environment. However, all somatic gene therapy applications currently being considered, actually are strategies for *gene augmentation*, i.e. the transfer of an additional normal copy of a gene into cells expressing a diseased copy, rather than the replacement of mutated copies.

The transfer of genes into humans has raised important ethical issues, which are currently being widely discussed (Morgan and Anderson, 1993; Wivel and Walters, 1993). In considering the transfer of genes directly into the brains of patients to cure disease, the scientific community and the public will have to evaluate carefully whether this will arouse unique ethical issues, different from those raised in response to trials for gene transfer into the bone marrow or other somatic tissues. Many disorders that affect the brain cause cognitive abnormalities, and gene therapy will tend to improve their intellectual abilities. Thus, it will be important to discuss whether gene therapy of cognitive disabilities is still best described as gene therapy, or whether it should be considered an application more likely described as gene enhancement. Either way, there is currently a need to explore the ethical dimensions of neurological gene therapy through an open and public debate, well in advance of the start of neurological gene therapy clinical trials.

16.2 CURRENT STATUS OF THERAPEUTIC INTERVENTION IN SELECTED NEUROLOGICAL DISORDERS

Although there is great variability in the time of onset of different types of neurological disease, inherited diseases in general have an earlier onset than non-inherited disorders. Inherited diseases account for 40% of infant mortality in developed countries (Connor, 1989). On the other hand, increased life expectancy in developed countries, has led to a constant increment in the total number of old people suffering from neurodegenerative disorders such as Parkinson's and Alzheimer's disease (Rocca *et al.*, 1991; Zhang and Roman, 1993; see Appendix 16.A for disease frequencies). In the absence of any curative treatments (Kopin, 1993; Schehr, 1994), the management of old-age neurodegenerative disorders represents a progressive burden for both families and the public health system.

Appendix 16.A presents a classification of both inherited and sporadic neurological disorders, with a brief description of their frequency, pathophysiology, and where known, molecular basis. Appendix 16.A has been divided into the following categories: (1) degenerative disorders of the central nervous system, including ataxias, and retinal degenerations; (2) tumors; (3) anterior horn motor neuron diseases; (4) inherited peripheral neuropathies; (5) muscle disorders; (6) metabolic disorders; and (7) mental retardation.

Unfortunately, shortcomings in our understanding of the pathophysiology of many CNS disorders constrain our capacity to develop effective therapies for many diseases. Surgical intervention is applicable to the resection of accessible tumors, while pharmacological treatments are successfully used to manage a wide range of afflictions, like infections, Parkinson's disease, headaches and pain syndromes (Kopin, 1993; Markert *et al.*, 1993; Takamiya *et al.*, 1993). Unfortunately, only the treatments of some infectious disorders are truly curative; most other therapies are palliative. Many opportunities are thus available to improve neurological therapeutics, given that even benign tumors can remain untreatable because of their location deep within the brain, and infections such as AIDS, remain incurable.

In diseases where the pathophysiology is relatively well understood, e.g. Parkinson's disease, many different pharmacological approaches have been developed with different degrees of success (Kopin, 1993). Treatment for Parkinson's disease is certainly palliative, but it has done much to improve the quality of life of patients. What palliative treatments cannot offer though, is to cure or halt the progression of the disorder. Unfortunately, even gene therapy cannot offer the prospect of a potential cure in the absence of an understanding of the ultimate cause of the disease. Nevertheless, gene therapy could offer another line of treatment for Parkinson's and Alzheimer's disease, after classical pharmacological ones have ceased to serve individual patients. Since gene therapy can also be defined broadly as 'the transfer of new genetic material to the cells of an individual with resulting therapeutic benefit to the individual' (Morgan and Anderson, 1993), any form of gene therapy can be regarded as a novel pharmacological agent, with all the power and limitations of pharmacological treatments.

The following factors complicate the design of novel therapeutic strategies for non-inherited degenerative disorders of the brain: (i) their time of onset cannot be predicted exactly; (ii) clinical diagnosis is often the only type of diagnosis available, and is not 100% accurate; (iii) there is incomplete information regarding their ultimate causes; (iv) it is unclear whether the natural evolution can be halted by effective palliative treatments; and (v) their long-term natural evolution can be erratic and unpredictable.

One factor which influences the time of onset of symptoms, their manifestations, and their response to treatment is the redundancy of neuronal networks present in the brain. Such redundancy provides functional flexibility and resilience to damage caused by the loss of component neurons. While neuronal network redundancy and functional adaptability are important evolutionary advantages of the mammalian brain, they determine that, in many diseases, symptoms will only become apparent after a majority (up to 80–90%) of neurons have been lost. This is thought to occur in Parkinson's disease, and stroke. As neurodegeneration proceeds, increased activity of remaining neurons can also compensate functionally for the reduction in absolute neuron numbers. This delays the overt onset of symptoms until the functional capability of remaining neurons cannot compensate for the amount of neuronal loss. The pathologically induced increase in the functional activity of remaining neurons also probably accounts for the failure of some treatments that attempt to further increase the activity of remaining cells. In slowly progressive diseases, early diagnostic methods could lead to significant clinical benefit by allowing prompt therapeutic intervention, e.g. the administration of growth factors, to halt neuronal death at initial stages of disease progression (Jinnah *et al.*, 1993; Tomec *et al.*, 1995).

In comparison with idiopathic neurological disease, the pathophysiology of many inherited metabolic disorders affecting brain function is well understood, and a variety of therapies have been developed for their treatment (Baranger and Ginns, 1989; Kolodny, 1989; O'Brien, 1989; Sandhoff *et al.*, 1989; Scriver *et al.*, 1989; Spence and Callahan, 1989; Suzuki and Suzuki, 1989; Brady and Barton, 1991; Emery, 1991; Krivit and Shapiro, 1991; Moser *et al.*, 1991; Bundey, 1992; Sege-Peterson *et al.*, 1993). In these disorders the presence of a defective enzyme within a multi-enzymatic metabolic pathway leads to an accumulation of toxic precursors (Conzelmann and Sandhoff, 1983/4; Leinekugel *et al.*, 1992). The accumulation of neurotoxic metabolites causes neuronal disease or neuronal death, and leads to the slowly progressive CNS dysfunction encountered in most, if not all, of the gangliosidoses (O'Brien, 1989; Sandhoff *et al.*, 1989), mucopolysaccharidoses, and leukodystrophies (Kolodny, 1989; Moser and Moser, 1989; Suzuki and Suzuki, 1989) (excluding Pelizaeus–Merzbacher disease, which causes CNS demyelination).

Importantly, the availability of newborn screening programs has led to early detection of many diseases and the prompt institution of treatments has drastically improved the long-term prognosis for these patients. For example, a deficiency of the enzyme phenylalanine hydroxylase causes the disease phenylketonuria (Table 16.1). Withdrawal of phenylalanine from the diet of patients will avoid the most severe manifestations of this disease. Nevertheless, controversy still exists regarding when to stop the treatment, and whether adult patients would suffer disease symptoms if they reintroduce phenylalanine into their diets. Dietary control of metabolic disorders has also been applied to other disorders including adrenoleukodystrophy (Moser *et al.*, 1991). This experimental therapy requires close long-term supervision of patient compliance and treatment performance, throughout the life of the affected individual, and, while slowing disease progression, it does not revert the pathological phenotype to normal (Scriver *et al.*, 1989).

Interestingly, in some disorders, such as the Lesch–Nyhan syndrome, the metabolic derangement perturbs neuronal physiology, but does not result in neuronal degeneration. Nevertheless, the relationship between the deficiency of the purine salvage pathway enzyme hypoxanthine–guanine phosphoribosyl transferase (HPRT), to the characteristic triad of mental retardation, self-mutilation, and neurological dysfunction of Lesch–Nyhan patients, remains to be elucidated (Sege-Peterson *et al.*, 1993). Thus, it is important to consider carefully the pathophysiology and anatomical expression of each disease, since this will determine which therapeutical options should be considered further.

Other treatments which have been made available for the management of inherited metabolic diseases, which affect brain function, are enzyme replacement by direct intravenous infusion (Sandhoff *et al.*, 1989; Brady and Barton, 1991) and bone marrow transplantation (Hoogerbrugge *et al.*, 1988; Krivit and Shapiro, 1991; Moser *et al.*, 1991; Shapiro *et al.*, 1991). Enzyme replacement has provided some clinical benefit mainly for non-neurological symptoms, but is useful for the treatment of only some metabolic disorders (e.g. Gaucher's disease, or the immune disorder caused by adenosine deaminase deficiency). Nevertheless, direct intravenous enzyme infusion reduces the level of circulating toxic metabolites, but the effect is necessarily transient. In addition, such treatments are extremely expensive, costing up to US$ 100 000 per year for the treatment of Gaucher's disease.

Microglial cells, or brain macrophages, originate from bone marrow macrophages. Thus, bone marrow transplantation has been attempted for the treatment of metabolic brain disorders in which the intracerebral accumulation of toxic metabolites could be reversed, at least theoretically, by expression of enzyme activity within brain microglial cells (Hoogerbrugge *et al.*, 1988; Suzuki and Suzuki, 1989). Studies using an experimental animal model for Krabbe's disease (the *twitcher* mouse) have shown, that the entry of macrophages originating from the transplanted bone marrow into the brain produces an increase in the level of

Table 16.1 Relationship between enzyme level (as % of control) and clinical phenotype in inherited metabolic disorders

1. Degradation of ganglioside G_{M2} (pmol/[h·mg·activator unit]) by extracts of fibroblasts from patients with G_{M2} gangliosidosis (Tay–Sachs and Sandhoff disease)[a]

	Mean	*Range*	*% of controls*	*Onset age*
Control	535	296–762	100	
Heterozygote	285	121–395	53.3	
Tay–Sachs disease				
Infantile	2.4	0.8–3.8	0.45	3–5 months
Juvenile	15.8	13.6–18.0	2.95	2–6 years
Adult	19.1	13.1–32.8	3.57	>14 years
Sandhoff disease				
Infantile	6.6	3.6–9.5	1.23	3–5 months
Juvenile	23.3	9.5–39.4	4.29	3–10 years
Healthy adults	105	75–143	19.6	

2. Glucocerebrosidase activity (nmol/mg protein/h) in leukocytes of Gaucher's disease patients[b]

	Mean	*Range*	*% of controls*	*Onset age*
Control	13.5	4.0–20.2	100	
Heterozygotes	9.4	2.9–24.6	69.6	
Type 1 (chronic, non-neuronal)	4.2	0.0–9.0	31.1	infancy–adulthood
Type 2 (acute, neuronal)	3.3	0.4–7.3	24.4	3 months
Type 3 (subacute, neuronal)	2.7	1.8–4.0	20.0	childhood, variable

[a] Modified from Sandhoff *et al.* (1989); [b] modified from Baranger and Ginns (1989); [c] modified from Kolodny (1989); [d] modified from Sege-Peterson *et al.* (1993); [e] modified from Scriver *et al.* (1989).

PAH, phenyalanine hydroxylase; Non-PKU (A and B), two types of hyperphenylalaninemias without phenylketonuria; HPRT, hypoxanthine-guanine; phosphoribosyl transferase; n.d., not determined

Table 16.1 *Continued*

3. Leukocyte sulfatidase activity (nmoles of sulfatide hydrolyzed/mg protein/2 hours), cerebral white matter-sulfatides and sulfatide excretion in metachromatic leukodystrophy[c]

	Mean	*Range*	*% of controls*	
Controls	12.92	10–24	100	
Heterozygotes 1–11 (5±3)	5.08	0.5–11	39.3	
Patients	0	0	0.0	
	White-matter sulfatides	*Sulfatide catabolism (%)*	*Excretion*	*Onset age (years)*
Control	1.7–4.1	Normal	Normal	
Late infantile	12.8–15.8	0–10	Increased	1–2
Juvenile	n.d.	15–30	Increased	4–12
Adult	6.5	30–50	Increased	15–62

4. Phenotype of HPRT deficiency[d]

	% control HPRT activity
Classical Lesch–Nyhan (mental retardation, self-mutilation, choreoathetosis)	≤1.4
Intelligent Lesch–Nyhan (self-mutilation, choreoathetosis)	1.4–1.6
Neurological (choreoathetosis)	1.6–8
Neurologically normals (hyperuricemic)	8–60
Neurologically normals (normal uricemia)	≥ 60

5. Phenylketonuria (PKU)[e]

	Mean	*Range*	*% of controls*
In vitro phenylalanine hydroxylase activity (μmoles tyrosine/g protein/h)			
Controls	100	60–120	100
Non-PKU		2–40	2–40
PKU	≤ 2	≤ 2	≤ 2
In vivo residual phenylalanine hydroxylase activity			
Controls	100		100±10
Non-PKU (A)	25		25±10
Non-PKU (B)	7		7±5
PKU	0.5		0.5±2

missing galactocerebroside enzyme activity in the CNS (Hoogerbrugge *et al.*, 1988). Bone marrow transplantation could thus be of therapeutic value in conditions like Krabbe's (Shapiro *et al.*, 1991), and Gaucher's disease. However, its clinical applicability could be limited by the inherent problems normally associated with bone marrow transplantation, and by the lack of direct transfer of enzyme activity into neurons. In many inherited diseases of metabolism in which neuronal physiology is specifically affected, enzymatic activity is required within affected neurons. Whether an elevation of intraneuronal enzyme activity could be achieved by either enzyme replacement or bone marrow transplantation has not yet been conclusively determined. Nevertheless, very recent data (Walkely *et al.*, 1994) indicate that intraneuronal α-mannosidase activity can be detected in neurons in a feline model of α-mannosidosis after heterologous bone marrow transplantation. How far these results could be extrapolated to other experimental paradigms remains to be determined. Thus, in the future, autologous bone marrow transplantation after *in vitro* transduction could be applicable to a number of inherited and possibly sporadic neurological disorders, as a method to deliver new genes and gene products to the brain, and possibly into neurons.

16.3 CANDIDATE DISEASES FOR NEUROLOGICAL GENE THERAPY

16.3.1 Non-inherited degenerative disorders, tumors, and viral infections

Alzheimer's and Parkinson's disease are two degenerative disorders of the brain whose ultimate causes remain unknown in the majority of cases (Appendix 16.A). Both conditions affect older patients and their incidence increases with increasing age (Rocca *et al.*, 1991; Zhang and Roman, 1993). More than 10% of the population of Europe over 80 years of age suffers from Alzheimer's disease (Rocca *et al.*, 1991; Appendix 16.A). Thus, the burden on families and society posed by these diseases is enormous, and this has created pressures to develop novel and effective therapeutic approaches (Lowenstein, 1994).

It is well established that the main pathophysiological manifestations of Parkinson's disease originate from the degeneration of nigrostriatal dopaminergic neurons. To restore the levels of dopamine in the basal ganglia of patients suffering from Parkinson's disease, several therapeutic options have been developed and exploited during the last 20 years, including the administration of dopamine precursors, dopamine agonists, inhibitors of amine transmitter metabolizing enzymes, or inhibitors of dopamine re-uptake (Jinnah *et al.*, 1993; Kopin, 1993). Nonetheless, while most of these pharmacological treatments greatly improve the patients' quality of life, it is unclear whether they also alter the long-term course of deterioration caused by the disease.

A further obstacle for developing new therapeutic approaches to Parkinson's disease is the lack of faithful animal models of the naturally occurring disease state. Although the nigrostriatal dopaminergic pathway can be lesioned in both rodents and monkeys using a variety of surgical or pharmacological approaches to create useful animal models of Parkinson's disease, such lesioned animals do not display the progressive nature and widespread brain involvement seen during the natural evolution of human Parkinson's disease. Interestingly the neurotoxin MPTP, a by-product synthesized during the illegal production of heroin, causes parkinsonian symptoms in humans and can be used to produce experimental parkinsonism in monkeys. In this case, the animal model does faithfully reproduce the human condition (Kopin, 1993), and such monkeys constitute a useful model for the study of both the human disease, and its response to novel treatments such as neuronal transplantation or gene therapy (Jinnah *et al.*, 1993).

Unfortunately, possible avenues for treating Alzheimer's disease remain unclear. The primary defect underlying this disease is unknown. However, brains of affected patients display an increased accumulation of β-amyloid peptide, and mutations in the β-amyloid peptide precursor have been detected in families affected by familial Alzheimer's disease. Overall, no unique pathophysiological hypothesis has yet been put forward to explain all cases of this devastating infirmity. Nevertheless, different therapeutic strategies have been proposed, such as the administration of inhibitors of acetylcholine-metabolizing enzymes (this neurotransmitter is decreased in Alzheimer's brains, and is also known to be involved in memory processes), the direct delivery to the brain of either nerve growth factor, or other growth factors, to halt neuronal death, and also the administration of compounds which either inhibit the accumulation of β-amyloid in the brain, or protect the brain from chronic inflammation (Schehr, 1994). Whether any of these strategies will prove successful in the long term remains to be determined.

Several approaches to the production of animal models of Alzheimer's disease are being pursued. Rats with lesions of their basal forebrain cholinergic neurons innervating the cortex, a deficit also seen in human Alzheimer's patients, display impaired memory, and this can be treated by either increasing their brain level of acetylcholine pharmacologically or through the use of neuronal transplants. The long-term, progressively deteriorating course of Alzheimer's, however, is not mimicked by these animal models. Also, transgenic animal models, expressing either additional copies or mutated copies of the β-amyloid gene are being developed (Gomes *et al.*, 1995). Yeast artificial chromosomes are being used to transfer large genomic sequences and their endogenous long-range regulatory elements into the germ-line of mice to develop transgenic animals models for a variety of both inherited and non-inherited neurological disorders, including Alzheimer's disease.

A theoretical drawback faced when designing new therapeutic approaches to treat degenerative disorders of the brain in which the ultimate cause remains unknown, is that such treatments, while providing much benefit to patients, remain necessarily palliative, not curative. Proposed novel therapies involve either restoring a missing neurotransmitter (e.g. dopamine in Parkinson's, acetylcholine in Alzheimer's), or replacing the missing neurons (Gage *et al.*, 1987; Fisher and Gage, 1993) [e.g. transplantation of fetal substantia nigra neurons in Parkinson's (Freed *et al.*, 1992; Spencer *et al.*, 1992), or basal forebrain neurons in Alzheimer's (Gomes *et al.*, 1995)]. Whether deficits seen in these patients are primary causes of disease, or secondary consequences, remains to be determined. Ideally, gene therapy will offer the possibility of *curing* the manifestations of those disorders for which the molecular cause is known (gene mutation), by either replacing the defective gene (gene transduction/augmentation), or inhibiting the expression of disease alleles (gene inhibition).

Although tumors can be classified as acquired genetic disorders, their treatment aims to eliminate the tumor cells. These are the only diseases of the CNS, which are currently being targeted using gene therapy. Specifically, the gene therapy strategies being developed are of three types; (i) selective killing of tumor cells transduced with retroviruses encoding potentially cytotoxic genes (Culver and Blaese, 1994); (ii) inhibition of growth factor production; and (iii) immunostimulation against tumor antigens (Summers, 1993; Anderson, 1994). An alternative strategy which is being used experimentally, but has not yet been proposed for human trials of gene therapy is the use of neurotoxic HSV-1 vectors (Markert *et al.*, 1993).

The selective killing of tumor cells transduced with toxic genes encoded by a retroviral vector, takes advantage of the fact that dividing tumor cells, but not normal post-mitotic neurons, are transduced by retroviral vectors. This therapeutic strategy consists of injecting retrovirus vectors or retroviral-producing cell lines directly into brain tumors using stereotactic surgery. The retroviral vectors encode the herpes simplex virus thymidine kinase gene, which will be expressed by transduced cells. When patients are later injected with ganciclovir, the herpes simplex thymidine kinase will convert it into a cytotoxic compound. Since only tumor or dividing cells, but not post-mitotic neurons, will express the herpes thymidine kinase, they should be killed upon ganciclovir challenge. This strategy is effective in significantly reducing or eliminating tumors from animals transplanted with human tumors (Takamiya *et al.*, 1993). Clinical trials are underway, and preliminary information indicates that this strategy can be effective in reducing tumor size in certain cases, although the responsible mechanisms (e.g. direct cytotoxicity, important bystander effect, or destruction of tumor blood vessels) remain unclear. The other strategies described above are based on the hypothesis that reducing the amount of available growth factors or inducing a strong immune response to tumor antigens, will decrease tumor growth and size. These strategies constitute effectively the first clinical trials for CNS disease (Culver and Blaese, 1994), and as such, their outcome will be important in dictating the future pathways to be followed by human neurological gene therapy.

Viral infections have also been considered as acquired disorders amenable to treatment using gene therapy protocols (Morgan and Anderson, 1993). Two such brain infections which should certainly become a target for gene therapy are the most severe manifestations of herpes simplex virus infection, namely, encephalitis caused by HSV-1 (in adults) or HSV-2 (in neonates), as well as the variety of neurological manifestations due to HIV infection of the brain. Antisense technology, as well as genetic suppressor elements, can certainly be exploited to inhibit the replication of these viruses in either the infected neurons (HSV-1, HSV-2) or brain microglial cells (HIV).

16.3.2 Inherited disorders

Are there any advantages in considering inherited brain disorders as candidate diseases for gene therapy? This question is often asked when evaluating whether it would be worth the effort for either the public health system or industry to develop treatments for disorders of relatively lower incidence? (Emery, 1991; Bundey, 1992; Martin, 1993; McKusick, 1994; Appendix 16.A)

In view of the efforts which have already been dedicated to unravel the causes of neurological disease (Bundey, 1992; Martin, 1993; McKusick, 1994), I believe that such disorders constitute both, proper targets for gene therapy in their own right, as well as test models to demonstrate the feasibility and applicability of gene therapy to neurological disease in general.

First, inherited neurological disorders for which the responsible gene has been cloned provide ideal *test* diseases to determine whether the molecular replacement of the missing function due to a mutated gene can achieve phenotypic reversion, and thus whether neurological gene therapy can be proposed as a *curative* treatment.

Second, using animal models with homologous lesions in corresponding genes (Smithies, 1993), we should be able to predict when, during a patient's lifetime, normal gene function, provided by effective gene therapy, is required to revert, or prevent, the appearance of a given pathological phenotype. This knowledge will be an important adjunct to molecular genetics, which can be used to predict, in families suffering from characterized inherited diseases and carrying identified mutations, which family members are likely to become ill (e.g. dominant disorders), or to diagnose diseases where family history is unknown (e.g. recessive disorders). Thus, experiments on animal models will be extremely useful in addressing the abovementioned questions, even if the symptoms shown by these models do not exactly mimic the human disease phenotype.

Early prediction afforded by molecular genetics will also be important within the future therapeutic management of inherited disorders, especially those presenting in early childhood, and with a rapidly deteriorating clinical evolution. Such early diagnosis (Martin, 1993; MacMillan and Harper, 1994) will allow prompt initiation of preventive or restorative treatment, as soon as the altered gene is detected, possibly even *in utero* (Ekhterae *et al.*, 1990). Thus, early diagnosis will offer the possibility of starting therapeutic interventions before any substantial neuronal loss has occurred. It is illustrative to compare this example with the idiopathic neurodegeneration of Parkinson's disease, in which the disease can only be detected after 90% of dopaminergic nigrostriatal neurons have been lost to the disorder.

Third, in how many neurons and at what level of expression will therapeutic transgenes need to be expressed to revert a pathological phenotype? Given the high degree of redundancy in the structure and possibly the function of the nervous system, it is likely that even a low percentage of transduced neurons (5–10%) could make a major difference in terms of clinical outcome. While it would be difficult to predict accurately the level of transgene expression required in all disorders, the study of patients suffering from recessively inherited metabolic disorders affecting brain function can provide quite accurate estimates of the target levels of transgene expression gene therapy should achieve.

Table 16.1 lists the levels of enzyme activity in controls, heterozygotes, and patients suffering from different inherited metabolic disorders. In many of these as little as 2% of remaining enzyme activity appears to be sufficient for humans to remain free of at least the most severe neurological symptoms. It will be important to ascertain whether the 2% actually refers to the *number* of neurons expressing enzyme activity, or whether it refers to the *amount of activity* needed in all neurons. The exact target percentages of both transduced neurons and enzyme levels needed to revert pathological phenotypes to normal are likely to depend on the particular disease examined, but useful approximations could be determined in advance of any therapeutic interventions based on animal and *in vitro* models. In Lesch–Nyhan, where no neurodegeneration has been described, 10% of basal ganglia neurons might be the required target population needed to be transduced, to restore CNS function to normal, with each transduced neuron expressing 20% of normal HPRT levels. On the other hand, in Tay–Sachs, we might need to restore intraneuronal hexosaminidase activity to 10% of normal in most neurons of the CNS, since it is still possible that non-transduced cells would continue to die, because of the continued intracellular accumulation of gangliosides (Table 16.1) (Friedmann and Jinnah, 1993; Jinnah *et al.*, 1993). Nevertheless, the phenomenon of metabolic cooperation, e.g. due to intercellular transfer of small molecules, might affect the absolute percentage of transduced neurons needed to revert the pathological phenotype of CNS dysfunction in diseases like Tay–Sachs.

To aid the development of future therapeutic approaches, Appendix 16.C lists criteria that could provide guidance in the designing and planning of future neurological gene therapy clinical protocols, and Table 16.2 outlines strategies that could be applied to more than one disease state. It is important to realize that experimental and clinical tools have now become available to pursue the development and implementation of neurological gene therapy. Not only are we in a position to ask questions about how to achieve the treatment of neurological disease by gene therapy, but we can also determine the relevant answers needed for the implementation of future clinical trials.

Huntington's disease is a neurodegenerative disorder mainly affecting basal ganglia neurons, which can manifest itself anytime during the patients' life. Its clinical manifestations evolve over 10–20 years. Importantly, the particular type of molecular lesion responsible for Huntington's disease (expansion of an unstable trinucleotide repeat) is common to other disorders also mainly affecting nervous system function, e.g. myotonic dystrophy, Fragile-XA and Fragile-XE, Kennedy's disease, dentatorubral pallidoluysian syndrome and spinocerebellar ataxia type I (SCAI) Appendix 16.A). Treatment for Huntington's disease will certainly need to focus on long-term expression of therapeutic gene products in presymptomatic individuals, since neurological lesions have been demonstrated in such cases.

It is also likely that other neurological diseases will share this kind of molecular lesion. Many of these disorders (Huntington's disease, myotonic dystrophy, DRPLA and SCA1) display anticipation, the earlier onset of the disease in subsequent generations, and this is related to the progressive expansion of the repeat sequence in transmission between generations; they also present a statistically significant positive correlation between triplet repeat expansion size and age of onset. The effects of trinucleotide repeat expansion, nevertheless, are disease specific; in Huntington's disease the expansion causes a gain of function, in Fragile X the FMR1 gene containing the trinucleotide repeat expansion is not transcribed due to hypermethylation, and in Kennedy's disease, the transcriptional activity of the mutated androgen receptor is reduced.

Thus, long-term strategies could be developed to inhibit the expression of the mutant, disease-causing allele as treatment for some inherited neurodegenerative disorders, where the molecular lesion is an expansion of unstable trinucleotide repeats resulting in a gain of function, and are phenotypically dominant, like Huntington's disease . This could be achieved through the use of vectors expressing antisense RNA (Murray and Crockett, 1992), genetic suppressor elements (Holzmayer *et al.*, 1992; Summers, 1993), or allele-specific ribozymes (Beaudry and Joyce, 1992). Pathophysiological considerations might also lead to the inclusion of other

Table 16.2 Strategies for human neurological gene therapy

Strategy	*Disease*	*Gene/gene product*	*Consequence of mutation*	*Target tissue*	*Vectors*
Gene augmentation	Lesch–Nyhan	HPRT	Loss of activity	Basal ganglia	HSV-1; Adenovirus, AAV
	Tay–Sachs	Hexosaminidase A	Loss of activity	Forebrain	Bone marrow, HSV1 Adenovirus, AAV
	Gaucher's	Glucocerebrosidase	Loss of activity	Forebrain	Bone marrow, HSV1 Adenovirus, AAV
	Kennedy's disease	Androgen receptor	Loss of function?	Anterior horn, motor neurons	HSV-1, Adenovirus, AAV
	Fragile X	FMR1	Loss of function	CNS	HSV-1, Adenovirus, AAV
	Familial dysautonomia	Unidentified	Unknown	Dorsal root ganglion neurons; posterior horn neurons	HSV-1, Adenovirus, AAV
Gene inhibition	Huntington's disease	IT15 (huntingtin)	Gain of function?	Basal ganglia, cerebral cortex	AAV, HSV-1, Adenovirus, Genetic suppressor elements
	Creutzfeld–Jakob disease	Prion protein gene	Accumulation of gene product	Forebrain	AAV, HSV-1, Adenovirus, Genetic suppressor elements
	Spinocerebellar ataxia I	Trinucleotide expanded allele (gene unidentified)	Unknown	Dorsal root ganglion neurons Cerebellum	HSV-1, Adenovirus, Genetic suppressor elements
	CNS infections: HSV-1, HIV.	Viral genes	Gain of function	Neurons, macrophages	AAV, Retroviruses, Genetic suppressor elements

Table 16.2 *Continued*

Strategy	*Disease*	*Gene/gene product*	*Consequence of mutation*	*Target tissue*	*Vectors*
Cell replacement	Parkinson's	Tyrosine hydroxylase	n/a	Basal ganglia neurons; alternatively fibroblasts, myoblasts, neuronal cell progenitors	Retroviruses and cell transplantation Direct gene transfer into neurons (e.g. HSV1)
	Alzheimer's	Neuronal growth factors? Choline acetyltransferase?	n/a	Neuronal cell progenitors, embryonic neurons, fibroblasts, etc., for transplantation Cortical neurons	Retroviruses and cell transplantation Direct gene transfer into neurons (e.g. HSV 1)
	Charcot–Marie–Tooth IA	PMP-22	Increase of function (duplication)? Dominant negative mutation (point mutations)?	Schwann cells	Retroviruses, liposomes, adenovirus particle-mediated gene transfer and cell transplantation
	Charcot–Marie–Tooth X1	Connexin 32	Loss of function	Schwann cells	Retroviruses, liposomes, adenovirus particle-mediated gene transfer and cell transplantation
	Pelizaeus–Merzbacher disease	PLP	Increase of function (duplication)? Dominant negative mutation (point mutations)?	Oligodendrocytes	Retroviruses, liposomes, adenovirus particle-mediated gene transfer and cell transplantation
	Multiple sclerosis	?	?	Oligodendrocytes	Retroviruses, liposomes, adenovirus particle-mediated gene transfer and cell transplantation

Table 16.2 *Continued*

Strategy	*Disease*	*Gene/gene product*	*Consequence of mutation*	*Target tissue*	*Vectors*
Cell killing, tumors	Neurofibromatosis	HSV-1 TK, other toxic genes;	NF1: decreased amount of protein NF2: loss of function?	Tumors	Retrovirus, HSV-1, Adeno-particle-mediated gene transfer, liposomes, antisense or genetic suppressor elements for growth factors, etc.
	von-Hippel Lindau	HSV-1 TK, other toxic genes.	Loss of function	Cerebellar tumors	Retrovirus, HSV-1, Adeno-particle-mediated gene transfer, liposomes, antisense or genetic suppressor elements for growth factors, etc.
	Tuberous sclerosis	HSV-1 TK, other toxic genes.	Loss of function	Brain tumors	Retrovirus, HSV-1, Adeno-particle-mediated gene transfer, liposomes, antisense or genetic suppressor elements for growth factors, etc.
	Neuroblastoma	HSV-1 TK, other toxic genes		Tumors	Retrovirus, HSV-1, Adeno-particle-mediated gene transfer, liposomes, antisense or genetic suppressor elements for growth factors, etc.

diseases into this therapeutic category, such as Creutzfeld–Jakob disease, which is likely to benefit from a decrease in expression of the mutated prion protein gene over a long period.

Avenues for the development of neurological gene therapy can also be forecasted for Charcot–Marie–Tooth and Pelizaeus–Merzbacher disease, two disorders affecting the myelinating cells of the peripheral and central nervous system, respectively. In both, the relevant disease genes have been recently cloned, and the molecular lesions identified. In either disease, normal oligodendrocytes or Schwann cells could either be transplanted, or patients' own cells could be engineered *in vitro* using retroviral vectors and returned to the patients (Appendix 16.A and Table 16.2). Alternatively, suitable vectors could be injected directly into the peripheral nerves of patients suffering from Charcot–Marie–Tooth, to target Schwann cells.

Thus, it may be proposed that various target diseases and possible strategies can already be identified, and that the precise tactics followed by neurological gene therapy should be adapted to the particular characteristics of each individual disease, e.g. the individual mutation, specific therapeutic requirements, disease progression, available vectors, the number of neurons to be transduced, and levels of transgene expression required. Having reviewed the possible target diseases for neurological gene therapy, the available strategies for transfering genes into the human brain, and directly into neurons will be examined. (For further in-depth reviews of individual diseases, see Rosenberg *et al.*, 1993.)

16.4 TARGETING FOREIGN GENES TO THE BRAIN

16.4.1 Transplantation of normal or genetically engineered cells

Both neurons and non-neuronal cells have been used in transplantation paradigms as platforms for gene transfer into the brains of experimental animals and human patients (Gage *et al.*, 1987; Freed *et al.*, 1992; Spencer *et al.*, 1992; Fisher and Gage, 1993). The cellular target of foreign gene transduction and transplantation will depend on the disorder under consideration. For the treatment of degenerative disorders, in which the anatomical locations of brain pathology are known, like Parkinson's or Alzheimer's disease, cells engineered to release growth factors together with neuronal cell transplantation might be the appropriate therapeutic strategy to halt ongoing neuronal death and attempt to replace the lost neurons (Jinnah *et al.*, 1993; Tomec *et al.*, 1995).

In Parkinson's disease more than 90% of the dopaminergic nigrostriatal neurons innervating the caudate and putamen will die before clinical parkinsonism is detected. In response to the long-term shortcomings of the pharmacological treatment of Parkinson's disease (Kopin, 1993), new therapeutic tools have been developed, the foremost of them, neural transplantation (Fisher and Gage, 1993). The main hypothesis is to transplant either dopaminergic neurons themselves (fetal midbrain transplants) or cells that produce dopamine or catecholamines (either from the patients' adrenal medulla, or genetically engineered ones). Much animal work supports the basic tenets of these approaches, and thus cell transplantation has already been attempted in Parkinson's disease patients (Freed *et al.*, 1992; Spencer *et al.*, 1992; Fisher and Gage, 1993).

However, because of the limited supply of human cells available for transplantation, other cellular alternatives have been developed (Gage *et al.*, 1987). Non-neuronal cells, e.g. fibroblasts, myoblasts, or neuroblasts, can be genetically engineered with the tyrosine hydroxylase gene (Fisher and Gage, 1993). Myoblasts lipofected with the gene encoding tyrosine hydroxylase release dopamine, rather than dopa, given that myoblasts already express the enzyme dopa decarboxylase (Jiao *et al.*, 1993). Since the missing metabolite is dopamine, myoblasts might represent a better cellular target than fibroblasts, which produce dopa. Alternatively, the genes encoding for both enzymes could be transferred into fibroblasts. Using animal models described above different types of engineered cells have been transplanted to the brain and shown to reverse the effects of experimental parkinsonism (Fisher and Gage, 1993). The extent to which transplanted neurons will integrate into the adult host's neuronal networks remains to be determined. This is likely to depend on the cell type employed, as well as environmental conditions within the host brain. Importantly, the long-term benefit to patients of transplantation paradigms is currently being evaluated (Freed *et al.*, 1992; Spencer *et al.*, 1992; Fisher and Gage, 1993).

Oligodendrocytes, Schwann cells, astrocytes, or neuronal progenitor cells are other nervous tissue-derived cells available for transplantation. Re-myelination of CNS demyelinating lesions has been achieved by transplanting oligodendrocyte precursors into demyelinating lesions (Groves *et al.*, 1993; Vignais *et al.*, 1993). It is envisaged that similar effects could be achieved in the peripheral nervous system by transplanting Schwann cells to treat demyelinating peripheral neuropathies. Myelinating cells could be engineered *ex vivo* and used towards the treatment of multiple sclerosis, or inherited disorders affecting myelination, such as Charcot–Marie–Tooth and Pelizaeus–Merzbacher disease (Appendix 16.A and Table 16.2). Also, the capacity of oligodendrocytes and astrocytes to migrate over long distances after transplantation into the CNS (Vignais *et al.*, 1993) could be harnessed to deliver gene products to anatomically widespread areas of the brain.

Neuronal progenitor cells can be grown in culture and transplanted back into the CNS of experimental animals. These precursor cell lines, when transplanted into the mouse CNS, differentiate into cell types similar to those normally found within the site of transplantation (Snyder *et al.*,

1992). Thus, precursor cell lines could be transduced with genes of interest before their transplantation into the brain. If they then were to differentiate into appropriate neuronal cell types, they would provide a phenotypically and genotypically normal replacement for the affected or dead neurons. Procedures to direct and predict the differentiation programme to be followed by neuronal progenitor cells *in vivo* have not yet been determined, but will be useful in their future application to neuronal gene therapy.

Further strategies are also being developed to deliver neuronal growth factors, e.g. nerve growth factor, ciliary neurotrophic factor, glial cell line derived neurotrophic factor or brain-derived neurotrophic factor (Thoenen *et al.*, 1993; Tomec *et al.*, 1995), directly into the brain to counteract pathologically induced neuronal death. This strategy has already been successful in a number of experimental animal models. Motorneurons' cell death occurring in a mutant mouse (*pmn/pmn*) suffering from progressive anterior horn motor neuropathy, a model for human spinal motor neuron disease, can be halted by administration of exogenous growth factors (Thoenen *et al.*, 1993).

Also, bone marrow transplantation can provide genetically engineered or normal macrophages to enter the brain as microglial cells. This provides an alternative option for gene delivery to the brain in metabolic disorders such as Krabbe's disease (section 16.2) (Shapiro *et al.*, 1991) or infectious diseases like AIDS, which mainly target the CNS through infected macrophages.

While transplantation represents a powerful strategy to deliver genes into the brain, this technique is limited to applications where the missing metabolite is active extracellularly, e.g. a neurotransmitter (dopamine), or a growth factor (NGF). It can be envisaged, however, that many disorders will require the direct *intracellular* delivery of missing genes and/or their products, since the missing proteins are required by intracellular metabolic pathways, and cannot be taken up from the extracellular milieu with high enough efficiency. A case in point are the metabolic storage diseases, where the lack of a particular enzyme induces a massive intracellular accumulation of toxic metabolic products. It is unlikely that extracellular delivery of enzymes, or other proteins could revert to a significant extent the pathological neuronal phenotype that occurs in such diseases.

Thus, three main applications have been proposed for gene transfer into the brain through cell transplantation: (i) to deliver substances to the brain which can act in the extracellular space, such as neurotransmitters (dopamine in Parkinson's; acetylcholine in Alzheimer's), or growth factors (NGF in Alzheimer's, GDNF in Parkinson's); (ii) to replace lost neurons, in Huntington's disease, stroke, or other disorders producing neuronal degeneration; and (iii) to re-myelinate demyelinating lesions in either inherited, or non-inherited disorders like multiple sclerosis. While uses of the first application have already undergone clinical trials testing, the second and third type of applications are still at the experimental stage.

16.4.2 Direct neuronal gene transfer: vectors for gene therapy of neurological disorders

Most vector systems in use, including retroviruses, adenoviruses, herpesviruses, liposomes, adeno-associated viruses or receptor-mediated gene transfer, could be applied to neurological gene therapy strategies, since target cells for gene transduction range from dividing myoblasts, neuronal precursor cells, glial cells, and microglial cells to post-mitotic mature neurons. Long-term expression of transgenes in a cell lineage (e.g. fibroblasts or oligodendrocytes) requires stable gene integration into the cells' chromosomes, while short-term expression does not. Importantly it has yet to be determined for how long vectors such as those derived from HSV-1 and adenovirus, which transfer genes into post-mitotic neurons, but remain episomal and do not integrate into the genome, will remain transcriptionally active.

Dividing cells can be targeted efficiently using either retroviruses or adeno-associated virus vectors which insert their genome into the host cell's DNA (Boris-Lawrie and Temin, 1993; Samulski, 1993); or non-viral systems such as liposomes, adenovirus particle-mediated transfection, or receptor-mediated gene transfer methods. Retroviral vectors have been improved over the last years, and new vectors display increased safety and efficiency. Their main drawback remains their stringent size limitation, ~9 kbp, their low titers, and the possibility of causing insertional mutagenesis upon insertion into host cell DNA. Insertional mutagenesis is a theoretical possibility which has not been substantiated by any *in vivo* experimental protocol using replication-deficient vectors in either experimental animals or human patients. Their size limitation restricts the range of promoter and genomic sequences which can be employed, and thus limits the transgenes to be transduced to their cDNAs. Their low titers restrict their use to *in vitro*/*ex vivo* applications. In spite of such limitations retrovirus vectors are already being used in human gene therapy (e.g. for severe combined immunodeficiency due to lack of adenosine deaminase), so far with good therapeutic outcome, and without major clinical problems. However, novel pseudotyped vectors (Burns *et al.*, 1993), in which the retroviral coat protein is replaced by the G protein of vesicular stomatitis virus, are likely to be increasingly useful, since these can be concentrated by centrifugation to titers in the range of 10^{10-12} pfu/ml. Also, pseudotyped vectors offer an expanded host range provided by the new coat protein used in the production of the virus, and could eventually be used in *in vivo* paradigms.

Retroviral vectors could be used in several neurological applications. Most of these involve the transduction of cells in preparation for transplantation, as described above. Disorders affecting Schwann cells or oligodendrocytes could be treated by *ex vivo* gene therapy. Patients' cells could be genetically engineered *in vitro*, expanded, and transplanted back into patients. Such strategies should be useful not only in the treatment of inherited disorders affecting Schwann cells (Charcot–Marie–Tooth disease), or CNS oligodendrocytes (Pelizaeus–Merzbacher disease), but also in the treatment of central demyelinating lesions such as those appearing in multiple sclerosis. In addition, retroviruses could also be used to transduce oligodendrocytes or astrocytes, *in vitro* to deliver gene products to widespread brain regions upon transplantation and intracerebral migration (Vignais *et al.*, 1993).

In addition to retroviruses, other gene transfer techniques have been developed, which provide useful alternatives for transducing cells *in vitro* before transplantation. These are intracellular injection, physical transfection methods (e.g. calcium phosphate precipitate-mediated, particle bombardment, and electroporation), lipofection and adenovirus- and receptor-mediated gene transfer (for a recent review of vectors for gene transfer, see Wolff, 1994). Additionally, liposomes can also be used as gene transfer vectors *in vivo*. Interestingly, Jiao *et al.* (1993) used lipofection to transduce muscle cells with a gene encoding tyrosine hydroxylase, thereby allowing the detection of gene expression for up to 6 months. Usually, shorter periods of expression are detected after transduction of other cells with retroviral vectors. Whether this represents a phenomenon of promoter shut-off, when acting from within viral vectors, or whether active expression can be detected in muscle cells for longer periods of time than in other target cells remains to be determined. It is known, however, that the immediate genomic environment surrounding the inserted gene can have important influences on the activity of a promoter–gene construct inserted into the host cell's genome.

Much work has been done over the last decade on herpes simplex virus 1 (HSV-1) and its possible application as a human gene therapy vector (Glorioso *et al.*, 1991; Breakfield *et al.*, 1992; Leib and Olivo, 1993; Lowenstein *et al.*, 1994a, b) (Figure 16.1). Herpes simplex provides the advantages of its neurotropism and its capacity to establish latency in neurons. Both during lytic infection and latency the double-stranded DNA genome remains episomal. Its neurotropism provides not only easy access and entry into neurons, but also a capacity to spread from one neuron to the next in synaptically linked networks (Enquist, 1994). Once it has established latency it will only express two to three species of mRNA during the lifetime of the infected neurons, without compromising their survival (Steiner and Kennedy, 1993). An important advantage of HSV-1 as vector for neuronal gene transfer and gene therapy would be to combine both of these characteristics, namely, the capacity to infect several neurons within a synaptic network (Enquist, 1994), and then establish latency in them (Steiner and Kennedy, 1993). Unfortunately, the full molecular understanding of latency has not yet been achieved.

What has been achieved is the production of severely disabled vectors which cannot complete the HSV-1 lytic cycle, and thus display greatly

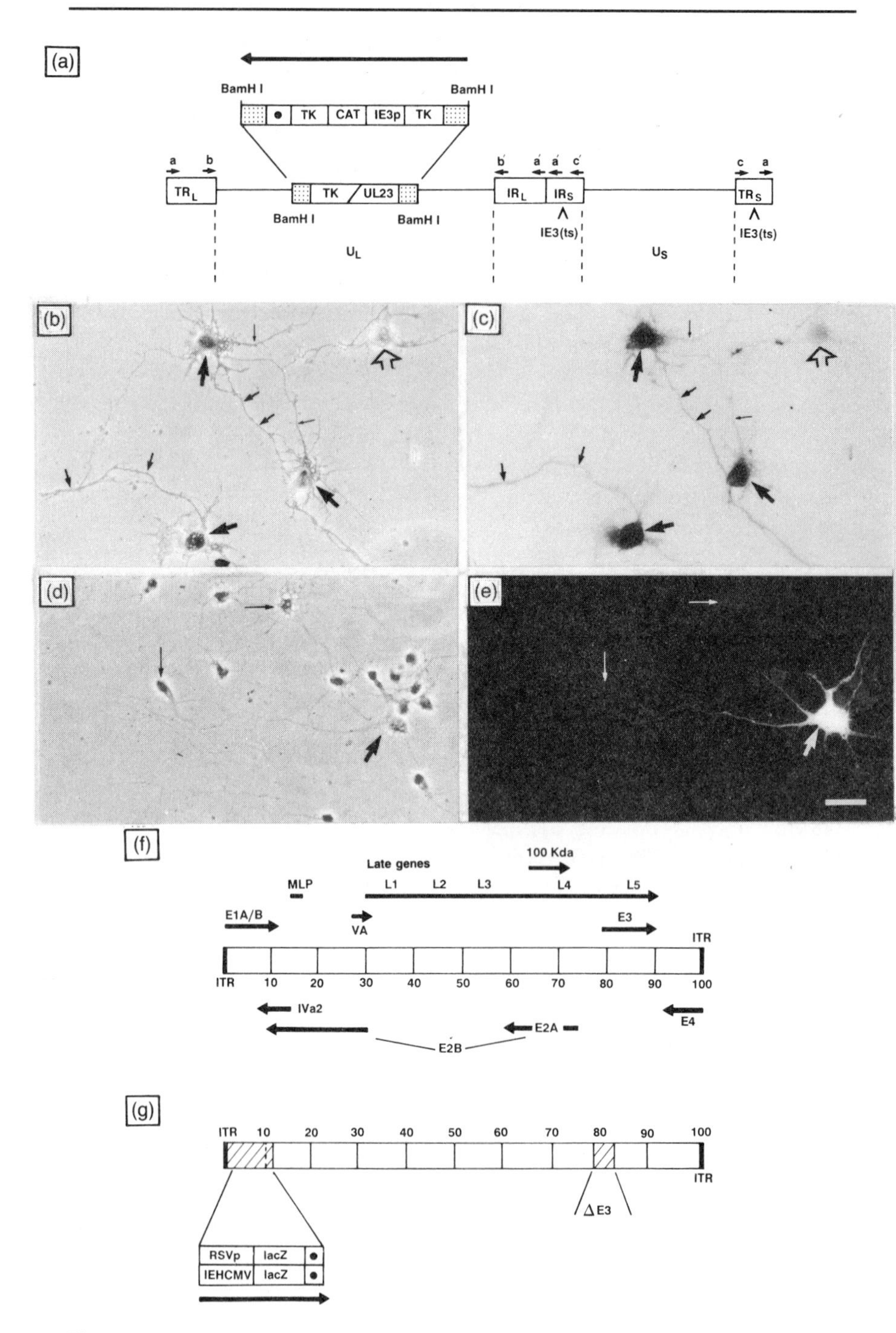
(a)
BamH I
BamH I
TK
CAT
IE3p
TK
a
b
b′
a′
a′
c′
c
a
TR$_L$
TK
UL23
IR$_L$
IR$_S$
TR$_S$
BamH I
BamH I
IE3(ts)
IE3(ts)
U$_L$
U$_S$
(b)
(c)
(d)
(e)
(f)
Late genes
100 Kda
MLP
L1
L2
L3
L4
L5
E1A/B
VA
E3
ITR
ITR
10
20
30
40
50
60
70
80
90
100
IVa2
E2A
E4
E2B
(g)
ITR
10
20
30
40
50
60
70
80
90
100
ITR
ΔE3
RSVp
lacZ
IEHCMV
lacZ

Figure 16.1

Figure 16.1 Gene transfer into primary cultures of post-mitotic neocortical neurons using viral vectors: herpes simplex type 1 and adenovirus vectors. (a) Schematic diagram of a vector derived from the herpes simplex virus type 1 temperature-sensitive mutant *ts*K (Lowenstein *et al.*, 1994). The 152-kbp of viral genome are shown diagramatically: it contains two unique sequences, a long one (U_L) and a shorter one (U_S), each flanked by either long terminal (TR_L) and long inverted (IR_L) repeats, or short terminal (TR_S) and the short inverted (IR_S) repeats. Genes located within the repeat sequences are found in two copies per genome. *ts*K contains a temperature-sensitive mutation in the immediate early 3 gene (IE3ts; permissive temperature for replication = 31°C), which is located within the repeat sequences flanking U_S. IE3 is an immediate early gene whose transcription is stimulated by the HSV-1 transactivating factor VP16 during the initiation of the viral lytic cycle in nuclei of infected cells. Its function is necessary for the virus to proceed through further stages of viral lytic cycle and further viral production. The repeat sequences also encode a cis-acting origin of replication (b, b′) and packaging signals (a, a′). The approximate location of the herpes simplex thymidine kinase (TK) gene (UL23) is shown. The stippled boxes surrounding the TK locus indicate flanking HSV-1 genomic sequences contained within a BamHI fragment, also present within the plasmid vector used to transfer genes into *ts*K by homologous recombination. The plasmid used is shown in diagrammatic form above the HSV1 genome. The TK coding sequence is interrupted by the HSV-1 IE3 promoter (IE3p), followed by the bacterial reporter gene encoding chloramphenicol acetyltransferase (CAT). The arrow overlying the foreign DNA sequences indicates the direction of transcription. Viral vectors which contain the foreign IE3p-CAT are also TK deficient. Details on the construction of these vectors are given in Lowenstein *et al.* (1994). (b, Phase contrast; c, brightfield) show neurons infected with the vector shown in (a); CAT activity was detected by immunocytochemistry. Three infected neurons are indicated by thick black arrows in both panels. An uninfected neuron is shown by a black arrow outline. Fine long arrows indicate dendrites, and short thicker arrows axons, displaying CAT immunoreactivity. Notice also the high concentration of CAT immunoreactivity over the nuclei of infected neurons. Neurons infected with wild type adenovirus 5 are shown in (d, phase contrasts; e, immunofluorescence). An infected neuron expressing the late 100 kDa scaffolding non-structural protein is shown (thick short arrow). Other non-infected cells are indicated by long fine arrows. The genomic organization of adenovirus is shown in (f). The double-stranded genome comprises 36 kbp and is divided into 100 map units of 0.36 kbp each. The inverted terminal repeats (ITR) contain the cis-acting elements conveying viral origin of replication and packaging sequences. The genomic location, approximate size direction of transcription (both DNA strands are transcribed) of non-structural early (E1, E2, E3, and E4), and structural late (L1–L5, IVa2) genes is shown. VA is transcribed by RNA polymerase III. Only the approximate location of genes L1–L5 is given by the long arrow; they are however transcribed as complex transcription units, each under the control of the major late promoter (MLP). Widely used available adenoviral vectors containing deletions in E3 to increase the capacity for transgenes, and the E1 region to render the virus replication deficient are shown in (g). E1 deleted virus can be grown on transcomplementing 293 cells, which constitutively express E1 adenoviral functions. Two different promoter/reporter gene cassettes are shown in (g), one encoding the Rous sarcoma virus promoter (RSVp) and the other the immediate early human cytomegalovirus promoter (IEHCMV; for details of constructions of this virus see Wilkinson and Akrigg, 1992) driving expression of the gene for β-galactosidase. No other genes are expressed when E1 deleted virus is used to infect non-transcomplementing cells with a very low virus input. Pseudo-adenoviral vectors only encode the viral ITR, and contain a coding capacity for transgene sequences of up to 30 kbp. Scale bar = 20μm.

reduced neuronal toxicity. Also, promoter cassettes have been developed which allow for long-term, high-level expression in many different neuronal systems. Both viral and cellular promoters have been used to drive the expression of a variety of transgenes within HSV-1 vectors (Dobson *et al.*, 1990). Importantly, the neuronal enolase promoter has been shown to remain neuronal specific when inserted into defective HSV-1 vectors (Breakfield *et al.*, 1992). Given the large size of the HSV-1 genome (153 kbp) even some foreign genomic sequences might be inserted into HSV-1 vectors, up to a theoretical insert of 30–35 kbp. Potential hazards of HSV-1-mediated gene transfer into the brain are a possible reactivation of latent herpes simplex virus, and thus the recombination and complementation of defective vectors with wild-type virus. Nevertheless, important advances are constantly being made in this field and new generations of vectors are likely to prove useful for neurological gene therapy (Dobson *et al.*, 1990; Glorioso *et al.*, 1991; Chapter 3).

Which diseases could be treated using HSV-1 vectors? Theoretically, many CNS disorders which would require transfer of genes directly into neurons could be treated with these vectors. Long-term expression (several months to years) can already be achieved in dorsal root ganglion neurons; it could also be achieved in many other neuronal systems using more effective promoter cassettes, and more definitely once latency is fully harnessed and understood. Short- to medium-term expression (weeks) can already be accomplished with currently available vectors. HSV-1 vectors have already been shown capable of transferring genes into neurons of different brain areas, including the brainstem and hippocampus, for up to 6 months (Wolfe *et al.*, 1992; Bloom *et al.*, 1994), albeit in a restricted number of neurons. Also, HSV-1 vectors could be developed into an excellent vehicle for gene therapy of diseases affecting dorsal root ganglion neurons, since such neurons are the site of HSV-1 latency during natural infection (Steiner and Kennedy, 1993). Many ataxias, as well as familial dysautonomia and chronic pain, affect dorsal root ganglion neuron function and survival. Thus these appear as excellent diseases to be targeted using HSV-1 vectors (Appendix 16.A and Table 16.2).

Adenoviruses on the other hand were not known, until recently, to infect neurons, and only infrequently cause clinically relevant brain infections. It was thus unexpected, when adenovirus vectors were shown to be active in expressing genes in infected neurons (Figure 16.1) (Akli *et al.*, 1993; Bajocchi *et al.*, 1993; Davidson *et al.*, 1993; Le Gal La Salle *et al.*, 1993). Adenovirus vectors can be made completely replication-deficient (Wilkinson and Akrigg, 1992; Vincent *et al.*, 1993; Wilkinson *et al.*, 1994; Chapter 2), although much of their toxicity is still provided by adenovirus coat proteins or endogenous proteins expressed at a very low level (Engelhardt *et al.*, 1994; Yang *et al.*, 1994). Their double-stranded DNA viral genome is smaller (36 kbp) than HSV-1, implying that less regions are dispensable for growth in tissue culture if helper-free vectors are being constructed. Up to 8 kbp of foreign DNA can be inserted into currently available first generation adenovirus vectors. This only allows for a limited amount of foreign promoters and coding sequences to be inserted. Although cell-specific promoters have been shown to maintain cell type specificity when expressed from within an adenovirus vector, currently available, as well as clinically used adenovirus gene therapy vectors all contain viral promoter sequences.

New generations of adenovirus vectors only encoding the inverted terminal repeats, containing the cis-acting packaging signals (pseudoadenoviral vectors) should provide up to 30 kbp of capacity for foreign DNA. However, such viruses will need to be grown up in the presence of helper virus, a system similar to HSV-1 amplicons (Leib and Olivo, 1993). The neurotoxicity of adenovirus vectors remains to be fully evaluated before these vectors are used in more widespread applications. However, adenovirus vectors have already been approved as vectors in human gene therapy clinical trials of cystic fibrosis (Anderson, 1994).

Due to the presence of the blood–brain barrier most vectors used to target neurons are delivered into the brain by stereotactic injection. This procedure is routinely used in both experimental animals and human patients, and is the route through which retrovirus has been delivered directly into brain tumors. However, a technique that would target brain neurons across the blood–brain barrier following a peripheral site of vector administration (Neuwelt *et al.*, 1991, 1944) would enhance the general availability and safety of neurological gene therapy.

Adenovirus particle-mediated transfection takes advantage of the activity of the adenovirus capsid proteins (including the penton protein). These contribute to the release of virus from endosomes to the intracellular cytoplasm (Curiel, 1993). Thus, by coupling transgene-containing plasmids to inactivated adenovirus, the efficiency of gene transfer is increased several-fold. There are several variants of this basic idea, and in the future, isolated adenovirus proteins could be complexed with either liposomes or other non-viral vectors to increase the efficiency of gene delivery, avoiding completely the introduction of further possible pathogenic particles. This is especially important in the case of adenovirus, where capsid proteins (penton) are also known to be cytotoxic. Until now, this technique has not been applied to *in vivo* paradigms of disease, although this remains a promising gene transfer system. Nevertheless, this technique might offer important advantages in both *ex vivo* protocols, or in some specific *in vivo* protocols, e.g. gene transfer into lung epithelia in cystic fibrosis using liposomes complexed to adenoproteins or particles. The technique can be made even more selective by adding specific ligands for cell surface receptors to these complex gene transfer vectors. Receptors would increase the efficiency of cell entry by directing the vector complex to the receptor-mediated endocytotic pathway. Individual proteins which aid in the release of vector particles from endosomes, or target the complexes to nuclei, could also be incorporated into such complex chimeric vectors.

Recent developments in the exploitation of adeno-associated viruses (AAV) as gene transfer vectors are also promising for both *in vivo* and *ex vivo* gene therapy protocols (Samulski, 1993; Kaplitt *et al.*, 1994). Adeno-associated viruses only grow in the presence of helper adenovirus or HSV-1 virus, but can be physically separated from helper virus. Helper-free vectors can thus be prepared to infect cells both *in vivo* and *in vitro*. AAV, like retroviruses, integrate into the host cell genome. Specificity in the chromosomal integration site (chromosome 19) has been proposed for AAV, but has yet to be confirmed for AAV-derived vectors. AAV vectors only contain the inverted terminal repeat regions of the virus, including the cis-acting packaging sequences, while all other viral sequences are removed. Unfortunately, the packaging capacity of AAV limits the size of transgene insert to approximately 5 kbp, a smaller size than retroviruses (~9 kbp). Nevertheless, the host range of AAV is wider than the retroviral host range, and AAV could be used to infect postmitotic neurons (Samulski, 1993; Kaplitt *et al.*, 1994).

Direct naked DNA transfer has been shown to be useful to transfer genes and antisense oligonucleotides to muscle cells, and recently this has been shown to be effective also in the transfer of genes into neurons, both *in vitro* and *in vivo* (Jiao *et al.*, 1992). Whether correct regulation of gene expression can be achieved from such minimal gene transfer systems, remains to be determined.

Also, yeast and mammalian artificial chromosomes are being developed as vectors for both somatic, and germ-line, gene therapy (Huxley, 1994). The advantage of artificial chromosomes is the virtual absence of size limitations of foreign DNA sequences to be delivered, thus allowing the insertion of both genomic and complete endogenous long-range regulatory sequences into such a vector. While YACs have been used to transfer genes into germ-line cells, methods to use these for somatic cell gene transfer are currently being set up. These methods will prove very useful for the development of transgenic animals using human genomic sequences, which should also allow the discovery of the long-range genomic sequences needed for adequate transcriptional regulation of transgenes.

16.5 GERM-LINE GENE THERAPY

In considering diseases suitable for gene therapy, this chapter has been limited to the consideration of effective somatic cell gene therapy. However, many missing gene products have now been shown to be expressed very early in development, throughout the CNS, and also throughout the whole body. Thus, somatic gene therapy might not achieve its aims in certain diseases in which the main pathological effect of the missing gene product occurs prenatally, and/or the gene is needed in more neurons than can be transduced with current gene transfer methods. Thus, if somatic gene therapy fails in achieving some of its therapeutic goals, germ-line gene therapy might be one of few available therapeutic alternatives left. For this to proceed, techniques will need to be developed for genes to be inserted into specific genomic sites, to come under adequate developmental transcriptional control, as well as to achieve adequate levels of transgene expression. Thus, it is important that the ethical dimensions of neurological germ-line gene therapy are addressed well in advance of the techniques being developed and proposed for clinical trials. If open discussion is maintained, ethical issues that might otherwise be seen as imposing insurmountable dilemmas, might eventually be resolved (Wivel and Walters, 1993).

Nevertheless, in the UK, the Committee on the Ethics of Gene Therapy (Clothier *et al.*, 1992) has clearly expressed that

> The purpose of gene modification of sperm or ova or cells that produce them would be to prevent the transmission of defective genes to subsequent generations. Gene modification at an early stage of embryonic development might be a way of correcting gene defects in both the germ line and somatic cells. However, we have concluded that there is insufficient knowledge to evaluate the risks to future generations. We recommend, therefore, that gene modification of the (human) germ line should not yet be attempted.

However, it is possible that when somatic gene therapy becomes routine, and shown to be safe for patients and perceived as advantageous by society, germ-line gene therapy will be considered more favorably. In view of the current debate on this issue, it is interesting to note that during early 1994, a patent request has been filed at the European Patent Office concerning a technique that could be applicable to human germ-line gene therapy. The procedure proposes to remove affected spermatogonia, the cells giving rise to sperm, from the patient, engineer and expand them *in vitro* to correct the gene defect, expand them in culture, and then replace them into 'purged' seminiferous tubules to re-populate the testis with genetically engineered spermatogonia. This will certainly stimulate the on-going ethical debate.

16.6 FUTURE PROSPECTS AND COMMERCIAL DEVELOPMENT

Having discussed the possible uses of available vectors for gene transfer into neurons it remains to be decided whether any of such vectors could be developed for use in clinical gene therapy for human neurological disorders. One issue that remains to be determined is how specific cell types will be targeted using currently available vectors. In principle, cell type-specific promoters could be used, and these have already been shown to work within the environment of HSV-1 or adenovirus vectors. Retroviruses have been exploited to selectively deliver genes encoding cytotoxic proteins into dividing tumor cells located within the CNS, since they are very unlikely to integrate into non-dividing mature neurons. Also, bone marrow transplantation of *ex vivo* transduced cells might represent a method to deliver genes into microglial cells which will eventually find their way into the CNS. As discussed above, glial cells could be transduced *ex vivo* and transplanted into the brain where they can migrate long distances to deliver genes throughout the CNS.

A particular challenging step that remains to be solved is whether genes can be transferred to a large enough number of post-mitotic neurons distributed throughout the whole brain. Present delivery systems involve the use of stereotactic injections of vectors. While useful, they allow transduction of only small numbers of cells. It is important to keep in mind the problem presented by the size of the human brain when considering human neurological gene therapy. The human substantia nigra is as large as a whole mouse brain, and available techniques are still unable to transduce a very high percentage of neurons in a mouse brain. Whether this will ultimately be necessary or not will depend on our understanding of the disease, its pathophysiology and our knowledge of the minimum percentage both of transgene activity required, as well as the number of transduced cells necessary to achieve reversion of the pathological phenotype to normal.

What will the future vector for human neurological gene therapy look like? Many of the available vectors will certainly be used in the experimental and initial clinical trial stages of human neurological gene therapy. However, it is likely that future vectors will be custom-designed to be administered by a physician on an outpatient basis, just like most medicines currently are. To achieve this, peripherally administered vectors should be able to cross the blood–brain barrier and target neurons throughout the brain. It is likely that much efforts will concentrate in future years on how to deliver gene transfer vectors to widespread regions of the brain across the blood–brain barrier. Also, new generation vectors will be modular, and likely to contain specific ligands to target them efficiently to various neuronal cell types. They will take advantage of specific proteins to mediate efficient intracellular uptake, endosomal lysis, and nuclear targeting, functions provided by individual proteins, nowadays only found in intact virus particles. It is also likely that such vectors will accommodate genomic sequences with their endogenous long-range regulatory elements to provide developmentally, and cell type-specific expression at an adequate level. The achievement of such chimeric and modular vectors available for peripheral administration will certainly provide a system with enough flexibility to apply gene therapy to more neurological diseases than is conceivable at present. At present, the most important hurdle remaining is still to demonstrate, in animal models of human neurological diseases, the effectiveness of gene transfer to achieve phenotype reversion to normal, and thus efficient and clinically relevant gene therapy.

While many questions relating to the practical implementation of human gene therapy remain to be tackled, I am confident these will be addressed as human neurological gene therapy proper will enter the clinical trials stage within the next to 5 to 10 years. In spite of current methodological limitations, there is much excitement highlighted by the fact that we now have tools to address many of the crucial questions that will need to be answered for neuronal gene therapy to become a clinical reality. Many animal models exist (Smithies, 1993) to be exploited in the development of gene therapy for both inherited diseases: e.g. Lesch–Nyhan syndrome (HPRT−mouse), Krabbe's disease (twitcher mouse), Charcot–Marie–Tooth disease (trembler mouse), Duchenne muscular dystrophy (*mdx* mouse), as well as for neurodegenerative diseases like Parkinson's disease (6-OHDA-lesioned rat; MPTP lesioned non-human primates), Alzheimer's disease (cholinergically denervated hippocampus; transgenic animals containing mutant forms of the amyloid precursor protein responsible for familial Alzheimer's disease). Also further transgenic animal models for both types of disorder will soon become available for a wider choice of target diseases, as our knowledge concerning the molecular basis of human neurological disease increases. In addition, the availability of selective neurotoxic agents, is being exploited to determine systematically which growth factors are necessary for the survival of identified neuronal populations (Thoenen *et al.*, 1993). Animal models will allow us to determine whether gene transduction needs to target 100% of cells, or whether 5% of transduced cells are sufficient for phenotypic reversion, whether cells will have to express 80%, 50% or 1% of transgene activity, for what period will transgenes need to be expressed before phenotypic reversion occurs, and at what stage of development will gene replacement be of therapeutic value.

For the first time, we now have the tools to test whether gene therapy of human neurological disorders will be feasible, how to identify and overcome current technical limitations, and thus how to proceed for neurological gene therapy to become a reality. Thus, in summarizing, I would like to propose that, in principle, a wide variety of neurological diseases, both inherited and idiopathic, should be considered suitable targets for gene therapy (Appendix 16.A and Table 16.2). This could be achieved by either correcting the genetic defect itself in the case of the inherited disorders, or by therapeutic intervention designed to modify the pathophysiological processes leading to non-inherited disease. Finally, I would also like to suggest that in view of (a) the high incidence and prevalence of neurological disease, (b) the fact that human gene therapy was already proposed in 1972 for a metabolic disorder with devastating neurological symptomatology, the Lesch–Nyhan syndrome, (c) the efforts that have already been invested in the understanding of the molecular causes of neurological disease, and

(d) the interest and contribution of pharmaceutical and biotechnological companies to the development, production and commercialization of novel therapies, I believe there exists a timely opportunity to develop new academic–industrial ventures, in combined efforts to provide new and powerful treatments for a diverse group of neurological diseases.

ACKNOWLEDGEMENTS

This paper is dedicated to the memory of Dr Terry Parsons who showed me earthly ways to pursue some of my dreams.

I apologise to many researchers concerned that the citation of a large number of original papers in this review has been limited by space. P.R.L. is a Research Fellow of The Lister Institute of Preventive Medicine. I wish to acknowledge the support my work receives from MRC, The Wellcome Trust, SERC, The Royal Society, and The Welsh Scheme for the Development of Health and Social Research. I wish to thank Derek Bain, Sarah Bundey, Maria Castro, Vincenzo Crunelli, Neva Haites, Tim Jacob, Stephen Jones, John MacMillan, Afshan Malik, Chris Preston, Julian Sampson, Andy Shering, Nevin Summers, Gavin Wilkinson, for challenging and enlightening discussions on various aspects of neurological gene therapy, its prospects and vectors, as well as their provision of critical revisions and suggestions for this review and its content.

REFERENCES

Akli, S., Cailland, C., Vigne, E. *et al.* (1993) Transfer of a foreign gene into the brain using adenovirus vectors. *Nature Genet.*, 3, 224–8.

Anderson, W.F. (1985) Human gene therapy: scientific and ethical considerations. *J. Med. Philos.*, **10**, 275–91.

Anderson, W.F. (1994) Human gene marker/therapy clinical protocols listing. *Human Gene Ther.*, 5, 553–63.

Bajocchi, G., Feldman, S.H., Crystal, R.G. *et al.* (1993) Direct *in vivo* gene transfer to ependymal cells in the central nervous system using recombinant adenovirus vectors. *Nature Genet.*, 3, 229–34.

Baranger, J.A. and Ginns, E.I. (1989) Glucosylceramide lipidoses: Gaucher disease, in *The Metabolic Basis of Inherited Disease*, (eds C.R. Scriver, A.L. Beaudet, W.S. Sly and D. Valle), 6th edition, McGraw Hill, New York, pp. 1677–98.

Beaudry, A.A. and Joyce, G.F. (1992) Directed evolution of an RNA enzyme. *Science*, **257**, 635–41.

Bloom, D.C., Lokensgard, J.R., Maidment, N.T. *et al.* (1994) Long term, expression of genes *in vivo* using non-replicating HSV vectors. *Gene Therapy*, **1** (Suppl. 1), S36–8.

Boris-Lawrie, K.A. and Temin, H.M. (1993) Recent advances in retrovirus vector technology. *Curr. Opin. Genet. Dev.*, 3, 102–9.

Brady, R.O. and Barton, N.W. (1991) Enzyme replacement therapy for type 1 Gaucher disease, in *Treatment of Genetic Disorders*, (ed. R.J. Desnick), Churchill Livingstone, New York, pp. 153–68.

Breakfield, X.O., Huang, Q, Andersen, J.K. *et al.* (1992) Gene transfer into the nervous system using recombinant herpes virus vectors, in *Gene Transfer and Therapy in the Nervous System*, (eds F.H. Gage and Y. Christen) Springer-Verlag, Berlin, pp. 118–32.

Bundey, S. (1992) *Genetics and Neurology*, 2nd edn, Churchill Livingstone, Edinburgh.

Burns, J.C., Friedmann, T., Driever, W. *et al.* (1993) Vesicular stomatitis virus G glycoprotein pseudotypes retroviral vectors: concentration to very high titre and efficient gene transfer into mammalian and non-mammalian cells. *Proc. Natl Acad. Sci. USA*, **90**, 8033–7.

Clothier, C., Auld, M., Bobrow, M. *et al.* (1992) *Report of the Committee on the Ethics of Gene Therapy*, HMSO, London, 41 pp.

Connor, J.M. (1989) Genetic aspects of prenatal diagnosis. *J. Inher. Metab. Dis.*, **12** (suppl. 1), 89–96.

Conzelmann, E. and Sandhoff, K. (1983/4) Partial enzyme deficiencies: residual activities and the development of neurological disorders. *Dev. Neurosci.*, **6**, 58–71.

Culver, K.W. and Blaese, R.M. (1994) Gene therapy for adenosine deaminase deficiency and malignant solid tumours, in *Gene Therapeutics*, (ed. J.A. Wolff), Birkhauser, Boston, USA, pp 263–80.

Curiel, D. (1993) Adenovirus facilitation of molecular conjugate-mediated gene transfer. *Prog. Med. Virol.*, **40**, 1–18.

Davidson, B.L., Allen, E.D., Kogovsky, K.F. *et al.*, (1993) A model system for *in vivo* gene transfer into the central nervous system using an adenoviral vector. *Nature Genet.*, **3**, 219–23.

Dobson, A.T., Margolis, T.P., Sedarati, F. *et al.* (1990) A latent, nonpathogenic HSV-1 derived vector stably expresses beta-galactosidase in mouse neurons. *Neuron*, **5**, 353–60.

Dodet, B. (1993) Commercial prospects for gene therapy – a company survey. *Trends Biotechnol.*, **11**, 182–9.

Ekhterae, D., Crumbleholme, T., Karson, E. *et al.* (1990) Retroviral vector-mediated transfer of the bacterial neomycin resistance gene into fetal and adult sheep and human hematopoietic progenitors *in vitro*. *Blood*, **75**, 365–9.

Emery, A.E.H. (1991) Population frequencies of inherited neuromuscular diseases – a world survey. *Neuromusc. Disord.*, **1**, 19–29.

Engelhardt, J.F., Ye, X., Doranz, B. *et al.* (1994) Ablation of EZA in recombinant adenoviruses improves transgene persistence and decreases inflammatory response in mouse liver. *Proc. Natl Acad. Sci. USA*, **91**, 6196–200.

Enquist, L.W. (1994) Infection of the mammalian nervous system by pseudorabies virus (PRV). *Semin. Virol.*, **5**, 221–31.

Fisher, L.J. and Gage, F.H. (1993) Grafting in the mammalian central nervous system. *Physiol. Rev.*, **73**, 583–616.

Freed, C.P., Breeze, R.E., Rosenberg, N.L. *et al.* (1992) Survival of implanted fetal dopamine cells and neurologic improvement 12 to 46 months after transplantation for Parkinson's disease. *N. Engl. J. Med.*, **327**, 1549–55.

Friedmann, T. and Jinnah, H.A. (1993) Gene therapy for disorders of the nervous system. *Trends Biotechnol.*, **11**, 192–7.

Friedmann, T. and Roblin, R. (1972) Gene therapy for human genetic disease? *Science*, **175**, 949–55.

Gage, F.H., Wolff, J.A., Rosenberg, M.B. *et al.* (1987) Grafting genetically modified cells to the brain: possibilities for the future. *Neuroscience*, **23**, 795–807.

Glorioso, J.C., Goins, W.F. and Fink, D.J. (1991) Herpes simplex virus-based vectors. *Semin. Virol.*, **3**, 265–76.

Gomes, D., Adams, D., Alessandrini, R. *et al.* (1995) Alzheimer type neuropathology in transgenic mice overexpressing V717F β-amyloid precursor protein. *Nature*, **373**, 523–7.

Groves, A.K., Barnett, S.C., Franklin, R.J.M. *et al.* (1993) Repair of demyelinated lesions by transplantation of purified 0–2A progenitor cells. *Nature*, **362**, 453–5.

Holzmayer, T.A., Pestov, D.G. and Roninson, I.B. (1992) Isolation of dominant negative mutants and inhibitory antisense RNA sequences by expression selection of random DNA fragments. *Nucleic Acids Res.*, **20**, 711–17.

Hoogerbrugge, P M., Suzuki, K., Poorthuis, B.J.H.M. *et al.* (1988) Donor-derived cells in the central nervous system of twitcher mice after bone marrow transplantation. *Science*, **239**, 1035–38.

Huxley, C. (1994) Mammalian artificial chromosomes: a new tool for gene therapy. *Gene Therapy*, **1**, 7–12.

Jiao, S.S., Acsadi, G., Jani, A. *et al.* (1992) Persistence of plasmid DNA and expression in rat brain cells *in vivo*. *Exp. Neurol.*, **115**, 400–13.

Jiao, S.S., Gurevich, V. and Wolff, J.A. (1993) Long term correction of a rat model of Parkinson's disease by gene therapy. *Nature*, **362**, 450–3.

Jinnah, H.A., Gage, F.H. and Friedmann, T. (1993) Gene therapy and neurological disease, in *The Molecular and Genetic Basis of Neurological Disease*, (eds S.B. Prusiner, R.N. Rosenberg, S. DiMauro, R.L. Barchi and L.M. Kunkel), Butterworth-Heinemann, Boston, pp. 969–76.

Kaplitt, M.C., Leone, P., Samulski, R.J. *et al.* (1994) Long term gene expression and phenotypic connection using adeno-associated virus vectors in the mammalian brain. *Nature Genet.*, **8**, 148–54.

Kolodny, E.H. (1989) Metachromatic leukodystrophy and multiple sulfatase deficiency: sulfatide lipidosis, in *The Metabolic Basis of Inherited Disease*, (eds C.R. Scriver, A.L. Beaudet, W.S. Sly and D. Valle), 6th edn, McGraw Hill, New York, pp. 1721–50.

Kopin, I.J. (1993) The pharmacology of Parkinson's disease therapy: an update. *Annu. Rev. Pharmacol. Toxicol.*, **32**, 467–95.

Krivit, W. and Shapiro, E.G. (1991) Bone marrow transplantation for storage diseases, in *Treatment of Genetic Disorders*, (ed. R.J. Desnick), Churchill Livingstone, New York, pp. 203–21.

Le Gal La Salle, G., Robert, J.J., Benard, S. *et al.* (1993) Adenovirus as a potent vector for gene transfer in neurons and glia in the brain. *Science*, **259**, 988–90.

Leib, D.A. and Olivo, P.D. (1993) Gene delivery to neurons: is herpes simplex virus the right tool for the job? *Bioessays*, **15**, 547–54.

Leinekugel, P., Michel, S., Conzelmann, E. *et al.* (1992) Quantitative correlation between the residual activity of β-hexosaminidase A and arylsulfatase A and the severity of the resulting lysosomal sotrage disease. *Hum. Genet.*, **88**, 513–23.

Lowenstein, P.R. (1994) Molecular neurosurgery: mending the brain's software and hardware. Gene therapy for neurological disease? *Bio/Technology*, **12**, 1075–80.

Lowenstein, P.R., Fournel, S., Bain, D. *et al.* (1994a) Herpes simplex virus 1 (HSV-1) helper co-infection affects the distribution of an amplicon encoded protein in glia. *NeuroReport*, **5**, 1625–30.

Lowenstein, P.R., Morrison, E.E., Bain, D. *et al.* (1994b) Use of recombinant vectors derived from herpes simplex virus 1 mutant *ts*K for short term expression of transgenes encoding cytoplasmic and membrane anchored proteins in post-mitotic polarized cortical neurons and glial cells *in vitro*. *Neuroscience*, **60**, 1059–77.

MacMillan, J.C. and Harper, P.S. (1994) Clinical genetics in neurological disease. *J. Neurol. Neurosurg. Psychiatry*, **57**, 7–15.

Markert, J.M., Malick, A., Coen, D.M. *et al.* (1993) Reduction and elimination of encephalitis in an experimental glioma therapy model with attenuated herpes simplex mutants that retain susceptibility to acyclovir. *Neurosurgery*, **32**, 597–603.

Martin, J.G. (1993) Molecular genetics in neurology. *Ann. Neurol.*, **34**, 757–73.

McKusick, V.A. (1995) *Genome data base and online mendelian inheritance in man.* Baltimore, Johns Hopkins University and Howard Hughes Medical Institute, 1994.

Morgan, R.A. and Anderson, W.F. (1993) Human gene therapy. *Annu. Rev. Biochem.*, **62**, 191–217.

Moser, H.W. and Moser, A.B. (1989) Adrenoleukodystrophy (X-linked), in *The Metabolic Basis of Inherited Disease*, (eds C.R. Scriver, A.L. Beaudet, W.S. Sly and D. Valle), 6th edn, McGraw Hill, New York, pp. 1511–32.

Moser, H.W., Auborg, P., Cornblath, D. *et al.* (1991) Therapy for X-linked adreno-leukodystrophy, in *Treatment of Genetic Disorders*, (ed. R.J. Desnick), Churchill Livingstone, New York, pp. 111–29.

Murray, J.A.H. and Crockett, N. (1992) Antisense techniques: an overview, in *Antisense RNA and DNA*, (ed. J.A.H. Murray), Wiley-Liss, New York, pp. 1–49.

Neuwelt, E.A., Pagel, M.A. and Dix, R.D. (1991) Delivery of ultraviolet-inactivated ^{35}S-herpesvirus across an osmotically modified blood–brain barrier. *J. Neurosurg.*, **74**, 475–9.

Neuwelt, E.A., Weissleder, R., Nilaver, G. *et al.* (1994) Delivery of virus-sized iron oxide particles to rodent CNS neurons. *Neurosurgery*, **34**, 777–84.

O'Brien, J.S. (1989) β-galactosidase deficiency (G_{M1} gangliosidosis, galactosialidosis, and Morquio syndrome type B); ganglioside sialidase deficiency (Mucolipidosis IV), in *The Metabolic Basis of Inherited Disease*, (eds C.R. Scriver, A.L. Beaudet, W.S. Sly and D. Valle), 6th edn, McGraw Hill, New York, pp. 1797–1806.

Rocca, W.A., Hofman, A., Brayne, C. *et al.* (1991) Frequency and distribution of Alzheimer's Disease in Europe: a collaborative study of 1980–1990 prevalence findings. *Ann. Neurol.*, **30**, 381–90.

Roemer, K. and Friedmann, T. (1992) Concepts and strategies for human gene therapy. *Eur. J. Biochem.*, **208**, 211–25.

Rosenberg, R.N., Prusiner, S.B., Di Mauro, S. *et al.* (eds) (1993) *The Molecular and Genetic Basis of Neurological Disease*, Butterworth-Heinemann, Boston.

Samulski, R.J. (1993) Adeno-associated virus: integration at a specific chromosomal locus. *Curr. Opin. Genet. Dev.*, **3**, 74–80.

Sandhoff, K., Conzelmann, E., Neufeld, E.F. *et al.* (1989) The G_{M2} gangliosidoses, in *The Metabolic Basis of Inherited Disease*, (eds C.R. Scriver, A.L. Beaudet, W.S. Sly and D. Valle), 6th edn, McGraw Hill, New York, pp. 1807–39.

Saunders, A.M., Strittmatter, W.J., Schmiechel, D. *et al.* (1993) Association of apolipoprotein E allele epsilon 4 with late onset familial and sporadic Alzheimer's disease. *Neurology*, **43**, 1467–72.

Schehr, R.S. (1994) Therapeutic approaches to Alzheimer's disease. An informal survey of promising drug discovery strategies. *Bio/Technology*, **12**, 140–4.

Scriver, C.R., Kaufman, S. and Woo, S.L.C. (1989) The hyperphenylalaninemias, in *The Metabolic Basis of Inherited Disease*, (eds C.R. Scriver, A.L. Beaudet, W.S. Sly and D. Valle), 6th edn, McGraw Hill, New York, pp. 495–546.

Sege-Peterson, K., Nyhan, W.L. and Page, T. (1993) Lesch–Nyhan disease and HPRT deficiency, in *The Molecular and Genetic Basis of Neurological Disease*, (eds R.N. Rosenberg, S.B. Prusiner, S. DiMauro, R.L. Barchi and L.M. Kunkel), Butterworth-Heinemann, Boston, pp. 241–59.

Shapiro, E.G., Lockman, L., Kennedy, W. *et al.* (1991) Bone marrow transplantation as treatment for globoid cell leukodystrophy, in *Treatment of Genetic Disorders*, (ed. R.J. Desnick), Churchill Livingstone, New York, pp. 223–38.

Smithies, O. (1993) Animal models of human genetic diseases. *Trends Genet.*, **9**, 112–16.

Snyder, E.Y., Deitcher, D.L., Walsh, C. *et al.* (1992) Multipotent neural cell-lines can engraft and participate in development of mouse cerebellum. *Cell*, **68**, 33–51.

Spence, M.W. and Callahan, J.W. (1989) Sphingomyelin-cholesterol lipidoses: the Niemann–Pick group of diseases, in *The Metabolic Basis of Inherited Disease*, (eds C.R. Scriver, A.L. Beaudet, W.S. Sly and D. Valle), 6th edn, McGraw Hill, New York, pp. 1655–76.

Spencer, D.D., Robbins, R.J., Naftolin, F. *et al.* (1992) Unilateral transplantation of human fetal mesencephalic tissue into the caudate nucleus of patients with Parkinson's disease. *N. Engl. J. Med.*, **327**, 1541–8.

Steiner, I. and Kennedy, P.G.E. (1993) Molecular biology of herpes simplex virus type 1 latency in the nervous system. *Mol. Neurobiol.*, **7**, 137–59.

Summers, N.M. (1993) Commercial prospects for gene therapy and the newly emerging human gene estate industry. MSc Thesis (Management of Technology, MIT).

Suzuki, K. and Suzuki, Y. (1989) Galactosylceramide lipidosis: globoid-cell leukodystrophy (Krabbe disease), in *The Metabolic Basis of Inherited Disease*, (eds C.R. Scriver, A.L. Beaudet, W.S. Sly and D. Valle), 6th edn, McGraw Hill, New York, pp. 1699–720.

Takamiya, Y., Short, P., Moolten, F.L. *et al.* (1993) An experimental model of retrovirus gene therapy for malignant brain tumours. *J. Neurosurg.*, **79**, 104–10.

Thoenen, H., Hughes, R.A. and Sendtner, M. (1993) Trophic support of motoneurons. Physiological, pathophysiological and therapeutic implications. *Exp. Neurol.*, **124**, 47–55.

Tomec, A., Lindquist, E., Liu, L.F.H. *et al.* (1995) Protection and repair of the nigrostriatal dopaminergic system by GDNF *in vivo*. *Nature*, **373**, 335–9.

Vignais, L., Nair Oumesmar, B.N., Mellouk, F. *et al.* (1993) Transplantation of oligodendrocyte precursors in the adult demyelinated spinal cord: migration and remyelination. *Int. J. Dev. Neurosci.*, **11**, 603–12.

Vincent, N., Ragot, T., Gilgenkrantz, H. *et al.* (1993) Long term correction of mouse dystrophic degeneration by adenovirus-mediated transfer of a minidystrophin gene. *Nature Genet.*, **5**, 130–4.

Walkley, S.U., Thrall, M.A., Dobrenis, K. *et al.* (1994) Bone marrow transplantation corrects the enzyme defect in neurons of the central nervous system in a lysosomal storage disease. *Proc. Natl Acad. Sci. USA*, **91**, 2970–4.

Wilkie, A.O.M. (1994) The molecular basis of genetic dominance. *J. Med. Genet.*, **31**, 89–98.

Wilkinson, G.W.G. and Akrigg, A. (1992) Constitutive and enhanced expression from a defective adenovirus vector. *Nucleic Acids. Res.*, **20**, 2233–9.

Wilkinson, G.W.G., Darley, R.L. and Lowenstein, P.R. (1994) Viral vectors for gene therapy, in *From Genetics to Gene Therapy*, (ed. D.S. Latchman), βIOS Scientific Publishers, Oxford, pp. 161–92.

Wivel, N.A. and Walters L. (1993) Germ-line gene modification and disease prevention: some medical and ethical perspectives. *Science*, **262**, 533–8.

Wolfe, J.H., Deshmane, S.L. and Fraser, N.W. (1992) Herpesvirus vector gene transfer and expression of β-glucuronidase in the central nervous system of MPS VII mice. *Nature Genet.*, **1**, 379–84.

Wolff, J.A. (1994) *Gene Therapeutics*, Birkhauser, Boston.

Wolff, J.A. and Lederberg, J. (1994) An early history of gene transfer and therapy. *Hum. Gene Ther.*, **5**, 469–80.

Yang, Y., Nunes, F.A., Berencsi, K. *et al.* (1994) Cellular immunity to viral antigens limits E1-deleted adenoviruses gene therapy. *Proc. Natl Acad. Sci. USA*, **91**, 4407–11.

Zhang, Z.X. and Roman, G.C. (1993) Worldwide occurrence of Parkinson's disease: an updated review. *Neuroepidemiology*, **12**, 195–208.

APPENDIX 16.A

NEUROLOGICAL DISORDERS: OVERVIEW OF THEIR PATHOLOGY, MOLECULAR LESIONS AND FREQUENCY

Disease	*Pathology/ pathophysiology/ symptomatology*	*Gene locus*	*Gene/product*	*Mutations*	*Disease mechanism*	*Incidence/prevalence (per 100 000)*	*Inheritance*
1. *Degenerative disorders of the central nervous system, including ataxias and retinal degenerations*							
Alzheimer's disease (AZD) (all cases)	Dementia Neurodegeneration, plaques and tangles throughout the CNS	–	–	–	Accumulation of neurotoxic products?	30–59 y: 20 [P] 60–69 y: 300 [P] 70–79 y: 3 200 [P] 80–89 y: 10 800 [P]	
Familial AZD, early onset (104300)	As sporadic	21q21.3-q22.05 (Rare) 14q24.3 (70% of FAD)	21: b-amyloid precursor protein; cloned 14: Not cloned	Point mutations	Accumulation of β-amyloid precursor protein	Rare	AD
Familial AZD, late onset	As sporadic	19q	Unknown Not cloned	Unknown	Unknown	Rare	AD
Parkinson's disease (PD) (168600)	Tremor, rigidity, bradikinesia. Degeneration of SN neurons. Symptoms with ⩾80% neurons	–	–	–		40–59 y: 100 [P] 60–69 y: 400 [P] 70–79 y: 800 [P] ⩾ 80 y: 1000 [P]	
Familial PD (168600)	As sporadic	Unknown	Mitochondrial respiratory chain enzymes? Not cloned	Unknown	Increased neuronal vulnerability to toxic factors?	Rare	AD
Familial Creutzfeld–Jakob disease (CJD) (123400)	CNS spongy degeneration	20pter-p12	Prion protein Cloned	Point mutations	Conformational change in protein; neurodegeneration	0.1–0.15 [P] (10–15% of total CJD)	AD
Huntington's disease (143100)	Chorea, dementia Degeneration of especially basal ganglia neurons	4p16.3	IT15/huntingtin Cloned	Trinucleotide repeat expansion (CAG)	Gain of function? Neurodegeneration	8 [P]	AD
Hereditary dentato-rubral and pallido-luysian atrophy (DRPLA) (125370)	Myoclonus, epilepsy, dementia, mental retardation, cerebellar ataxia, choreoathetosis	12p	DRPLA gene Gene not cloned; trinucleotide repeat expansion cloned.	Trinucleotide repeat expansion (CAG)	Unknown	Rare Common in Japan	AD
Torsion dystonia-1 (128100)	Involuntary movements	9q32-q34 (50%)	DYT1 Not cloned	Unknown	Unknown	6.7 [I] 16–50 in Ashkenazi Jews [P]	AD

APPENDIX 16.A *Continued*

Disease	*Pathology/pathophysiology/symptomatology*	*Gene locus*	*Gene/product*	*Mutations*	*Disease mechanism*	*Incidence/prevalence (per 100 000)*	*Inheritance*
Ataxias							
Ataxias (all)						2 [P]	
Ataxia Telangiectasia (208900)	Cerebellar ataxia, telangiectasias, immune defects, tumors	11q23	Unknown Not cloned	Unknown	Unknown	0.3 [P]	AR
Spinocerebellar, ataxia 1(SCA-1) (164400)	Cerebellar ataxia, chorea	6p23-p24	Unknown Gene not cloned; trinucleotide repeat expansion cloned	Trinucleotide repeat expansion (CAG)	Unknown	Rare	AD
Spinocerebellar ataxia 2 (SCA2) (183090)	Cerebellar ataxia, chorea	12q23-q24.1	Unknown Not cloned	Possible trinucleotide repeat expansion?	Unknown	Rare	AD
Machado–Joseph disease (109150)	Cerebellar ataxia	14q24-q32	Unknown Not cloned	Unknown	Unknown	Rare	AD
Friedrich's ataxia (229300)	Spinal and cerebellar ataxia	9q12-q21.1	Unknown Not cloned	Unknown	Neurodegeneration?	1.5 [P]	AR
Retinal degenerations							
Retinitis pigmentosa (RP); (all)	Constriction of the visual fields, night blindness, fundus changes				Defective photoreceptor function; retinal degeneration	20–33 [P]	
RP2 (312600)		Xp11.4-p11.2	Not cloned	Unknown	Unknown	All XL = 15% of total	XL
RP3 (312610)		Xp21.1-p11.4	Not cloned	Unknown	Unknown		XL
RP6 (312612)		Xp21.3-p21.2	Not cloned	Unknown	Unknown		XL
RP1 (180100)		8p11-q21	Not cloned	Unknown	Unknown	All AD = 50% of total	AD
RP4 (180380)		3q21-q24	Rhodopsin (cloned)	Point mutations	Unknown		AD
RP7 (179605)		6p21.1-cen	rds/peripherin	Deletions, point mutations	Unknown		AD
RP9 (180104)		7p15.1-p13	Unknown	Unknown	Unknown		AD
RP10 (180105)		7q	Unknown	Unknown	Unknown		AD
RP-CNGC (123825)		4p14-q13	Rod CGHP cloned protein	Point mutations	Unknown	All AR = 25% of total	AR
RP (180072; human homolog of mouse Rd)		4p16.3	Phosphodiesterase β cGHP	Point mutations	Unknown		AR

APPENDIX 16.A *Continued*

Disease	*Pathology/ pathophysiology/ symptomatology*	*Gene locus*	*Gene/product*	*Mutations*	*Disease mechanism*	*Incidence/prevalence (per 100 000)*	*Inheri-tance*
2. *Tumors and infections*							
Tuberous sclerosis (TSC)	Various tumors, skin lesions					10 [I]	
Tuberous sclerosis 1 (191100)		9q34	Unknown; not cloned	Unknown	Unknown		AD
Tuberous sclerosis 2 (191092)		16p13.3	TSC2/tuberin (homology with GTPase activating proteins) Cloned	Deletions	Tumor suppressor Loss of heterozygosity		AD
Neurofibromatosis type 1(162200)	Neurofibromas, skin lesions	17q11.2	Neurofibromin (member of the GTPase activating protein family) Cloned	Macro and microdeletions Insertions Point mutations	Tumor suppressor Loss of heterozygosity	28.6	AD
Neurofibromatosis type 2 (101000)	Acoustic neuromas, bilateral meningiomas,	22q12.2	Merlin (cytoskeletal protein) Cloned	Point mutations. Deletions	Tumor suppressor Loss of heterozygosity		AD
von Hippel Lindau disease (193300)	Cerebellar hemangioblastomas	3p25	VHL gene Cloned	Frameshifts. Deletions	Tumor suppressor Loss of heterozygosity	2.8	AD
Neuroblastoma (256700)	Tumors of the CNS	1p36.2-p36.1 14q32-qter	Unknown Not cloned	Deletions	Loss of heterozygosity Amplification of N-myc?		Sporadic
Glioblastoma (137800)	Glial tumours of the CNS	Many loci	Unknown	Deletions	Tumor suppressor Loss of heterozygosity		AD
Retinoblastoma (inherited [5–10%]; sporadic [60–70%]; germ-line mutation [20–30%]) (180200)	Retinal and other tumors	13q14.1-q14.2	Rb gene (member of the GTPase activating protein family) Cloned	Deletions	Tumor suppressor Loss of heterozygosity	7 (inherited) 45 (sporadic)	AD
Herpes simplex virus 1 infection	Encephalitis	–	Expresses up to 70 viral genes	–	–	0.5	–
Acquired immunodeficiency syndrome (AIDS)	Encephalitis. Opportunistic infections	–	Expresses viral genes	–	–	High incidence among hemophiliacs, drug abusers, homosexuals; lower in heterosexuals	–

APPENDIX 16.A *Continued*

Disease	*Pathology/ pathophysiology/ symptomatology*	*Gene locus*	*Gene/product*	*Mutations*	*Disease mechanism*	*Incidence/prevalence (per 100 000)*	*Inheri-tance*
3. *Anterior horn (motor-neuron) diseases*							
Spinal muscular atrophy (SMA)	Degeneration of anterior horn motor-neurons					16–66 [I] 2–40 [P]	
SMA (Werding–Hoffman [infantile, (253300)], Kugelberg Welander [juvenile, (253400)], Intermediate forms, (253550)	Neuronal apoptasis inhibitory protein and/or survival motoneuron protein	5q12.2-q13	Neuronal apoptosis inhibitory protein and/or survival motorneuron protein	Deletions Point mutations	Unknown	4–10 [I]	AR
SMA (chronic – adult) (158590)		Unknown	Unknown Not cloned.	Unknown	Unknown	4.2 [I]	AD/AR/XL
Familial amyotrophic lateral sclerosis (ALS) (105400)	Spinal and bulbar motor-neuron degeneration	21q22.1-q22.2 (some)	Cu, Zn superoxide dismutase Cloned	Point mutations	Free radical-mediated neurotoxicity?	0.1 [I] (10% of total cases)	AD
X-linked bulbo-spinal muscular atrophy (Kennedy's disease) (313200)	Degeneration of bulbar and spinal motor-neurons	Xq12	Androgen receptor Cloned	Trinucleotide repeat expansion (CAG)	Decreased androgen receptor function?	Rare	XL

APPENDIX 16.A *Continued*

Disease	*Pathology/ pathophysiology/ symptomatology*	*Gene locus*	*Gene/product*	*Mutations*	*Disease mechanism*	*Incidence/prevalence (per 100 000)*	*Inheritance*
4. *Inherited peripheral neuropathies*							
Hereditary motor and sensory neuropathy (HMSN) [all]						14–280 [P]	
Charcot–Marie–Tooth type 1A (CMT1A) [HMSN IA] (118220)	Peripheral nerve demyelination	17p11.2	Peripheral myelin protein-22 (PMP-22) Cloned	Duplications, point mutations,	Decreased myelinating capacity of Schwann cells		AD
CMT1B (118200) [HMSN IB]	Peripheral nerve demyelination	1q21.2-q23	Peripheral myelin protein P_0 Cloned	Point mutations	Decreased myelinating capacity of Schwann cells		AD
CMT1C [HMSN IC]	Peripheral nerve demyelination	Unknown	Unknown Not cloned	Unknown	Decreased myelinating capacity of Schwann cells		AR/AD
CMTX1 (302800)	Peripheral nerve demyelination	Xq13	Connexin-32. Cloned	Point mutations	Decreased myelinating capacity of Schwann cells		XL
Hereditary neuropathy with liability to pressure palsies (162500)	Peripheral nerve demyelination	17p11.2	PMP-22 Cloned	Deletions	Impaired myelinating capacity of Schwann cells		AD

APPENDIX 16.A *Continued*

Disease	*Pathology/ pathophysiology/ symptomatology*	*Gene locus*	*Gene/product*	*Mutations*	*Disease mechanism*	*Incidence/prevalence (per 100 000)*	*Inheri-tance*
5. *Muscle disorders*							
Myotonic dystrophy (DM) (160900)	Myotonia, muscle wasting, hypogonadism	19q13.3	Myotonin (DM protein kinase) Cloned	Trinucleotide repeat expansion (CTG)	Unknown	9–96 [P]	AD
Duchenne muscular dystrophy (DMD) (310200)	Muscle weakness; mental retardation (some)	Xp21.2	Dystrophin (cytoskeletal protein) Cloned	Frameshift mutations	Unknown	17–244 [P] 97–390 [I] (in males)	XL
Becker muscular dystrophy (310200)	Muscle weakness	Xp21.2	Dystrophin (cytoskeletal protein)	Deletions, duplications, point mutations	Unknown	12–17 [P] 16–55 [I]	XL
Facio-scapulo humeral muscular dystrophy (158900)	Muscle weakness – wasting of face and girdle muscles	4q35	Unknown Not cloned	Unknown	Unknown	2–67 [P]	AD
Periodic paralysis II (Paramyotonia congenita) (170500)	Episodic myotonic symptoms	17q23.1–25.3	adult muscle sodium channel a subunit Cloned	Point mutations	Alterations in sodium channel function		AD
Malignant hyperthermia (145600)	Idiosynchratic reaction to anesthetics: anesthetic death; myopathy.	I:19q13.1-q13..2 (50%) II: 17q11.2-q24 III: Unknown	I: skeletal muscle Ca^{2+} release channel (ryanodine receptor) II-III Unknown	I: Point mutations	Increased muscle sensitivity to anesthetics, rise in intracellular Ca^{2+}, fever, muscular rigidity	Not determined, important cause of anesthetic deaths	AD

APPENDIX 16.A *Continued*

Disease	*Pathology/ pathophysiology/ symptomatology*	*Gene locus*	*Gene/product*	*Mutations*	*Disease mechanism*	*Incidence/prevalence (per 100 000)*	*Inheritance*
6. *Autonomic nervous system disorders*							
Familial dysautonomia (223900)	Altered sensory and autonomic function	9q31-q33	Unknown	Unknown	Unknown	5–33 (Ashkenazi jews)	AR
7. *Metabolic disorders*							
Purine metabolism							
Lesch–Nyhan (308000)	Basal ganglia dysfunction; no neuronal cell loss.	Xq2.6-q2.7	Hypoxanthine Guanine Phosphoribosyl transferase Cloned	Point mutations Deletions	Impaired dopamine neurotransmission?	0.25–1 200/year USA	XL
Amino acid metabolism							
Phenylketonuria (PKU)(261600)	Mental retardation	12q22-q24.1	Phenylalanine hydroxylase Cloned	Point mutations	Defect in myelin formation	12.5	AR
Neuronal storage diseases/lysosomal disorders							
Gangliosidoses							
G_{M1}-gangliosidoses Infantile (230500); juvenile (230600); and adult variants (230650)	Mental retardation	3p21.33	β-galactosidase Cloned	Point mutations Duplications	Intraneuronal accumulation of G_{M1}-ganglioside; neurodegeneration	Rare	AR
G_{M2}-gangliosidoses Tay–Sachs disease (272800)	Mental retardation	15q23-q24	Hexosaminidase A Cloned	4bp insertion Point mutations Deletions	Intraneuronal accumulation of G_{M2}-ganglioside; neurodegeneration	25 jewish [I] 1 non-jewish [I]	AR
G_{M2}-gangliosidoses Sandhoff's disease (268800)	Mental retardation	5q13	Hexosaminidase A+B; β-subunit mutation	Deletions Insertions Duplications Point mutations	Intraneuronal accumulation of G_{M2}-ganglioside; neurodegeneration	0.1 (jewish) 0.3 (non-jewish)	AR

APPENDIX 16.A *Continued*

Disease	*Pathology/ pathophysiology/ symptomatology*	*Gene locus*	*Gene/product*	*Mutations*	*Disease mechanism*	*Incidence/prevalence (per 100 000)*	*Inheritance*
Lipidoses							
Gaucher's disease, (230800); Type II: acute; subacute neuronopathic	Mental retardation	1q21	Glucocerebrosidase Cloned	Point mutations Microdeletions	Intraneuronal accumulation of glucosyl-ceramide; neurodegeneration	Rare	AR
Niemann Pick disease (257200)	Mental retardation	11p15	Deficiency of acid sphingomyelinase Cloned	Microdeletions Point mutations	Intraneuronal accumulation of sphingomyelin; neurodegeneration	Rare	AR
Neuronal ceroid lipofuscinosis							
Batten's disease Infantile, late infantile, juvenile, Adult forms (204200)	Mental retardation	1p32: infantile; 16p12.1: juvenile; unknown: others	Unknown Not cloned	Unknown	Intraneuronal accumulation of subunit C of mitochondrial ATP synthase?	1 [I]	AR
Mucopolysaccharidoses (MPS)							
MPS I (Hurler's disease) (252800)	Mental retardation	4p16.3	α-L-iduronidase Cloned	Point mutations Microdeletions Deletion/ insertion	Severe enzyme deficiency Intracellular accumulation of dermatan sulfate and heparan sulfate	Rare	AR
MPS II (Hunter's syndrome) (309900)	Mental retardation	Xq28	Idurono-sulfatase Cloned	Point mutations Macrodeletions	Severe enzyme deficiency Intracellular accumulation of dermatan sulfate and heparan sulfate	1–3 [I]	XL
MPS III (Sanfilippo syndromes), type: A (252900), B (252920), C (252930), D (252940)	Mental retardation	Unknown Unknown Unknown 12q14	A: Heparan N-sulfatase B: N-acetyl-α-D-glucosaminidase C: acetylCoA: α-glucosaminidase N-acetyl-transferase D: N-acetyl-glucosamine-6-sulfate sulfatase Cloned	Unknown Unknown Unknown Unknown	Intracellular accumulation of heparan sulfate	A: 4 [I] B: rare C: rare D: rare	AR

APPENDIX 16.A *Continued*

Disease	*Pathology/ pathophysiology/ symptomatology*	*Gene locus*	*Gene/product*	*Mutations*	*Disease mechanism*	*Incidence/prevalence (per 100 000)*	*Inheritance*
Mucolipidoses							
I cell disease (Mucolipidosis II) (252500)	Mental retardation	4q21-q23	UDP-N-acetylglucosamine: N-acetyl-glucosaminyl-1-phospho-transferase Not cloned	Unknown	Lack of addition of mannose-6-phosphate lysosomal targeting signal onto lysosomal enzymes	Rare	AR
Leukodystrophies							
Metachromatic leukodystrophy (Late infantile, early juvenile, juvenile, and adult forms) (250100)	CNS demyelination; segmental demyelination in PNS	22q13-ter	Deficiency of arylsulfatase A Cloned	Point mutations Microdeletions	Intracellular accumulation of sulfatides	1 [I]	AR
Adrenoleukodystrophy (ADL) Adrenomyelo-neuropathy (AMN) (300100)	CNS demyelination; adrenal insufficiency	Xq28	ALD protein (ALDP) – peroxisomal transporter protein – (member of the superfamily of ABC transporters) Function: peroxisomal import of lignoceroyl-CoA synthetase?) Cloned	Partial deletions Point mutations	Intracellular accumulation of saturated very long chain fatty acids with C_{25} and C_{26}	1–5 [I]	XL
Krabbe's disease (Infantile; late infantile; juvenile; Adult) (245200)	CNS demyelination; segmental demyelination in PNS	14q24-q32	Galactocerebrosidase Cloned	Unknown	Intracellular accumulation of galactosylceramide	0.5–1 [I] 600 (Druze Israelis) [I]	AR

APPENDIX 16.A *Continued*

Disease	*Pathology/ pathophysiology/ symptomatology*	*Gene locus*	*Gene/product*	*Mutations*	*Disease mechanism*	*Incidence/prevalence (per 100 000)*	*Inheri-tance*
Canavan disease (271900)	Neurodegeneration	17pter	Aspartoacylase Cloned	Point mutations	Intraneuronal accumulation of N-acetyl-aspartic acid	Rare	AR
Pelizaeus-Merzbacher disease (312080)	CNS demyelination	Xq22	Proteolipid protein Cloned	Point mutations, duplications	Decreased myelinating capacity of oligodendrocytes	Rare	XL
8. *Mental retardation, non-metabolic*							
Fragile X, A (309550)	Mental retardation	Xq27	FMR1 Cloned	Trinucleotide repeat expansion (CGG), gene and promoter deletions	Decreased expression of FMRI	57 (males) 36 (females)	XL
Fragile X, E (309548)	Mental retardation	Xq28	Not cloned	Trinucleotide repeat expansion (CGG)	?		XL
Down's syndrome	Mental retardation	21	Various genes Some cloned	Gene dosage (trisomy)	Unknown	60–4100 [I] (depending on maternal age)	

This table does not provide a list of all human inherited neurological disorders. For complete listings and details of these, please consult the references from which data for this table were taken.
Gene column indicates the name of the gene, and whether it has been cloned or not. The Mendelian Inheritance in Man number (McKusick, 1994) for each disease, if available, is provided after its name.
y, years; AD, autosomal dominant; AR, autosomal recessive; XL, X-linked; [P], prevalence; SN, substantia nigra; [I], incidence; microdeletions, less than 20 bp deletions; macrodeletions, more than 20 bp deletions.
Data for this table are based on the following references: Emery, 1991; Rocca *et al.*, 1991; Bundey, 1992; Martin, 1993; Rosenberg *et al.*, 1993; Zhang and Roman, 1993; McKusick, 1995.

APPENDIX B
GENERAL CRITERIA FOR SOMATIC GENE THERAPY

General criteria (Anderson, 1985)

1. The new gene can be inserted into the correct target cells and will remain there long enough to be effective.
2. The new gene will be expressed in the cells at an appropriate level.
3. The new gene will not harm the cell or by extension, the animal.

Criteria (requirements) set by the Committee on the Ethics of Gene Therapy (UK) for Somatic Cell Gene Therapy (Clothier *et al.*, 1992)

1. The gene must have been isolated and be available for therapeutic use.
2. Something must be known of the function of the gene, so that treatment can be sensibly designed.
3. It must be known which tissues and cell types need to express the gene and when, during development, its expression is required.
4. The genetic sequences that control the function of the gene, for example by switching it on and off, must be known and have been isolated. If gene therapy is to work it is clearly important that the product of the gene is made in the right cells, in the right amounts and at the right time. Too much or too little of the gene product, or its production in the wrong cells, or at the wrong time during development, could be harmful.
5. There must be means available for inserting the gene, and its controlling elements, into the right cells and under the right control. Conversely, the gene must be prevented from getting into the wrong cells, or to the wrong place within a cell, or from spreading to other tissues, or even to other individuals, any of which might cause harm.
6. Means must be available for monitoring the efficacy and safety of the treatment, for a long time.

Special care should also be taken *vis à vis* the following issues:

1. Questions of safety, hightened by the possibility of inadvertent and unpredictable consequences of gene therapy to the patient, and the possible long-term consequences.
2. The need for long-term surveillance and follow-up.
3. The matter of consent.
4. The probability that children will be among the first candidates for therapy.
5. Confidentiality, and disclosure of genetic information important to kindred.
6. Somatic cell gene therapy should be governed by the exacting requirements which already apply in the United Kingdom to other research involving human subjects.
7. Gene therapy treatment should be limited to patients in whom the

potential for benefit is greatest in relation to possible inadvertent harm. First candidates for gene therapy should be patients in whom the disorder is:

(a) life threatening or causes serious handicap;
(b) for which treatment is at present unavailable or is unsatisfactory; and
(c) directed to alleviating disease, not to change human traits not associated with disease.

APPENDIX 16.C
CRITERIA DEVELOPED FOR SPECIFIC APPLICATIONS OF SOMATIC CELL GENE THERAPY

Ideal criteria to be fulfilled by neurological disorders to be treated by gene therapy

1. Single gene defect.
2. Gene should be cloned and mutations should be identified.
3. Natural history of disease history should be known.
4. Regulatory mechanisms of gene expression should be understood.
5. Target diseases should be correctable by direct intraneuronal gene transfer or by transplantation.
6. What level of expression is needed? Known target level of gene expression required to correct defect should have been determined.
7. For how long? The length of gene expression required to achieve therapeutic benefit should be known.
8. In how many neurons? The extent of transgene expression required to achieve therapeutic benefit should be known.
9. What is the acceptable risk levels for each disease/patient?
10. When, during disease evolution in any individual patient, should gene therapy be started in order to be of therapeutic benefit?

17

Mucopolysaccharidosis

OLIVIER DANOS and JEAN-MICHEL HEARD

17.1 INTRODUCTION

Most of the digestion process in the living cell takes place in the lysosome. These organelles are formed in the trans Golgi network, from vesicles in which more than 40 different enzymes, mostly acid hydrolases, are selectively packed. The mature lysosome results from a fusion between these enzyme-containing vesicles and late endosomes where macromolecules awaiting disposal are entrapped. The acid pH maintained within the lysosome activates the enzymes such that proteins, nucleic acids and complex sugars are degraded. A deficiency in one of these digestive enzymes results in the toxic accumulation of its undegraded substrate. This 'cellular dyspepsia' (De Duve, 1964) results in a lysosomal storage which can severely impair the physiology of the cell.

Glycosaminoglycans (GAG, also termed mucopolysaccharides) are continuously produced by the normal turnover of proteoglycans in connective tissues. Their stepwise degradation in the lysosome involves four glycosidases, five sulfatases and one non-hydrolytic transferase. In mucopolysaccharidosis (MPS), one of these enzymes does not function properly because of a genetic defect and, as a consequence, most cells in the organism are clogged with undigested intermediates. This results in a wide spectrum of clinical symptoms which include mild to severe bone and joint abnormalities, hepatosplenomegaly, corneal clouding, nerve entrapment syndromes and mental retardation (Neufeld and Muenzer, 1989; Hopwood and Morris, 1990). The discovery of the mechanism for enzyme routing to the lysosome has suggested that these deficiencies could be complemented by infusing the purified enzyme.

Molecular and Cell Biology of Human Gene Therapeutics
Edited by George Dickson
Published in 1995 by Chapman & Hall. ISBN 0 412 62550 4

In a classic series of experiments, Neufeld and collaborators (1977) have shown that fibroblasts from MPS patients could be corrected by factors secreted by normal fibroblasts or present in urine concentrates. These 'corrective factors' were identified as the normal enzymes themselves, which were taken up by the mutant cells and targeted to the lysosomes (Neufeld *et al.*, 1977). It is now known that, like most lysosomal enzymes, the ten enzymes involved in MPS follow the secretion/biosynthesis pathway. The glycosylated pro-enzyme is specifically modified in the Golgi compartment by phosphorylation on its mannose residues and becomes a ligand for the mannose-6-phosphate receptors. These membrane receptors cycle between the lysosome and the plasma membrane and direct the phosphorylated enzyme precursors to the organelles, either by selectively packing them into pre-lysosomal vesicles, or by capturing molecules secreted in the extracellular media (Pfeffer, 1991). The secretion/capture mechanism has served as a rationale for the design of therapies for MPS. The main treatment is bone marrow transplantation, in which it is expected that a large number of normal cells from the graft will infiltrate the diseased tissues and provide a source of enzyme for other cells and/or degrade the accumulated GAG themselves.

Several different animal models of MPS have been characterized providing a truly unique situation for the testing of experimental therapies including gene therapy (Table 17.1). Small laboratory animals like the MPS VII mouse (Birkenmeier *et al.*, 1989) or the MPS VI rat (Yoshida *et al.*, 1993) are very valuable because they are easily accessible and the mutation can be bred on an homogenous genetic background. Pre-clinical studies can be performed on colonies of MPS I dogs (Shull *et al.*, 1982) or cats (Haskins *et al.*, 1979), MPS VI cats (Jezyk *et al.*, 1977) or MPS VII dogs (Haskins *et al.*, 1984). A goat model of MPS IIID has also been described (Thompson *et al.*, 1992). Although considerable difficulties must be surmounted, MPS present a favorable situation for gene therapy. Low and unregulated levels of enzyme activity should be sufficient for correction and various gene transfer strategies can be thoroughly evaluated in animal models.

17.2 MOLECULAR BASIS, PHYSIOPATHOLOGY AND CURRENT TREATMENTS

17.2.1 General description

The prevalence of MPS is approximately 1 in 15 000. MPS I, II and III are equally frequent and together represent more than 80% of the cases (Table 17.2). Clinical features of variable intensities are shared by all MPS and some are specific for certain diseases (Neufeld and Muenzer, 1989; Hopwood and Morris, 1990). In most cases, symptoms affecting multiple organs appear in early infancy, and the disease has a chronic and progress-

Table 17.1 Classification of the mucopolysaccharidosis (MPS)

	Eponym	*Enzyme deficiency*	*Animal models*	*Reference*
MPS I	Hurler/Scheie	α-L-iduronidase[a]	Cat, dog	Scott *et al.* (1991)
MPS II	Hunter	Iduronate sulfatase[a]	–	Wilson *et al.* (1990)
MPS IIIA	Sanfilippo	Heparan-*N*-sulfatase	–	
B		*N*-Acetylglucosaminidase	–	
C		*N*-Acetyltransferase	–	
D		*N*-Acetylglucosamine-6-sulphatase[a]	Goat	Robertson *et al.* (1992)
MPS IVA	Morquio	Galactose-6-sulphatase[a]	–	Tomatsu *et al.* (1991)
B		β-Galactosidase[a]	–	Oshima *et al.* (1987)
MPS VI	Maroteaux–Lamy	Arylsulfatase B[a]	Cat, rat	Peters *et al.* (1990)
MPS VII	Sly	β-Glucuronidase[a]	Mouse, dog	Oshima *et al.* (1987)

[a] Cloned cDNAs

Table 17.2 Relative geographical frequencies (%) of mucopolysaccharidosis (MPS)

	MPS I	*MPS II*	*MPS III*	*MPS IV*	*MPS VI*	*MPS VII*	
Europe	28	20	30	14	5	3	n = 562[a]
USA	26	27	31.5	10	4.4	1.1	n = 89[b]
Japan	11	46	27.5	13.5	0.5	1.5	n = 259[c]

[a] Data from the European Registry of *Diagnostic Laboratories for Inborn Errors of Metabolism*; [b] Partial records from the United States *Metabolic Information Network*; [c] K. Sukegawa, personal communication

ive course, frequently leading to death before adulthood. The association of organomegaly, dysostosis multiplex, corneal clouding, and abnormal facies is a distinguishing feature of MPS. Except for MPS IV and VI and MPS IS (Scheie's disease), neurological manifestations and mental retardation are also present. These symptoms are predominant in MPS II and III where they appear early and in severe forms of MPS IH (Hurler syndrome), where they are delayed until after skeletal abnormalities become obvious.

The characteristic biochemical feature common to all MPS patients is elevated urinary GAG excretion. However, since the magnitude of this abnormal excretion is highly variable, the diagnosis is more reliably made on fibroblast or leukocyte extracts, using simple and sensitive assays to identify the enzyme deficiency (Hopwood and Morris, 1990).

Because of the polymorphic nature of these diseases and of the wide spectrum of clinical severity associated with each enzyme deficiency, an accurate prognosis is often very difficult, hindering the interpretation of therapeutic trials. Attempts to correlate the course of a disease with the residual level of enzyme activity measured in patient cells have been unsuccessful (Neufeld and Muenzer, 1989). The advancement of our knowledge on the genetic basis of MPS has provided hopes that accurate correlations could be established between the genetic lesions found in patients and the severity of their diseases. Unfortunately, the data available so far indicate again that MPS are very heterogeneous (Wilson *et al.*, 1991; Fukuda *et al.*, 1992; Jin *et al.*, 1992; Bunge *et al.*, 1993; Shipley *et al.*, 1993; Yamada *et al.*, 1993). As shown in Table 17.3, a tentative genotype/phenotype correlation can only be made in MPS IH and MPS IS patients carrying one of the frequent mutations in the α-L-iduronidase gene (Clarke and Scott, 1993; Scott *et al.*, 1993), and in MPS II patients with a large deletion or rearrangement of the iduronate sulfatase gene.

17.2.2 Approaches for specific treatments

Enzyme replacement therapy through cell transplantation was initially attempted in MPS patients by whole-blood or leukocyte transfusions and

Table 17.3 Genetic abnormalities in mucopolysachoridosis patients

	Chromosomal location	*Point mutation or small deletion/insertion*			*Large deletion or rearrangement*	
		Nature	*Allele frequency*	*Phenotype*	*Frequency*	*Phenotype*
IH	4p16–3	W402X	31%	Severe	NR	–
		Q70X	15%	Severe		
		P533R	3%	Severe		
IS	4p16–3	R89Q/678–79≥a	40%	Mild	NR	–
II	Xq28	Heterogeneous	80%	Variable	20%	Severe
IV-a	16p24.3	1342ΔCA	1[a]	Severe	2/12[b]	Severe
		N204K	1[a]	Mild		
		W273L	1[a]	Mild		
VI	5q13.3	Heterogeneous	–	Variable	NR	–
VII	7q21	Heterogeneous	–	Variable	NR	–

NR, not reported; [a], one patient reported; [b] 2 of 12 patients reported.

resulted in a minimal transient effect. Long-term biochemical improvement was observed after the transplantation of allogeneic normal skin fibroblasts in MPS II patients (Dean, *et al.*, 1979). However, this treatment had no effect on the course of the disease and was not investigated on a large scale.

In contrast, the benefit of bone marrow transplantation (BMT) has been unambiguously demonstrated in animal models of MPS. Evaluation of therapeutic effects are indeed much easier in these models where the natural course of the disease is well documented, even though variations exist between offsprings. The β-glucuronidase-deficient mouse recapitulates the symptomatology observed in MPS VII patients, with skeletal deformities, corneal clouding, hearing defect, and a major lysosomal accumulation of GAG (Birkenmeier *et al.*,1989). In this animal, syngeneic BMT has been shown to normalize the histology of liver and spleen and to increase life span (Birkenmeier *et al.*, 1991). The treatment of adult animals did not affect skeletal deformities and brain lysosomal storage. In contrast, animals transplanted as newborns became less dysmorphic and displayed a better mobility and a decreased neuronal storage (Sands *et al.*, 1993). BMT has also been performed in MPS VII dogs. Two animals treated at 1 month showed an almost normal skeletal development and no obvious neurological problem in adulthood (M. Haskins, personal communication). Successful BMT was also reported in five MPS I dogs (Shull *et al.*, 1987). Five months after transplantation the GAG storage disappeared in various tissues including brain, and the animals had improved their joint mobility. Similar observations were made after BMT in one MPS VI cat (Gasper *et al.*, 1984).

BMT has now been performed on a significant number of MPS patients but there is still no general consensus about the indication of this treatment (Krivit *et al.*, 1984, 1992; Vellodi *et al.*, 1992). An HLA-matched relative is available as donor in 15% of the cases, and an unrelated HLA-matched donor can be found for 30%. Mortalities are 10% and 25% respectively in these contexts. Successful engraftment always results in increased enzyme levels in leukocytes, normalization of the size of liver and spleen, and decreased levels of GAG in urine and cerebrospinal fluid (CSF). In a series of 38 treated MPS I patients, followed for a period of up to 12 years, Krivit observed a longer survival (W. Krivit, personal communication). In most children a stabilization of skeletal symptoms occurs, but the improvement of pre-existing lesions is minimal. Gibbus formation and dysostosis multiplex often persist and the improvement of corneal clouding is partial. Although definitive conclusions about the intellectual development require a long-term follow-up of patients, the neurological symptoms can be prevented by early BMT and the intellectual development of treated children is stable. Severe neurological symptoms are not reverted and seriously affected children are usually not treated.

Both the studies in animal models and the results obtained in patients emphasize the importance of an early transplantation, performed before the appearance of severe skeletal and neurological symptoms. They also stress the need for very early diagnosis. When history of MPS is known in the family, prenatal diagnosis can be proposed. In families where there is no record of MPS, the early recognition of patients could only be achieved through a systematic screening of infants for lysosomal disorders.

Enzyme therapy is not currently available for the treatment of MPS, as it is for Gaucher's disease. Early attempts to provide enzyme through the infusion of serum from normal individuals have failed to show significant improvement, except for a transient decrease in the urinary GAG excretion (Di Ferrante *et al.*, 1971). Injection of purified recombinant β-glucuronidase preparations in newborn MPS VII mice have shown that enzyme activity can be measured for up to 4 days in various tissues, with up to 30% of the normal activity in the brain (Vogler *et al.*, 1993). Attempts to treat MPS I dogs with infusions of purified recombinant human α-L-iduronidase are currently in progress (Shull *et al.*, 1994). However, because only small numbers of patients will be eligible for enzyme therapy, the large-scale industrial production of purified recombinant enzymes would meet serious economical problems.

17.3 STRATEGIES FOR GENE THERAPY

The partial success of BMT, which can only be offered to one-half of MPS patients, and the economical obstacles for enzyme therapy, have stimulated the search for gene therapy approaches. As in the other therapeutic interventions, the aim of these strategies would be to provide each tissue with the minimal level of enzyme required to avoid GAG overload. This could be achieved by implanting genetically modified autologous cells delivering the enzyme systemically. The first benefit of a gene therapy approach over allogenic tissue transplantation would be to eliminate the requirement for an HLA-matched donor. Any cell type could be chosen as a source of enzyme, provided that efficient methods for *ex vivo* gene transfer and stable re-implantation exist. This includes fibroblasts, hematopoietic cells, muscle cells, hepatocytes, endothelial cells, glial progenitors and others. Certain tissues could also be directly targeted *in vivo* with an adenovirus- or a herpesvirus-derived vector.

The normal cDNA is available for seven out of the ten enzymes affected in MPS (Table 17.1). Significant amounts of these enzymes can be secreted after transfection into COS or CHO cells. The recombinant enzyme is produced in the cell culture medium as a precursor which can be taken up by other cells, targeted to the lysosome and processed correctly (Oshima *et al.*, 1987; Morreau *et al.*, 1989; Peters *et al.*, 1990; Scott *et al.*, 1991; Tomatsu *et al.*, 1991; Anson *et al.*, 1992a, b; Bielicki *et al.*, 1993). The secretion efficiency varies, however, from one enzyme to the other. For example, β-galactosidase and β-glucuronidase are efficiently recovered from the culture media of over-expressing cells, whereas arylsulfatase B secretion is mediocre and must be stimulated *in vitro* by the addition of ammonium chloride. In some cases, secretion may thus be the limiting factor of cell-mediated enzyme distribution *in vivo*. A possible solution could be to over-express the cation-dependent mannose-6-phosphate receptor along with the enzyme, since this receptor seems to favor the targeting of mannose-6-phosphate containing proteins to the extracellular medium (Chao *et al.*,1990).

Corrective cDNAs have been introduced using retroviral vectors in cultured cells from MPS patients or animal models, resulting in the correction of the storage phenotype (Smith *et al.*, 1990; Wolfe *et al.*, 1990; Peters *et al.*, 1991; Anson *et al.*, 1992a, b, 1993; Braun *et al.*, 1993). In some cases, however, unwanted side effects linked to over-expression were observed. A 250-fold increase in α-L-iduronidase was associated with a decreased activity in other GAG-processing enzymes, possibly because of a competition for the mannose-6-phosphate-targeting pathway (Anson *et al.*, 1992a). When arylsulfatase B was over-expressed 100-fold, lysosomal and microsomal sulfatases were inhibited (Anson *et al.*, 1993). It remains to be established whether this would compromise the function of an implant of modified cells. This secondary storage may well be corrected *in vivo* by the presence in the organism of normal levels of the down-regulated enzymes.

In addition to systemic enzyme distribution, the phenotypic correction would be enhanced if the modified cells were able to infiltrate the tissues and clean up the accumulated GAG locally. This could be achieved by bone marrow-derived tissue macrophages which can be corrected after gene transfer to hematopoietic stem cells. In this situation, an efficient secretion may not be necessary, since the corrected macrophages would scavenge the tissue themselves. Genetically-modified astroglial or neuronal precursors with migrating properties, could also act as local detoxifiers in the brain.

17.3.1 Fibroblasts

Fibroblasts are easily obtained from skin biopsies and grown in primary culture. Freshly explanted cells of murine, canine or human origin can be efficiently infected with retroviral vectors. Between 10^9 and 10^{10} phenotypically normal diploid cells expressing the new genetic information can be obtained after a couple of weeks in culture.

The inclusion of primary skin fibroblasts into collagen lattices has been shown to result in the formation of transplantable derm equivalent (Bell *et al.*, 1979). Moullier *et al.* (1993) have implanted into the peritoneal cavity this type of lattices, mixed with bFGF-coated polytetrafluoroethylene (PTFE) fibers as a support for vascularization. This led to the rapid formation of individualized neo-organs in which the genetically modified fibroblasts were metabolically active for months. The dense vascularization of these implants brought the enzyme-secreting fibroblasts into permanent close contact with the mesenteric circulation (Moullier *et al.*, 1993).

A recurrent problem associated with gene transfer into primary fibroblasts has been the disappearance of gene expression upon re-implantation *in vivo*. This has been observed with sequences expressed from strong promoter elements commonly used in retroviral vectors, such as the retroviral LTR, the SV40 or the cytomegalovirus immediate early promoters. The available data suggest that these promoters are turned off within a few days following re-implantation of the modified fibroblasts (Palmer *et al.*, 1991; Scharfmann *et al.*, 1991). In contrast, the promoters from dihydrofolate reductase (DHFR) and phosphoglycerate kinase (PGK) remain active after re-implantation (Scharfmann *et al.*, 1991; Moullier *et al.*, 1993). The reason for the sustained activity of these regulatory elements which both control house-keeping genes, and belong to the class of TATA-less promoters, is not understood.

The human β-glucuronidase cDNA under the control of the mouse PGK-1 promoter, has been transferred with a retroviral vector into primary fibroblasts prepared from MPS VII mouse skin. The modified cells secreting 100–200 units of β-glucuronidase/10^6 cells per 24 hours were included into collagen/PTFE lattices and implanted into the peritoneal cavity of 14 syngeneic MPS VII animals (Moullier *et al.*, 1993). Each animal received two lattices containing $10–15 \times 10^6$ fibroblasts, its urinary GAG levels being diminished within a few days. β-Glucuronidase activity was found in the liver, spleen, lung, brain, kidney, heart and bone marrow of the implanted animals which were sacrificed after 10 to 155 days. The highest activity levels (between 4% and 6% of normal) were found in the liver and spleen and a complete correction of the lysosomal storage lesions was observed in these organs (Figure 17.1). Although the enzyme activity in the liver appeared concentrated into tissue macrophages, the storage lesions were also corrected in hepatocytes. This suggested that corrected macrophages could act in locally clearing out substrates which were not degraded by the hepatocytes.

These experiments have shown that engineered fibroblasts, if re-implanted in a suitable environment can provide long-term therapeutic levels of enzyme in an MPS model. The cure was not complete, however, in those animals which displayed severe skeletal abnormalities when they were treated at the age of 6 to 8 weeks. Experiments are in progress to test whether implanting enzyme-secreting fibroblasts within the first days of life could facilitate the enzyme access to the developing bones and joints and to the central nervous system.

Intraperitonal neo-organs containing autologous cells secreting human β-glucuronidase have also been implanted in normal dogs. The procedure developed in mice has been scaled up in order to prepare implants made of

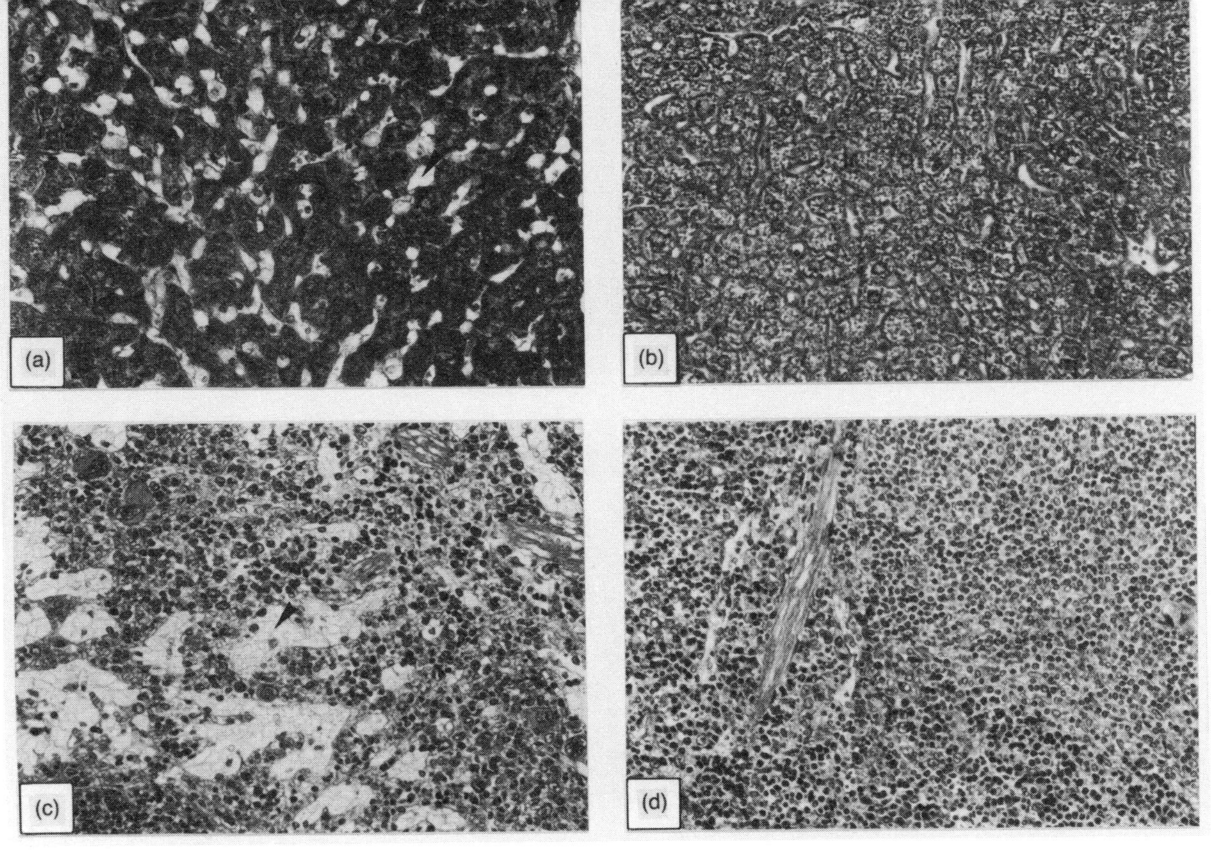

Figure 17.1 Correction of lysosomal storage in the liver and spleen of MPS VII mice. Sections (5 μm) from paraffin-embedded livers (a, b) and spleens (c, d), stained with Masson trichrome are shown. (a) Liver and (c) spleen of a 4-month-old control animal receiving fibroblasts expressing the β-galactosidase gene and showing abundant lysosomal storage within the dark and light pulp and in Küpffer cells (arrows). (b) Liver and (d) spleen of an experimental animal sacrificed 155 days after the implantation of a neo-organ secreting human β-glucuronidase, where the storage lesions have disappeared. (Original magnification ×75.)

10^9 cells and a minimally invasive surgical procedure was used for installing them into the peritoneal cavity. During the one year follow-up after implantation, the uptake of the secreted human enzyme was demonstrated in liver biopsies after heat inactivation of the endogenous activity (P. Moullier, unpublished results). These studies indicate that the procedure could be transferred to a clinical setting.

17.3.2 Hematopoietic cells

The relative success of histocompatible bone marrow transplantation in human and animal MPS indicates that the engraftment of normal hematopoietic stem cells can provide a long-term corrective effect. This correction can be mediated by the secretion of enzyme throughout the organism from the numerous and permanently renewed differentiated hematopoietic cells of normal donor origin, and its uptake by the deficient cells of the host. In addition, normal cells differentiating through the monocyte/macrophage lineage probably reduce the amount of storage in deficient cells by degrading the GAGs accumulated in their proximity.

Efficient procedures for gene transfer into hematopoietic stem cells have been developed in the mouse using retroviral vectors. Bone marrow cells from a donor animal are used for the hemato-lymphoid reconstitution of lethally irradiated syngeneic recipients, after a period of *in vitro* culture in the presence of fibroblasts producing a retroviral vector. If the new genetic information is transferred to a stem cell with long-term reconstituting capacity, it can be permanently amplified in the population of peripheral differentiated cells and it has been possible to obtain hematopoietic chimeras expressing a gene for long periods of time in most of the peripheral cells, in all lineages (Karlsson, 1991).

In the mouse model of MPS VII, Wolfe *et al.* (1992b) have used a retroviral vector to transfer the rat β-glucuronidase cDNA into bone marrow cells, under the control of a thymidine kinase promoter. Three treated animals were analyzed 6 months after bone marrow transplantation and two were found to be fully reconstituted with hematopoietic cells containing the cDNA. A complete disappearance of lysosomal storage was observed in the liver and spleen of treated animals (Wolfe *et al.*, 1992b).

In another study with the MPSVII mouse, Maréchal *et al.* (1993) have obtained partial hematopoietic chimeras using a low-dose irradiation treatment of the recipients. In 10 animals with less than 5% hematopoietic cells containing the human β-glucuronidase cDNA under the control of the PGK promoter, a significant improvement of the liver and spleen pathology was observed, suggesting that small amounts of enzyme delivered locally can be sufficient for correction (Maréchal *et al.* 1993). This observation is interesting in the prospect of clinical applications, since the current technology would not allow to obtain more than a few percent cells expressing the transgene in a treated patient (Bodine *et al.*, 1993).

17.3.3 Central nervous system

The enzyme delivered into the serum will not cross the blood–brain barrier under normal conditions (Pardridge, 1988). The barrier can be temporarily permeabilized by osmotic shock but this severe treatment cannot be applied regularly. BMT experiments in animal models of other lysosomal storage diseases have shown that hematopoietic cells from the monocyte/macrophage lineage naturally enter the central nervous system and can provide enzyme to the brain (Hoogerbrugge *et al.*, 1988). It is possible that the small amounts of β-glucuronidase found in the brain of MPS VII mice implanted with secreting fibroblasts correspond to enzyme molecules absorbed by monocytes in the periphery and transported across the barrier. Monocyte-mediated enzyme delivery into the brain is likely to be more efficient if the corrective cells arise from genetically modified bone marrow and continuously express the protein at high levels.

Intracerebral implants of genetically modified fibroblasts can locally deliver a therapeutic product (Gage *et al.*, 1991; Jiao *et al.*, 1993). In the case of MPS, however, enzyme delivery throughout the organ is needed and the modified cells should be able to migrate after implantation. We have observed that astroglial (O2A) progenitors isolated from newborn MPS VII mice secrete human β-glucuronidase after retrovirus-mediated gene transfer. After re-implantation in newborn mutants, they migrate to various sites of the brain where they can be detected after several months (unpublished data). This experimental system can be used to assess the capacity of a limited number of cells scattered in the brain to eliminate lysosomal storage. However, the difficulty to access the target cell for *ex vivo* gene transfer makes this procedure of little therapeutic relevance. Multipotent immortalized neural progenitor cell lines with high migration capacity have been described in the mouse (Snyder *et al.*, 1992) and used to obtain long-term diffuse engraftment in the brain of newborn MPS VII animals after the transfer of the β-glucuronidase cDNA (J. Wolfe, personal communication). The availability of such cells in humans could be of genuine interest for the treatment of CNS lesions in MPS.

Direct gene transfer into the CNS is feasible with herpesvirus or adenovirus vectors (Breakefield, 1993). Wolfe *et al.* (1990) have constructed an HSV-1 based vector where the rat gusB cDNA was placed under the control of the LAT promoter. The recombinant virus was used to infect MPS VII mice by corneal inoculation. After 9, 10 and 18 weeks, few positive neurons were detected by histochemical staining in the trigeminal ganglia and brainstem of 11 of 13 mice (Wolfe *et al.*, 1990). The disappearance of lysosomal storage in or around the positive cells was not studied. Although this has not been tested in MPS models, a more potent gene transfer can be obtained with adenovirus, by stereotactic injection of vector particles in the brain tissue or in the ventricular space. This second approach leads to the infection of the ependymal cells lining the ventricule and can be used to secrete a protein in the CSF (Bajocchi *et al.*, 1993). In MPS, this could directly reduce GAG levels in the CSF and might help enzyme diffusion to larger areas of the brain.

17.3.4 Other target tissues

The skeletal muscle has been proposed as a convenient organ for the systemic delivery of therapeutic proteins (Dhawan *et al.*, 1991). Myoblasts have been isolated from MPS VII dog skeletal muscle, grown in culture and infected with a rat β-glucuronidase cDNA-containing retroviral vector. Enzyme expression was documented in both myoblasts and myotubes (Smith *et al.*, 1990). Myoblasts from adult MPS VII mice were also isolated and engineered with a retroviral vector to secrete human β-glucuronidase (50–90 unit/10^6 cells per 24 hours). Eighteen MPS VII recipients were injected with the recombinant myoblasts, following a muscle injury. The secreting cells were observed by histochemical staining on muscle sections after 10 weeks, and were found to efficiently participate in the constitution of new muscle fibers (Naffakh *et al.*, 1994). Despite the high level of enzyme secretion measured *in vitro*, only trace amounts of activity were found in the liver and spleen of treated animals (unpublished results). This suggested that β-glucuronidase was blocked before it could access the bloodstream, possibly at the level of the muscle basal membrane.

The liver occupies a strategic position as a provider of proteins into the bloodstream. Retrovirus-mediated gene transfer *in situ* into the liver has been described in mice, rats and dogs (Ferry *et al.*, 1991; Cardoso *et al.*, 1993). Attempts at transferring the β-glucuronidase cDNA into the liver of MPS VII dogs are currently being made. The first results indicate that the fraction of hepatocyte which can be modified by this procedure may be too small to provide therapeutic levels of the enzyme (unpublished data).

17.4 PROSPECTS FOR CLINICAL TRIALS

So far, all *in vivo* gene transfer data have been obtained with MPS VII animals and it would seem logical to consider patients with Sly syndrome as the first candidates for a gene therapy trial. These patients are however exceedingly rare, with less than 20 known cases of live birth. For this reason, the inter-individual variations cannot be well documented in this disease and an adequate group of patients is difficult to define in view of a clinical trial.

Hurler syndrome is much more frequent than MPS VII (Table 17.2), with estimates varying between 1 : 50 000 and 1 : 100 000 live birth (Anson *et al.*,

1992a). As discussed above, genotype/phenotype correlation has begun to be established and pre- or peri-natal diagnosis is possible. The mechanisms of synthesis, processing, secretion and uptake of β-glucuronidase and α-L-iduronidase are closely related, and it is likely that many of the observations made in gene transfer experiments involving MPS VII models can be extrapolated to MPS I. Furthermore, the gene therapy approaches defined in MPS VII models can also be tested in MPS I dogs. Two types of intervention on MPS I patients could be proposed in the near future, involving retrovirus-mediated gene transfer to either hematopoietic (CD34+) cells or to skin fibroblasts re-implanted into the peritoneal cavity.

CD34+ hematopoietic cells can be prepared from bone marrow, from peripheral blood after treatment with granulocyte colony stimulating factor (G-CSF), or from cord blood if a prenatal diagnosis has been made. For the first trials, the procedures for infection with a retroviral vector and re-injection will be modeled after those used in ADA deficiency. In SCID patients however, corrected cells are expected to have a selective advantage, justifying the absence of conditioning regimen for endogenous bone marrow ablation. The situation might be different in MPS patients who could require conditioning to ensure even a low level of engraftment.

The graft of autologous skin fibroblasts secreting α-L-iduronidase from vascularized neo-organs could be performed using a minimally invasive surgical procedure. Phase I/II trials will have to assess the feasibility of the procedure, its tolerance by the patient, the efficiency and duration of enzyme secretion and the effect on the course of the disease.

Ultimately a combination of these two treatments may be required to obtain a significant improvement of the condition. The addition of a direct adenovirus-mediated gene transfer in the CNS and in joints (Roessler *et al.*, 1993) will require careful testing in the animal models.

ACKNOWLEDGEMENTS

We thank all our colleagues interested in gene therapy of MPS and members of our laboratory for discussions and access to unpublished results. This work was supported by grants from 'Vaincre les Maladies Lysosomales', the Association Française contre les Myopathies and the Pasteur Institute.

REFERENCES

Anson, D.S., Bielicki, J. and Hopwood, J.J. (1992a) Correction of mucopolysaccharidosis type I fibroblasts by retroviral-mediated transfer of the human α-L-iduronidase gene *Hum. Gene Ther.*, **3**, 371–9.

Anson, D.S., Taylor, J.A., Bielicki, J. *et al.* (1992b) Correction of human mucopolysaccharidosis type-VI fibroblasts with recombinant N-acetylgalactosamine-4-sulphatase. *Biochem. J.*, **284**, 789–94.

Anson, D.S., Muller, V., Bielicki, J. *et al.* (1993) Overexpression of N-acetylgalactosamine-4-sulphatase induces a multiple sulphatase deficiency in mucopolysaccharidosis-type VI fibroblasts. *Biochem. J.*, **294**, 657–62.

Bajocchi, S.H., Feldman, S., Crystal, R. and Mastrangeli, A. (1993) Direct *in vivo* gene transfer to ependymal cells in the central nervous system using recombinant adenovirus vectors. *Nature Genet.*, **3**, 229–34.

Bell, E., Ivarsson, B. and Merrill, C. (1979) Production of a tissue-like structure by contraction of collagen lattices by human fibroblasts of different proliferative potential *in vitro*. *Proc. Natl Acad. Sci. USA*, **76**, 1274–8.

Bielicki, J., Hopwood, J.J., Wilson, P. and Anson, D. (1993) Recombinant human iduronate-2-sulphatase: correction of mucopolysaccharidosis-type II fibroblasts and characterization of the purified enzyme. *Biochem. J.*, **289**, 241–6.

Birkenmeier, E.H., Davisson, M.T., Beamer, W.G. *et al.* (1989) Murine mucopolysaccharidosis type VII. Characterization of a mouse with β-glucuronidase deficiency. *J. Clin. Invest.*, **83**, 1258–66.

Birkenmeier, E.H., Barker, J.E., Vogler, C.A. *et al.* (1991) Increased life span and correction of metabolic defects in murine MPS VII following syngeneic bone marrow transplantation. *Blood*, **78**, 3081–92.

Bodine, D.M., Moritz, T., Donahue, R.E. *et al.* (1993) Long-term *in vivo* expression of a murine adenosine deaminase gene in rhesus monkey hematopoietic cells of multiple lineages after retroviral mediated gene transfer into CD34+ bone marrow cells. *Blood*, **82**, 1975–80.

Braun, S.E., Aronovich, E.L., Anderson, R.A. *et al.* (1993) Metabolic correction and cross-correction of mucopolysaccharidosis type II (Hunter syndrome) by retroviral-mediated gene transfer and expression of human iduronate-2-sulfatase. *Proc. Natl Acad. Sci. USA*, **90**, 11830–4.

Breakefield, X.O. (1993) Gene delivery into the brain using virus vectors. *Nature Genet.*, **3**, 187–8.

Bunge, S., Steglich, C., Zuther, C. *et al.* (1993) Iduronate-2-sulfatase gene mutations in 16 patients with mucopolysaccharidosis type II (Hunter syndrome). *Hum. Mol. Genet.*, **2**, 1871–5.

Cardoso, J.E., Branchereau, S., Prema Roy, J. *et al.* (1993) *In situ* retrovirus-mediated gene transfer into dog liver. *Hum. Gene Ther.*, **4**, 411–18.

Chao, H.H.J., Waheed, A., Pohlman, R. *et al.* (1990) Mannose-6-phosphate receptor dependent secretion of lysosomal enzymes. *EMBO J.*, **9**, 3507–13.

Clarke, L.A. and Scott, H.S. (1993) Two novel mutations causing mucopolysaccharidosis type I detected by single strand conformational analysis of the α-L-iduronidase gene. *Hum. Mol. Genet.*, **2**, 1311–12.

Dean, M.F., Stevens, R.L., Muir, H. *et al.* (1979) Enzyme replacement therapy by fibroblasts transplantation. *J. Clin. Invest.*, **63**, 138–45.

De Duve, C. (1964) From cytases to lysosomes. *Fed. Proc.*, **23**, 1045–9.

Dhawan, J., Pan, L.C., Pavlath, G.K. *et al.* (1991) Systemic delivery of human growth hormone by injection of genetically-modified myoblasts. *Science*, **254**, 1509–12.

Di Ferrante, N., Nichols, B.L., Donnelly, P.V. *et al.* (1971) Induced degradation of glycosaminoglycans in Hurler's and Hunter's syndromes by plasma infusions. *Proc. Natl Acad. Sci. USA*, **68**, 303–7.

Ferry, N., Duplessis, O., Houssin, D. *et al.* (1991) Retroviral-mediated gene transfer into hepatocytes *in vivo*. *Proc. Natl Acad. Sci. USA*, **88**, 8377–81.

Fukuda, S., Tomatsu, S., Masue, M. *et al.* (1992) Mucopolysaccharidosis type IVa. N-acetylgalactosamine-6-sulfate sulfatase exonic point mutations in classical Morquio and mild cases. *J. Clin. Invest.*, **90**, 1049–53.

Gage, F.H., Kawaja, M.D. and Fisher, L.J. (1991) Genetically modified cells: applications for intracerebral grafting. *Trends Neurosci.*, **14**, 328–33.

Gasper, P.W., Thrall, M.A., Wenger, D.A. *et al.* (1984) Correction of feline arylsulfatase B deficiency (mucopolysaccharidosis VI) by bone marrow transplantation. *Nature*, **312**, 467–9.

Haskins, M.E., Jezyk, P.F., Desnick, R.J. *et al.* (1979) Alpha-L-iduronidase deficiency in a cat: a model for mucopolysaccharidosis I. *Pediatr. Res.*, **13**, 1294–7.

Haskins, M.E., Desnick, R.J., DiFerrante, N. *et al.* (1984) Beta-glucuronidase deficiency in a dog: a model of mucopolysaccharidosis VII. *Pediatr. Res.*, **18**, 980–4.

Hoogerbrugge, P.M., Suzuki, K., Suzuki, K. *et al.* (1988) Donor-derived cells in the central nervous system of twitcher mice after bone marrow transplantation. *Science*, **239**, 1035–8.

Hopwood, J.J. and Morris, C.P. (1990) The mucopolysaccharidoses: diagnosis, molecular genetics and treatment. *Mol. Biol. Med.*, **7**, 381–404.

Jezyk, P.F., Haskins, M.E., Patterson, D.F. *et al.* (1977) Mucopolysaccharidosis in a cat with aryl-sulfatase B deficiency: a model of Maroteaux–Lamy syndrome. *Science*, **198**, 834–6.

Jiao, S., Gurevich, V. and Wolff, J.A. (1993) Long-term correction of rat model of Parkinson's disease by gene therapy. *Nature*, **362**, 450–3.

Jin, W.D., Desnick, R.J. and Schuchman, E.H. (1992) Mucopolysaccharidosis type VI: identification of three mutations in the arylsulfatase B gene of patients with the severe and mild phenotypes provides evidence for genetic heterogeneity. *Am. J. Hum. Genet.*, **50**, 795–800.

Karlsson, S. (1991) Treatment of genetic defects in hematopoietic cell function by gene transfer. *Blood*, **78**, 2481–92.

Krivit, W., Pierpont, M.E., Ayaz, K. *et al.* (1984) Bone marrow transplantation in the Maroteaux–Lamy syndrome (Mucopolysaccharidosis VI). *N. Engl. J. Med.*, **311**, 1601–11.

Krivit, W., Shapiro, E., Hoogerbrugge, P.M. and Moser, H.W. (1992) State of the art review: bone marrow transplantation treatment for storage diseases. *Bone Marrow Transplant.*, **10**, 87–97.

Maréchal, V., Naffakh, N., Danos, O. and Heard, J.M. (1993) Disappearance of lysosomal storage in spleen and liver of mucopolysaccharidosis VII mice after transplantation of genetically-modified bone marrow cells. *Blood*, **82**, 1358–65.

Morreau, H., Galjart, N.J., Gillemans, N. *et al.* (1989) Alternative splicing of β-galactosidase mRNA generates the classic lysosomal enzyme and a β-galactosidase-related protein. *J. Biol. Chem.*, **264**, 20655–63.

Moullier, P., Bohl, D., Heard, J.M. and Danos, O. (1993) Correction of lysosomal storage in the liver and spleen of MPS VII mice by implantation of genetically-modified skin fibroblasts. *Nature Genet.*, **4**, 154–9.

Naffakh, N., Pinset, C., Montarras, D. *et al.* (1994) Transplantation of adult-derived myoblasts in mice following gene transfer. *Neuromusc. Disord.*, **3**,

413–17.
Neufeld, E.F. and Muenzer, J. (1989) The mucopolysaccharidoses, in *The Metabolic Basis of Inherited Disease*, (eds C.R. Scriver, A.L. Beaudet, W.S. Sly and D. Valee), McGraw-Hill, New York, pp. 1565–87.
Neufeld, E.F., Lim, T.W. and Shapiro, L.J. (1977) Inherited disorders of lysosomal metabolism. *Annu. Rev. Biochem.*, **44**, 357–76.
Oshima, A., Kyle, J.W., Miller, R.D. *et al.* (1987) Cloning, sequencing, and expression of cDNA for human β-glucuronidase. *Proc. Natl Acad. Sci. USA*, **84**, 685–9.
Palmer, T.D., Rosman, G.J., Osborne, W.R.A. and Miller, A.D. (1991) Genetically modified skin fibroblasts persist long after transplantation but gradually inactivate introduced genes. *Proc. Natl Acad. Sci. USA*, **88**, 1330–4.
Pardridge, W.M. (1988) Recent advances in blood–brain barrier transport. *Annu. Rev. Pharmacol. Toxicol.*, **28**, 25–39.
Peters, C., Schmidt, B., Rommerskirch, W. *et al.* (1990) Phylogenetic conservation of arylsulfatases. cDNA cloning and expression of human arylsulfatase B. *J. Biol. Chem.*, **265**, 3374–81.
Peters, C., Rommerskirch, W., Modaressi, S. and von Figura, K. (1991) Restoration of arylsulphatase B activity in human mucopolysaccharidosis-type-VI fibroblasts by retroviral-vector-mediated gene transfer. *Biochem. J.*, **276**, 499–504.
Pfeffer, S.R. (1991) Targeting of proteins to the lysosome *Curr. Top. Microbiol. Immunol.*, **170**, 43–63.
Robertson, D.A., Freeman, C., Morris, C.P. and Hopwood, J.J. (1992) A cDNA clone for human glucosamine-6-sulphatase reveals differences between arylsulphatases and non-arylsulphatases. *Biochem. J.*, **288**, 539–44.
Roessler, B.J., Allen, E.D., Wilson, J.M. *et al.* (1993) Adenoviral-mediated gene transfer to rabbit synovium *in vivo*. *J. Clin. Invest.*, **92**, 1085–92.
Sands, M.S., Barker, J.E., Vogler, C. *et al.* (1993) Treatment of murine mucopolysaccharidosis type VII by syngeneic bone marrow transplantation in neonates. *Lab. Invest.*, **68**, 676–86.
Scharfmann, R., Axelrod, J.H. and Verma, I.M. (1991) Long-term *in vivo* expression of retrovirus-mediated gene transfer in mouse fibroblast implants. *Proc. Natl Acad. Sci. USA*, **88**, 4626–30.
Scott, H.S., Anson, D.S., Osborn, A.M. *et al.* (1991) Human alpha-L-iduronidase: cDNA isolation and expression. *Proc. Natl Acad. Sci. USA*, **88**, 9695–9.
Scott, H.S., Litjens, T., Nelson, P.V. *et al.* (1993) Identification of mutations in the α-L-iduronidase gene (IDUA) causing Hurler and Scheie syndromes. *Am. J. Hum. Genet.*, **53**, 973–86.
Shipley, J.M., Klinkenbreg, M., Wu, B.M. *et al.* (1993) Mutational analysis of a patient with mucopolysaccharidosis type VII, and identification of pseudogenes. *Am. J. Hum. Genet.*, **52**, 517–26.
Shull, R.M., Munger, R.J., Spellacy, E. *et al.* (1982) Canine α-L-iduronidase deficiency: a model of mucopolysaccharidosis I. *Am. J. Pathol.*, **109**, 244–8.
Shull, R.M., Hastings, N.E., Selcer, R.R. *et al.* (1987) Bone marrow transplantation in canine mucopolysaccharidosis I. *J. Clin. Invest.*, **79**, 435–43.
Shull, R.M., Kakkis, E.D., McEntee, M.F. *et al.* (1994) Enzyme replacement in a canine model of Hurler syndrome. *Proc. Natl Acad. Sci. USA*, **91**, 12937–41.
Smith, B.F., Hoffman, R.K., Giger, U. and Wolfe, J.H. (1990) Genes transferred by

retroviral vectors into normal and mutant myoblasts in primary cultures are expressed in myotubes. *Mol. Cell. Biol.*, **10**, 3268–71.

Snyder, E.Y., Deitcher, D.L., Walsh, C. *et al.* (1992) Multipotent neural cell lines can engraft and participate in development of mouse cerebellum. *Cell*, **68**, 33–51.

Thompson, J.-N., Jones, M.Z., Dawson, G. and Huffman, P.S. (1992) N-acetylglucosamine 6-sulphatase deficiency in a Nubian goat: a model of Sanfilippo syndrome type D (mucopolysaccharidosis IIID). *J. Inherit. Metab. Dis.*, **15**, 760–8.

Tomatsu, S., Fukuda, S., Masue, M. *et al.* (1991) Morquio disease: isolation, characterization and expression of full-length cDNA for human N-acetylgalactosamine-6-sulfate sulfatase. *Biochem. Biophys. Res. Commun.*, **2**, 677–83.

Vellodi, A., Young, E., New, M. *et al.* (1992) Bone marrow transplantation for Sanfilippo disease type B. *J. Inher. Metab. Dis.*, **15**, 911–18.

Vogler, C., Sands, M., Higgins, A. *et al.* (1993) Enzyme replacement with recombinant β-glucuronidase in the newborn mucopolysaccharidosis type VII mouse. *Pediatr. Res.*, **34**, 837–40.

Wilson, P.J., Morris, C.P., Anson, D.S. *et al.* (1990) Hunter syndrome: isolation of an iduronate-2-sulfatase cDNA clone and analysis of patient DNA. *Proc. Natl Acad. Sci. USA*, **87**, 8531–5.

Wilson, P.J., Suthers, G.K., Callen, D.F. *et al.* (1991) Frequent deletions at Xq28 indicate genetic hetrogeniety in Hunter syndrome. *Hum. Genet.*, **86**, 505–8.

Wolfe, J.H., Schuchman, E.H., Stramm, L.E. *et al.* (1990) Restoration of normal lysosomal function in mucopolysaccharidosis type VII cells by retroviral vector-mediated gene transfer. *Proc. Natl Acad. Sci. USA*, **87**, 2877–81.

Wolfe, J.H., Deshmane, S.L. and Fraser, N.W. (1992a) Herpesvirus vector gene transfer and expression of β-glucuronidase in the central nervous system of MPS VII mice. *Nature Genet.*, **1**, 379–84.

Wolfe, J.H., Sands, M.S., Barker, J.E. *et al.* (1992b) Reversal of pathology in murine mucopolysaccharidosis type VII by somatic cell gene transfer. *Nature*, **360**, 749–53.

Yamada, Y., Tomatsu, S., Sukegawa, K. *et al.* (1993) Mucopolysaccharidosis type II (Hunter Disease): 13 mutations in 52 Japanese patients and carrier detection in four families. *Hum. Genet.*, **92**, 110–14.

Yoshida, M., Noguchi, J., Ikadai, H. *et al.* (1993) Arylsulfatase B-deficient mucopolysaccharidosis in rats. *J. Clin. Invest.*, **91**, 1099–104.

18

DNA-based immunization

HEATHER L. DAVIS and ROBERT G. WHALEN

18.1 INTRODUCTION

Nucleic acid-mediated immunization refers to the induction of an immune response to a protein expressed *in vivo* subsequent to the introduction of its encoding polynucleotide (DNA or RNA). If proven to be safe, nucleic acid vaccines may one day be an attractive alternative to classical vaccines, which contain the antigen itself.

This is a new methodology which has frequently been referred to as 'genetic immunization' or 'genetic vaccination'. At a recent meeting held by the World Health Organization in Geneva (May 17–18, 1994), it was recommended that in the interest of clarity the term 'genetic vaccine' be discarded in favor of 'nucleic acid vaccine', which includes DNA-based and RNA-based immunization. Therefore, in the following text we will follow this recommendation. Additionally, we will restrict our discussion to DNA-based immunization since there are few descriptions of RNA-based vaccines (e.g. Martinon *et al.*, 1993).

Although gene transfer may be carried out by a variety of means (as discussed in other chapters of this book), only some of these are appropriate for the purpose of immunization. Since vaccination is almost always applied to a large population mass (either human or animal), only direct (*in vivo*) gene transfer could ever be envisioned for immunization purposes. Furthermore, although direct gene transfer may be carried out using either viral vectors or recombinant plasmid DNA, the use of viral vectors has the same limitations and disadvantages as classical vaccines based on attenuated viruses (see below), and are thus of limited interest for the purpose of immunization. Thus, for the purpose of this chapter, we will restrict the

Molecular and Cell Biology of Human Gene Therapeutics
Edited by George Dickson
Published in 1995 by Chapman & Hall. ISBN 0 412 62550 4

subject of 'nucleic acid vaccination' to the induction of an immune response to a protein expressed from a gene introduced directly in the form of pure plasmid DNA.

18.2 THE RESPONSE OF THE IMMUNE SYSTEM TO IMMUNIZATION

Successful vaccination leads to the induction of appropriate humoral and/or cell-mediated immunity (CMI) through effects on both B and T lymphocytes. Humoral immunity is the production of neutralizing antibodies by B cells, whereas CMI, as used here, involves the destruction of infected cells by cytotoxic T lymphocytes (abbreviated CTL or Tc). Both humoral and CMI responses are influenced by the action of T helper cells (Th).

B lymphocytes are involved solely with humoral immunity. When activated by binding circulating soluble antigen, most differentiate into plasma cells that produce and secrete neutralizing antibodies. Some activated B cells become memory cells which react quickly to a secondary challenge of antigen, with higher and more sustained levels of antibody.

T lymphocytes do not recognize circulating antigen (unlike B cells), but only recognize antigen presented on surfaces of antigen-presenting cells by major histocompatibility complex (MHC) glycoprotein surface molecules. Class I MHC presentation occurs on cells in which there is endogenous cytosolic synthesis of antigen (e.g. virus-infected cells), whereas class II MHC presentation is associated with phagocytosis and digestion of exogenous antigen (e.g. by macrophages). T cells may be divided into two more or less separate groups on the basis of their expression of two surface molecules, CD4+ and CD8+, which bind to class II and class I MHC molecules respectively. Most CD8+ cells, which when activated by binding to class I MHC presenting foreign antigen in the presence of appropriate secondary signals (e.g. cytokines, see below), become CTLs capable of lysing the antigen-presenting (i.e. infected) cells. In contrast, CD4+ T cells generally act as Th cells which secrete cytokines. Th cells may be divided into two subclasses (Th1, Th2) according to the cytokines secreted. Th1 and Th2 activities are mutually inhibitory and broadly speaking tend to favor CMI and humoral immunity respectively. Some CD4+ T lymphocytes may act as CTL, but these destroy target cells expressing antigen presented by MHC II.

In summary, both humoral and cytotoxic immune responses involve complex, simultaneous and sequential interactions between several cell types, through cell–cell contact and/or cellular factors. With humoral immunity, neutralizing antibody is produced by activated B cells, but this depends largely on the stimulatory action of Th2 cells, which themselves are activated by antigen-presenting cells. For CMI, a CD8+ cell recognizing MHC I antigen on an infected cell becomes a CTL, but full activation depends largely on cytokines secreted by a CD4+ T-helper cells, principally

Th1, which themselves were previously stimulated by antigen presentation in lymphoid tissue (Kuby, 1994).

18.3 LIMITATIONS OF CURRENT VACCINATION STRATEGIES

An ideal vaccine would have the following properties: (i) safe for all individuals, including those immunosuppressed; (ii) easily administered, preferably by oral means; (iii) induction of the full range of immune responses; (iv) long-lasting effect from a single dose (i.e. no boosters); (v) easy and inexpensive to manufacture; (vi) simplified yet rigorous quality assessment and control; and (vii) heat-stability.

Classical vaccination, which involves administration of an immunogen, may be subdivided into three general types: subunit, live attenuated and killed vaccines.

18.3.1 Subunit vaccines

Subunit vaccines involve administration of only selected components of the pathogen (virus or bacteria). Common examples include synthetic peptides, recombinant proteins purified from cell lines genetically engineered to produce the desired polypeptide or protein (e.g. hepatitis B surface protein), or purified macromolecules (e.g. non-infectious hepatitis B virus particles purified from the serum of chronically infected individuals). These vaccines are generally safe and effective for inducing humoral immunity. In some cases (e.g. where antigen is in the form of large particles) they may also induce CD4+ Th response, but their primary limitation is an inability to invoke a CD8+ CTL response. This is an important disadvantage since CMI is thought to be superior to humoral immunity alone and may even be essential for protection against certain pathogens. Another major disadvantage of subunit vaccines is that they induce only short-term immunity and thus there is a need for boosters. Other drawbacks may include limited supply (purified macromolecules), difficult and expensive production and purification (recombinant protein vaccines) and low immunogenicity (synthetic peptides).

18.3.2 Live attenuated vaccines

Live vaccines consist of viruses or bacteria grown for prolonged periods under abnormal conditions such that they become attenuated (i.e. non-pathogenic). Although this is possible for a only few viruses (e.g. measles, mumps, polio–Sabin), they usually provide excellent immunity due to prolonged exposure to the immunogen and synthesis in self-cells, and thus the induction of CD8+ CTL function. More recently, live viral vaccines have become applicable to a greater number of diseases by cloning recombinant antigens into established attenuated viral (or bacterial cell) lines (e.g. vaccinia virus) to make recombinant viral vectors (World Health Organization, 1990). Such vaccines, which may be considered a type of 'genetic' vaccine, are more effective, less expensive and easier to produce than recombinant protein vaccines. The most important disadvantage for live viral vaccines, whether or not they carry a recombinant gene, is the possibility of reversion to a virulent form, especially in immunodepressed (e.g. AIDS) or malnourished individuals, and in the presence of other viruses as contaminants.

18.3.3 Killed vaccines

Viruses or bacteria may be killed by treatment with heat or chemicals, and while this renders them safer as a vaccine there is a loss of CD8+ CTL immunity so that they essentially behave like a multicomponent subunit vaccine. Other disadvantages may include infection due to incomplete inactivation, loss of immunogenicity owing to denaturation and unexplained encephalitis reactions. Current examples of killed virus vaccines are those against rabies and polio (Salk).

Another disadvantage of both protein and live viral vaccines is that they are not heat-stable. This is an important factor for developing countries since the requirement for refrigeration, frequently referred to as the 'cold-chain', adds considerably to the cost of vaccines.

18.4 POTENTIAL ADVANTAGES OF NUCLEIC ACID VACCINATION

In contrast to classical vaccination where antigens are administered, nucleic acid vaccination involves the administration of genetic material (DNA or RNA) encoding the antigen, which is then synthesized *in vivo*. In a sense they are like subunit vaccines except that the antigen is produced within the cells of the individual to be immunized. There are two unique features of DNA-based vaccines that make them particularly interesting to pursue.

The first is the possibility of long-lived gene expression resulting in sustained presentation of the antigen at low levels to the immune system. This could obviate the need for booster injections, and may even result in a superior immunological response. However, it will be important to ensure that chronic antigen presentation does not result in deleterious consequences such as the induction of anergy, tolerance and auto-immunity.

The second unique feature of DNA-based immunization is due to the synthesis of antigen *in vivo* (but without infection) and presentation by class I MHC surface glycoproteins. As such, DNA-based vaccines should induce CMI by activation of CD8+ CTL, as do the live viral or bacterial vaccines, but without the risk of inadvertent infection, an important consideration for use in immunosuppressed individuals. In addition, if CD8+ CTL can be induced, nucleic acid-mediated immunization could also be applied in certain cases for the treatment of infected individuals.

The use of pure recombinant DNA offers several other technical, economical and logistical advantages over classical vaccines, especially for use in developing parts of the world. For example, once large-scale production is in place it will probably be easier and less expensive to produce and maintain quality control for DNA than for recombinant protein or attenuated viral vaccines, which may in turn facilitate transfer of the technology itself to developing countries. In addition, DNA is heat-stable so does not require refrigeration, which could result in considerably lower costs for delivery to underdeveloped areas of the world. Multi-disease vaccines could be created by mixing together different DNAs (this is often impossible with protein vaccines which can interfere with one another). Finally, owing to the ease by which the DNA sequences to be expressed can be modified, a single gene can include epitopes of more than one antigen, or of different or changing strains of pathogens.

Thus, DNA-based vaccination promises to overcome many of the shortcomings of classical vaccination provided that it does not present new safety or technical problems itself. DNA-based immunization has not yet been applied to humans; however, its potential clinical interest has been indicated by several different animal models. Although potential safety issues will likely delay use of DNA for large-scale vaccination of healthy people, there are plans for phase I clinical trials in the near future for DNA-based immunotherapy on cancer patients and HIV seropositive individuals.

18.5 ANIMAL MODELS OF DNA-BASED IMMUNIZATION

The possibility of using direct gene transfer with plasmid DNA to induce an immune response to the expressed protein was first indicated by Acsadi *et al.* (1991). They found that reporter gene expression after plasmid DNA transfer into rats' hearts was short-lived but showed that treatment with the immunosuppressant cyclosporin succeeded in prolonging expression. This was interpreted to mean that the foreign reporter gene products (e.g. luciferase, β-galactosidase and chloramphenicol acetyl transferase) were immunogenic in heart cells. Subsequently, it was demonstrated that plasmids encoding human growth hormone (Tang *et al.*, 1992; Eisenbraun *et al.*, 1993) and human α_1-antitrypsin (Tang *et al.*, 1992) could induce antibodies to these proteins. In these cases the DNA was coated onto gold particles and introduced by bombardment. In 1992, several meeting reports described the phenomenon of nucleic acid immunization against a known viral antigen; however, these communications were not generally available to the gene therapy community. Since the beginning of 1993, several models of DNA-based immunization have appeared in peer-reviewed journals, and

Table 18.1 Animal models of DNA-based immunization

Reference[a]	*Vaccine model*	*Proteins expressed*	*Animal model*	*Route*	*Immunization schedule*	*Ab induction*	*T-cell proliferation*	*Cytotoxic T-cell response*	*Protection against challenge*
Conry *et al.* (1994)	Tumor immunity	Carcino-embryonic antigen	Mouse	IM	200 μg at 0, 3 and 6 weeks	Yes	Yes	ND	ND
Cox *et al.* (1993)	Bovine herpes virus 1	Glycoproteins	Mouse, cattle	IM	*Mice*: 50 μg at 0 and 4 weeks or 25–100 μg at 0 and 6 weeks *Cattle*: 125–500 μg at each of 5 treatments over 5 months	Yes	ND	ND	Partial
Davis *et al.* (1993b)	Hepatitis B virus	Surface antigen	Mouse	IM	100–200 μg one time only	Yes	ND	ND	ND
Montgomery *et al.* (1993)[1] Ulmer *et al.* (1993)[2] Yankauckas *et al.* (1993)[3]	Influenza A virus	Nucleoprotein	Mouse	IM	200 μg at 0, 3 and 6 weeks[1] 200 μg one time only[2] 1–100 μg, 1+ times[3]	Yes	ND	Yes	Yes
Fynan *et al.* (1993a[1], b[2]) Montgomery *et al.* (1993)[3] Robinson *et al.* (1993)[4]	Influenza A virus	Hemagglutinin glycoprotein	Mouse[2,3] Chicken[1,2,4]	ID IM IN IV SC IV+IP+IM	300 μg at 0 and 4 weeks[1,4] 40 ng–300 μg at 0 and 4 weeks[2] 200 μg at 0, 3 and 6 weeks[3]	Yes	ND	ND	Yes
Geissler *et al.* (1994)	MHC class I immunity[b]	MHC class I antigen	Rat	IM	1–1000 μg one time only	No	ND	Yes	Partial[b]
Wang *et al.* (1993a,b)	HIV-1	gp160	Mouse, monkey	IM	100 μg at 0, 2 and 4 weeks	Yes	Yes	(Yes)[c]	ND
Xiang *et al.* (1994)	Rabies virus	Glycoprotein	Mouse	IM	10–250 μg, 1–3 × at 2–3-week intervals	Yes	Yes	Yes	Yes
Sedegah *et al.* (1994)	Malaria	Circumsporozoite protein	Mouse	IM	100 μg at 0, 5, 8 weeks	Yes	ND	Yes	Yes

[a] Superscript numbers against references relate to the immunization schedule indicated in same horizontal row.
[b] This challenge model involves the rejection of a cardiac allograft; [c] Cytotoxic T-cell responses are cited as unpublished, preliminary studies.

ID, intradermal or intraepidermal; IM, intramuscular; IN, intranasal; IP intraperitoneal; IV, intravenous; SC, subcutaneous; ND, not determined those documented up until May 1994 (the time of writing of this review) are summarized in Table 18.1.

18.5.1 DNA delivery methods

Intramuscular injection

The possibility of using pure plasmid DNA for direct gene transfer was originally demonstrated by intramuscular injection of an aqueous saline solution of DNA into hindlimb muscles of mice using a syringe and small-gauge needle (Wolff *et al.*, 1990). This method results in low efficiency of gene transfer and considerable variability of gene expression, problems which are likely to be due in large part to limitation of diffusion by physical factors such as the organization of the connective tissue and extent of the extracellular matrix (Jiao *et al.*, 1992; Davis *et al.*, 1993c). This is evidenced by the distribution of intramuscularly injected dye, which is highly variable and often restricted to interfascicular spaces. The diffusional barrier can be partly overcome by pretreating mature muscle with a hypertonic solution of sucrose, which presumably forces the myofibers apart and thus opens channels in the connective tissue. Such sucrose-treated muscles show better diffusion of injected substances and produce less variable expression of injected reporter genes (Davis *et al.*, 1993c). In a muscle such as the diaphragm which is not extensively subdivided by connective tissue, gene transfer of plasmid DNA is more efficient without pretreatment than that in sucrose-treated hindlimb muscles (Davis and Jasmin, 1993).

The efficiency of gene transfer using pure DNA in skeletal muscle is at least 10-fold better in regenerating muscle than in normal mature muscle (Davis *et al.*, 1993a; Danko *et al.*, 1994; Vitadello *et al.*, 1994). Regeneration may be induced by anoxia, crush or freezing, but these often give inconsistent results. The local anesthetic bupivicaine (Marcaine®) is also effective, although it frequently fails to induce complete degeneration. Controlled and complete regeneration can be obtained in muscle tissue by the injection of certain snake toxins such as cardiotoxin and notexin, which have been used in mice and rats (d'Albis *et al.*, 1989; Whalen *et al.*, 1990).

Although viral vectors might be considered for the purpose of vaccination, it is interesting to note that in skeletal muscle, plasmid DNA is superior to adenovirus for gene transfer in normal mature muscle and they are equally efficient in regenerating muscle. In contract, retrovirus works only poorly in regenerating muscle and not at all in mature muscle (Davis *et al.*, 1993a). While physical barriers surely affect diffusion of viral vectors in muscle, it is also likely that fully mature myofibers are somewhat refractory to infection by adenoviruses and completely so by retroviruses (Davis *et al.*, 1993a). Furthermore, it is clear that the use of viral vectors induces an immune response to the viral proteins themselves, which on the one hand, is detrimental to the state of the muscle tissue and additionally may prevent the use of the same viral vector for subsequent gene transfer. In contrast, plasmid DNA itself does not appear to be strongly immunogenic (Nabel *et al.*, 1992; Davis *et al.*, 1993a)

Jet injection systems

Needleless jet injection systems can also be used to deliver DNA to tissue. One example was described in 1992 by Furth *et al.* who demonstrated that mammary tissue could be transfected using the Ped-O-Jet system, which was designed largely for intradermal or subcutaneous injections. The Biojector® jet injection system (Bioject Inc., Portland, OR), which uses cartridges of compressed gas to expel a fine stream of liquid at high speed from variable-dose needleless syringes, is capable of a true intramuscular delivery. We have successfully used the Biojector® to carry out DNA-based immunization against the surface antigen of hepatitis B virus on rats and rabbits (Davis *et al.*, 1994).

Particle bombardment

DNA may be coated onto gold particles and introduced by a method known as particle bombardment or 'biolistics' (Pecorino and Lo, 1992). This has been shown to be effective for transferring reporter genes into epidermis, dermis, muscle, liver and pancreas (Tang *et al.*, 1992; Cheng *et al.*, 1993). It has also been used to introduce plasmid DNA encoding influenza hemagglutinin into the skin of mice for the purpose of immunization (Fynan *et al.*, 1993a,b). This method appears to be a highly efficient way to immunize, since a protective immune response, albeit in the absence of a strong humoral response, can be elicited with considerably less DNA (as little as 40 ng) than is typically used for intramuscular injection (10–400 μg). This is likely due to the fact that the particle bombardment method allows direct intracellular delivery and that the skin contains all the necessary cell types for eliciting both humoral and cytotoxic cellular immune responses (Eisenbraun *et al.*, 1993; Fynan *et al.*, 1993a,b).

Other methods of DNA delivery

Plasmid DNA may also be administered by intravenous injection as well as instillation or inhalation into the respiratory passages. None of these methods however results in highly efficient gene transfer unless the DNA is first incorporated into liposomes, a situation which might be preferable to avoid for the purposes of immunization.

18.5.2 Choice of tissue for gene transfer

There are many possible routes of administration of DNA for the purposes of immunization, many of which are the same as those commonly used for classical vaccination. As outlined in Table 18.1, successful immunization has been demonstrated with administration of plasmid DNA by intramuscular (IM), intranasal (IN), intradermal/intraepidermal (ID), intravenous (IV) and subcutaneous (SC) routes. Several factors may influence the choice of target tissue for nucleic acid immunization.

Efficiency of gene transfer

The most efficient route to transfer pure plasmid DNA is intramuscular when it is injected as an aqueous solution, or ID if it is coated onto gold particles. Other routes are generally inefficient, although this might be improved if the DNA were contained within liposomes. However, this solution is not desirable for several reasons including cost, practicality and immunological or other untoward reactions to lipid components.

Stability of gene expression

Genes transferred in the form of plasmid DNA remain non-integrated and therefore will be quickly lost or diluted out in dividing cells. Thus muscle fibers, which are permanently post-mitotic, afford the possibility of longer expression than do cells which have a rapid rate of turnover (e.g. respiratory epithelium, epidermis). Longer-lasting expression may even permit single-dose immunization (i.e. without boosters).

Ability to synthesize and secrete antigen

Different types of cells will have different capacities to synthesize and secrete proteins. For example, some antigens require special post-translational modifications and this may be less efficient in some cell types. Additionally, it is not clear whether all cells secrete efficiently. While it was thought that skeletal muscle fibers were not effective as secretory cells, this may not necessarily be the case judging by the release of hepatitis B surface antigen from transfected muscle fibers (H. Davis *et al.*, unpublished results).

Ease of administration and acceptability to patients

Oral administration is certainly the easiest route for vaccination, although to date there is no reason to believe that naked or liposome-complexed plasmid DNA could be transferred by this means. Another easily accessible route which might be more hospitable to DNA are the respiratory passages, which could be treated with inhaled aerosol or nose-drops. The IM and SC routes are not difficult but require the use of a syringe and needle or jet injection, whereas the IV route is definitely more cumbersome. The most inconvenient method is to use DNA-coated gold particles, which might require considerable effort to prepare. Additionally, these particles are introduced using a particle accelerator. Originally this was a large machine, but a hand-held model is now available which must nevertheless be connected to a large tank of compressed gas (Pecorino and Lo, 1992). Further developments will likely produce an even more convenient apparatus. Intraperitoneal (IP) administration, an easy technique for use on rodent animal models, was not effective for eliciting an immune response with injection of DNA encoding influenza virus hemagglutinin glycoproteins (Robinson *et al.*, 1993).

18.6 IMMUNE RESPONSE TO DNA-BASED IMMUNIZATION

18.6.1 Humoral responses

In most cases of immunization using plasmid expression vectors, antibodies have been found against the proteins produced. However, the success of the induction of a humoral response varies considerably between models. A single injection of the plasmid DNA expressing the hepatitis B surface antigen protein (HBsAg) resulted in levels of antibody which increased continuously over a period of at least 8 weeks (Davis *et al.*, 1993b). Antibodies to the HBsAg can be measured in milliInternational units per milliliter (mIU/ml). This allows a comparison to be made with the levels known to provide protection in humans, which are on the order of 10 mIU/ml. In C57BL/6 mice (a strain of mice which is considered a good responder) inoculated with the HBsAg expression vector, the mean antibody levels expressed in these terms were 100–300 mIU/ml. DNA-based immunization in BALB/c mice, which are considered highly responsive to HBsAg, gave rise to mean values of 1000 mIU/ml with some mice reaching 13 000 mIU/ml (unpublished results). Thus the humoral responses obtained in mice can achieve levels of 10 to 1000 times greater than those considered to be protective in humans. When expressed as antibody titers measured by a typical ELISA protocol, the humoral response in C57BL/6 mice reached values of between 2×10^4 and $>10^5$, and these titers were attained over a 4- to 8-week period following a single injection of DNA expression vectors (Michel *et al.*, 1995).

In most of the other studies of DNA-based immunization in which an antibody response was studied, multiple injections of DNA vectors were given at intervals of several weeks, although a single DNA injection was apparently sufficient in at least one case to produce significant antibody titers (Ulmer *et al.*, 1993). Owing to the widely varying immunization schedules, it is not possible to assess the relative efficacy of the individual expression systems and the immunogens used (Table 18.1). Nonetheless, three injections of a vector expressing influenza nucleoprotein (NP) at 0, 3 and 6 weeks resulted in antibody titers of nearly 10^5 (after background subtraction) (Ulmer *et al.*, 1993) which are comparable with those obtained with a single injection of the HBsAg vector (Davis *et al.*, 1993b).

Cox *et al.* (1993) obtained antibody to bovine herpes glycoprotein in both mice and calves, but in this case only after several DNA injections. Conry *et al.* (1994) have also demonstrated a humoral response in mice following immunization with a series of three IM injections of plasmid DNA expressing human carcinoembryonic antigen. Xiang *et al.* (1994) obtained neutralizing antibodies to rabies virus upon IM injection of a plasmid encoding the viral glycoprotein. After three DNA injections, the mice attained antibody titers of $>10^4$.

In studies using plasmids expressing the HIV-1 envelope protein gp160 (from HIV-1 IIIB or Z6) as an immunogen, antibody titers remained low in most mice even with several injections of DNA expression vectors (Wang *et al.*, 1993a,b; Williams *et al.*, 1993). Nonetheless, nearly 100% seroconversion (defined as a reading of 3 standard deviations above control sera) was achieved after three to four DNA injections. Cynomolgus monkeys also gave a humoral response to DNA inoculation. Despite low antibody titers, the immune response obtained in either mice or monkeys was sufficient to achieve neutralization of viral infection in cell culture and to prevent syncytium formation. Thus, a broadly effective immune response was obtained with the gp160 expression system, and low antibody titers are clearly not the only feature with which to judge the efficacy of DNA-based immunization.

In the studies of Fynan *et al.* (1993a,b) which used influenza hemagglutinin as the protein expressed from a plasmid DNA construct, antibodies were only occasionally present, and only after a second DNA injection. Nonetheless, when influenza virus was used to challenge mice or chickens, antibody levels seemed to increase more rapidly (although not to higher levels) in animals inoculated with DNA expression vectors than in those treated with control DNA plasmids.

Finally, a humoral response was also detected against the circumsporozoite protein of a malarial parasite following plasmid DNA-based immunization in mice (Sedegah *et al.*, 1994).

18.6.2 Cytotoxic T lymphocyte response

One of the direct consequences of immunization using DNA expression vectors is that the proteins are necessarily synthesized *in vivo*. Therefore, the polypeptides can be processed via the pathway leading to presentation of protein-derived peptides by MHC class I surface molecules. Such presentation can lead to the induction of CTLs, although the mechanism by which this might occur in the case where proteins are synthesized by skeletal muscle fibers is far from clear, since other specialized accessory molecules are thought to be required for activation of CD8+ cells. Nonetheless, the induction of CTL after intramuscular injection of DNA expression vectors has been demonstrated particularly convincingly for the influenza NP model (Montgomery *et al.*, 1993; Ulmer *et al.*, 1993; Yankauckas *et al.*, 1993). In order to reveal specific CTL activity, it normally is necessary to re-stimulate spleen cells by growing them in culture in the presence of antigen-presenting cells. A more stringent assay for CTL is to test lytic activity of cells derived from the animal without re-stimulation by antigen-presenting cells, but rather with non-specific stimulation by concanavalin A and IL-2. In both protocols, specific lysis of target cells was obtained by spleen cells of mice previously inoculated with NP-expressing DNA constructs. We have also demonstrated potent CTL activity in mice injected with HBgAg-expressing vectors (Davis *et al.*, unpublished results). Indeed by only 6 days after a single injection of DNA, spleen cells specifically restimulated *in vitro* for 5 days, exhibited 80–90% specific lysis at effector/target rates as low as 2.5 or 1.

In the work of Xiang *et al.* (1994) with a rabies virus glycoprotein expression vector, virus-specific CTL were also obtained. Their induction required at least two injections of DNA and after three injections some non-specific CTL activity was also evident.

A CTL response was also detected with DNA-based immunization to the circumsporozoite protein of *Plasmodium yoelii*, a malarial parasite which infects mice (Sedegah *et al.*, 1994).

An interesting approach has been taken by Geissler *et al.* (1994) who used a rat model of allograft rejection of heterotopic heart transplants. The effect of heterologous MHC class I molecules on graft rejection was evaluated after IM injection of plasmids coding either the membrane-bound form of a rat MHC molecule or the analogous secreted form. A single injection of either plasmid led to accelerated graft rejection by 1–2 days with 1 mg of plasmid expressing the membrane-bound form being the most effective. Even 1 μg of DNA encoding the membrane-bound form was sufficient to cause greater CTL activity 3 days post-transplant than was found with use of control plasmid or that expressing the soluble form. These results, while not demonstrating induction of CTL by the DNA injection, can be interpreted to suggest that priming of CD8+ T lymphocytes had occurred. In principle, the use of MHC class I molecules may represent a direct means to more efficiently induce CTL, although accessory molecules might still be needed, depending on the nature of the antigen-presenting cells.

18.6.3 T-helper response

T-cell proliferative responses have been identified in several of the studies on DNA-based immunization. Wang *et al.* (1993a) showed that T-cell proliferation could be obtained upon stimulation in cell culture with specific HIV-1 peptides after immunization with a gp160 expression vector. Conry *et al.* (1994) found that while both lymphocyte proliferative and humoral responses could be obtained to human carcinoembryonic antigen, the two responses could be dissociated in that some mice had antibodies but no proliferative response. It was suggested that this might be due to activation of different Th populations in the individual mice, and it will therefore be interesting to determine which parameters of nucleic acid immunization cause the induction of one or the other set of Th populations.

A very complete study was carried out on the immunological response in mice to DNA-based immunization against rabies virus glycoprotein (Xiang *et al.*, 1994). Lymphocytes derived from both the spleen and the lymph nodes demonstrated a Th cell response to the rabies virus. Interestingly, the level of production of IL-2 indicated that this response was of the Th1 type, which is similar to that obtained by immunization with inactivated rabies virus. This study provides a good model for the types of immunological analyses that should be carried out for the various (and future) DNA-based vaccines in order to better understand the general and specific features of the immune response they induce.

The presence of a lymphocyte proliferative response is perhaps of no surprise in those models where gene transfer was carried out in the dermis and epidermis where are found many cells specialized for class II MHC presentation of antigens to T-helper cells. This is especially true in those experiments where gene transfer was carried out with DNA-coated gold particles, in which case direct entry into professional antigen-presenting cells was likely.

Perhaps more difficult to explain is the presence of T-helper response in models using intramuscular gene transfer. Induction of CD4+ T cells requires presentation of antigen by surface class II MHC, which is normally associated only with professional antigen-presenting cells, and these molecules are likely to be absent on the surface of transfected muscle fibers. There are three possible explanations for the induction of T-helper responses with immunization based on direct injection of plasmid DNA in muscle.

First, if the transfected muscle fibers secrete the antigenic protein, it may be taken up and processed by local MHC class II cells (e.g. macrophages) which have been attracted to the site in response to the physical trauma of the intramuscular injection or to endotoxins present in the plasmid DNA preparation. This could be possible (although it has not been demonstrated) in the case of expression of hepatitis B surface antigen, where good secretion of antigen was found *in vitro* and was indicated *in vivo* (Davis *et al.*, 1993b). This cannot, however, explain the demonstrated T-cell proliferation seen with expression of rabies virus glycoprotein, which does not appear to have been secreted (Xiang *et al.*, 1994). Additionally, T-helper response to HIV-1 gp160 or carcinoembryonic antigen was detected despite undetectable levels of antibody, which may indicate lack of secretion of antigen (Wang *et al.*, 1993a; Conry *et al.*, 1994).

In the event that muscle fibers do not secrete the antigenic protein, a second possible explanation is that MHC class I presentation by transfected muscle fibers is able to induce a CD8+ CTL response without the assistance of T-cell help. The CTL could then destroy those same myofibers, releasing antigen which could then be processed and presented by class II MHC APCs. Although we have demonstrated destruction of antigen-expressing muscle fibers at about 2 weeks after DNA transfer (Watkins *et al.*, 1994, unpublished results), this possibility is also unlikely to explain the findings with the rabies DNA vaccine where the kinetics of the CTL and Th responses suggest simultaneous rather than sequential onset (Xiang *et al.*, 1994).

The third possibility is that intramuscular injection of plasmid DNA transfects non-muscle cells within the muscle tissue, such as dendritic cells which can present antigen by class I and class II MHC. This would allow simultaneous appearance of both T helper cells and CTL. Although this may well take place, there is no evidence to date that non-muscle cells within muscle tissue take up and express plasmid DNA, and this will indeed be difficult to demonstrate.

18.6.4 Protection against virulent challenge

In order for nucleic acid immunization to be considered a viable approach to vaccination, it must be demonstrated that the immune response obtained is sufficient to protect against infection by a pathogen. 'Challenge' studies on animals in several of the DNA-based immunization models, especially those using influenza virus proteins (Table 18.1), have shown that such protection can be obtained.

Protection against influenza is possible to study in mice since certain strains of influenza will infect them, causing severe disease and death. In the study by Ulmer *et al.* (1993), intramuscular injection of a plasmid encoding the influenza nucleoprotein (NP) was able to greatly diminish viral titers in the lung and to prevent weight loss after intranasal challenge of mice with heterologous virus. Moreover, DNA-vaccinated mice had greater survival rates than control mice when infected with either homologous or heterologous virus strains. Although both high-titer antibodies to NP and CTL were obtained, several lines of evidence suggested that anti-NP alone would not provide protection against disease. It was thus concluded that CTL induced by DNA immunization were responsible for the protection against viral challenge. Additionally, the cytotoxic immune response to the conserved NP protein provided cross-protection, a phenomenon of particular interest for vaccination against viral diseases such as influenza where sequence variability frequently diminishes the efficacy of a vaccine to one strain against a different strain.

Using several different routes of DNA introduction, including parenteral, mucosal and particle bombardment, Fynan *et al.* (1993b) found that plasmids expressing the influenza virus hemagglutinin induce a protective immune response in mice and chickens, even though very little antibody was induced. The most effective protection was obtained with epidermal delivery of DNA-coated gold particles, which led to 95% protection in mice using only 0.4 μg of DNA.

Certain other unusual and unexpectedly efficient routes of DNA introduction (intranasal mucosal and particle bombardment) which led to protective immunity (albeit a weak immune response overall) open new avenues to explore, and particularly suggest induction of mucosal immunity.

The study involving rabies glycoprotein in mice (Xiang *et al.*, 1994) also provides clear evidence for protection against viral infection after DNA-based vaccination. The immune response to the plasmid-encoded glycoprotein, which was both humoral and cellular in nature, was sufficient to protect the mice against a lethal challenge with rabies virus. Finally, Cox *et al.* (1993) showed that calves immunized against bovine herpesvirus developed a humoral response which was sufficient to confer some level of protection to virus challenge in that the immunized animals shed less virus than did a control calf. Although these results are not definitive with respect to protection, it is clearly interesting that such a large animal can be induced to produce protective antibodies with reasonably small quantities of plasmid DNA.

Finally, protection against infection by a malarial parasite (*Plasmodium yoelii*) was also demonstrated in mice immunized with plasmid DNA encoding the circumsporozoite protein (Sedegah *et al.*, 1994).

18.7 FUTURE DIRECTIONS FOR DNA-BASED IMMUNIZATION

Although progress over the last year has been impressive, it is important to realize that each of the different models reviewed in this chapter is a first description of DNA-based vaccination, and thus a great deal of optimization and refinement doubtlessly remains to be done. Fortunately, given the simplicity of the methods involved, there will be considerable incentive to develop and test new techniques and formulations to improve the DNA transfer, increase the efficiency and longevity of gene expression, and improve or alter the immune response induced by the expressed protein. For example, further work could be done to find stronger promoters, although the widely used CMV promoter is certainly adequate, and to associate tissue-specific regulatory elements with the promoters if tissue-restricted expression is particularly required. Other regulatory sequences (introns and the untranslated sequences of mRNA) can presumably be found which will increase transcript levels or enhance stability of the mRNAs (Whalen,

1994). The amino acid sequence signals required for protein secretion and/or modification can likewise be easily evaluated. Generally speaking, all manipulations concerning the coding and non-coding expressed sequences can be carried out and evaluated very rapidly, due to the use of standard molecular biological methods in combination with *in vivo* plasmid transfer.

As abundantly demonstrated by the examples discussed here, the value of plasmid-based gene transfer for basic research in immunology has now been definitively established. However, when considering vaccination of human populations, the one great uncertainty remaining today is the theoretical possibility of DNA integration, which could have pathogenic effects. Specifically, plasmid DNA might integrate into the chromosomal DNA of a somatic cell and this could lead to insertional mutagenesis by activation of a proto-oncogene or by inactivation of a tumor suppressor gene. The likelihood of this process is low, even though at present the amounts of plasmid injected in a typical experiment contain 10^{10} to 10^{12} molecules of DNA. First, most injected DNA is rapidly degraded after injection and only an extremely small fraction actually enters muscle fibers or other cells. Second, for cells that do take up plasmid DNA, the probability of integration is very low. In muscle fibers, which are permanently post-mitotic, integration is a particularly remote possibility, but even for transfected mononucleated cells in muscle or other tissues, the likelihood of integration is still low since most of the plasmid DNA would soon be lost during subsequent cell division. Finally, even if integration into the host genome did occur, it would occur randomly and thus only a small fraction of such events would have deleterious consequences.

Nonetheless, no matter how low the calculated final probability of an oncogenic or other pathogenic event, only long-term retrospective data will allow the assessment of any substantially increased risk of oncogenesis in human populations as a result of vaccination with DNA. The nature of medical progress is such that initial human trials will be done in ethically acceptable clinical settings where the potential benefits will clearly outweigh the theoretical risks. In the meantime, it is incumbent on researchers to design experiments to test the perceived risks using suitable and sensitive experimental models, not always for vaccination *per se*, but more specifically for oncogenesis. Finally, while different standards for medical care for different countries are to be avoided, it is indisputable that populations in certain areas of the world may be able to benefit from DNA-based vaccination in the near future, given the risks that they otherwise face from endemic and otherwise uncontrolled infectious disease. Likewise, other therapeutic approaches such as post-infection vaccination (e.g., for HIV-seropositive individuals or chronic carriers of hepatitis B virus) as well as immunotherapy for other conditions could conceivably arise from the expected rapid increase in knowledge that will come from studies on nucleic acid immunization.

18.8 SUMMARY

Over the past 2 years the principle of nucleic acid immunization has been demonstrated in several different animal models. Each of these was based on direct gene transfer using plasmid DNA into one or more tissues, with skeletal muscle being the preferred target. The expression vectors used to date have encoded antigens from several different viruses as well as a tumor-specific protein, an MHC class I molecule and a parasite antigen. It is clear that the induction of a broad range of immune responses is possible with DNA-based immunization, including the ability to confer protection against viral or parasitic challenge. In some cases, the immune response is superior to that obtained with traditional recombinant protein vaccination. For example, in the case of hepatitis B surface antigen, antibody appears earlier and levels rise more quickly with DNA-based than with recombinant protein vaccination (Davis *et al.*,1994).

Despite the very promising beginning of this new approach, many issues remain to be examined before application to humans, especially those concerned with safety. In the meantime however, DNA-based immunization offers an extremely powerful tool for molecular immunologists to study the immune system and with which to develop new vaccines and other immunotherapeutic approaches. One can easily and rapidly clone and modify genes in plasmid DNA expression vectors, allowing many new constructs to be produced and tested in a short period of time which can be on the order of weeks. In contrast, the preparation of viral vectors, or the production and purification of recombinant proteins from bacteria, yeast or stably transfected mammalian cell lines, can easily take months to develop. The DNA-mediated induction of an immune response to a protein produced *in situ* has therefore initiated a new era of vaccine research. There is now the possibility to dramatically modify the way one approaches prophylactic vaccination, and as a consequence public health care can potentially be improved in a highly cost-effective manner.

ACKNOWLEDGEMENTS

We particularly wish to thank Marie-Louise Michel, Simon Watkins, Maryline Mancini and Martin Schleef for many fruitful discussions regarding our on-going collaborative work on nucleic acid immunization. H.L. Davis is a Principal Investigator at the Loeb Medical Research Institute, Ottawa Civic Hospital and an Associate Professor in the Faculties of Health Sciences and Medicine at the University of Ottawa. Her laboratory is supported by grants from the University of Ottawa, the Medical Research Council of Canada (MRC) and the Muscular Dystrophy Association (USA). R.G. Whalen is at the Pasteur Institute and is Director of Research in the French Centre National de Recherche Scientifique (CNRS), both of which support his laboratory. His research is also financed by grants from the

Association Française contre les Myopathies, the Institut National de la Santé et de la Recherche Médicale and the French Ministry of Research. The authors have also received collaborative travel grants from NATO, MRC and CNRS.

REFERENCES

Acsadi, G., Dickson, G., Love, D.R. *et al.* (1991) Human dystrophin expression in *mdx* mice after intramuscular injection of DNA constructs. *Nature*, **352**, 815–18.

Cheng, L., Ziegelhoffer, P.R. and Yang, N.S. (1993) *In vivo* promoter activity and transgene expression in mammalian somatic tissues evaluated by using particle bombardment. *Proc. Natl Acad. Sci. USA*, **90**, 4455–9.

Conry, R.M., LoBuglio, A.F., Kantor, J. *et al.* (1994) Immune response to a carcinoembryonic antigen polynucleotide vaccine. *Cancer Res.*, **54**, 1164–8.

Cox, G.J., Zamb, T.J. and Babiuk, L.A. (1993) Bovine herpesvirus 1: immune responses in mice and cattle injected with plasmid DNA. *J. Virol.*, **67**, 5664–7.

d'Albis, A., Couteaux, R., Janmot, C. and Mira, J.-C. (1989) Myosin isoform transitions in regeneration of fast and slow muscles during postnatal development of the rat. *Dev. Biol.*, **135**, 320–5.

Danko, I., Fritz, J.D., Jiao, S. *et al.* (1994) Pharmacological enhancement of *in vivo* foreign gene expression in muscle. *Gene Ther.*, **1**, 114–21.

Davis, H.L. and Jasmin, B.J. (1993) Direct gene transfer into mouse diaphragm. *FEBS Lett.*, **333**, 146–50.

Davis, H.L., Demeneix, B.A., Quantin, B. *et al.* (1993a) Plasmid DNA is superior to viral vectors for direct gene transfer in adult mouse skeletal muscle. *Human Gene Ther.*, **4**, 733–40.

Davis, H.L., Michel, M.-L. and Whalen, R.G. (1993b) DNA based immunization for hepatitis B induces continuous secretion of antigen and high levels of circulating antibody. *Human Mol. Genet.*, **2**, 1847–51.

Davis, H.L., Whalen, R.G. and Demeneix, B.A. (1993c) Direct gene transfer into skeletal muscle *in vivo*: factors affecting efficiency of transfer and stability of expression. *Human Gene Ther.*, **4**, 151–9.

Davis, H.L., Michel, M.-L., Mancini, M. *et al.* (1994) Direct gene transfer in muscle with plasmid DNA for the purpose of nucleic acid immunization. *Vaccine*, **12**, 1503–9.

Eisenbraun, M.D., Fuller, D.H. and Haynes, J.R. (1993) Examination of parameters affecting the elicitation of humoral immune responses by particle bombardment-mediated genetic immunization. *DNA Cell Biol.*, **12**, 791–7.

Furth, P.A., Shamay, A., Wall, R.J. and Hennighausen, L. (1992) Gene transfer into somatic tissues by jet injection. *Anal. Biochem.*, **205**, 365–8.

Fynan, E.F., Robinson, H.L. and Webster, R.G. (1993a) Use of DNA encoding influenza hemagglutinin as an avian influenza vaccine. *DNA Cell Biol.*, **12**, 785–9.

Fynan, E.F., Webster, R.G., Fuller, D.H. *et al.* (1993b) DNA vaccines: protective immunizations parental, mucosal, and gene-gun inoculations. *Proc. Natl Acad. Sci. USA*, **90**, 11478–82.

Geissler, E.K., Wang, J., Fechner, J.H. *et al.* (1994) Immunity to MHC class I antigen after direct DNA transfer into skeletal muscle. *J. Immunol.*, **152**, 413–21.

Jiao, S., Williams, P., Berg, R.K. *et al.* (1992) Direct gene transfer into nonhuman primate myofibers *in vivo*. *Human Gene Ther.*, **3**, 21–33.

Kuby, J. (1994) *Immunology* W.H. Freeman & Co., New York, 1994.

Martinon, F., Krishnan, S., Lenzen, G. *et al.* (1993) Induction of virus-specific cytotoxic T lymphocytes *in vivo* by liposome-entrapped mRNA. *Eur. J. Immunol.*, **23**, 1719–22.

Michel, M.-L., Davis, H.L., Mancini, M. *et al.* (1995) DNA-based immunization to hepatitis B surface antigen in mice: humoral response mimics that with natural hepatitis B viral infection in humans. *Proc. Natl Acad. Sci. USA*, in press.

Montgomery, D.L., Shiver, J.W., Leander, K.R. *et al.* (1993) Heterologous and homologous protection against influenza A by DNA vaccination: optimization of DNA vectors. *DNA Cell Biol.*, **12**, 777–83.

Nabel, E.G., Plautz, G. and Nabel, G.J. (1992) Transduction of a foreign histocompatibility gene into the arterial wall induces vasculitis. *Med. Sci.* **89**, 5157–61.

Pecorino, L.T. and Lo, D.C. (1992) Gene transfections: having a blast with gene transfer. *Curr. Biol.*, **2**, 30–2.

Robinson, H.L., Hunt, L.A. and Webster, R.G. (1993) Protection against a lethal influenza virus challenge by immunization with a haemagglutinin-expressing plasmid DNA. *Vaccine*, **11**, 957–60.

Sedegah, M., Hedstrom, R., Hobart, P. and Hoffman, S.L. (1994) Protection against malaria by immunization with plasmid DNA encoding circumsporozoite protein. *Proc. Natl Acad. Sci. USA*, **91**, 9866–70.

Tang, D., DeVit, M. and Johnston, S.A. (1992) Genet immunization is a simple method for eliciting an immune response. *Nature*, **356**, 152–4.

Ulmer, J.B., Donnelly, J.J., Parker, S.E. *et al.* (1993) Heterologous protection against influenza by injection of DNA encoding a viral protein. *Science*, **259**, 1745–9.

Vitadello, M., Schiaffino, M.V., Picard, A. *et al.* (1994) Gene transfer in regenerating muscle. *Human Gene Ther.*, **5**, 11–17.

Wang, B., Boyer, J., Srikantan, V. *et al.* (1993a) DNA inoculation induces neutralizing immune responses against human immunodeficiency virus type 1 in mice and nonhuman primates. *DNA Cell. Biol.* **12**, 799–805.

Wang, B., Ugen, K.E., Srikantin, V. *et al.* (1993b) Gene inoculation generates immune responses against human immunodeficiency virus type I. *Proc. Natl Acad. Sci. USA*, **90**, 4156–60.

Whalen, R.G. (1994) Promoters, enhancers and inducible elements for gene therapy, in *Gene Therapeutics*, (ed. J.A. Wolff), Birkhäuser, Boston, pp. 60–79.

Whalen, R.G., Harris, J.B., Butler-Browne, G.S. and Sesodia, S. (1990) Expression of myosin isoforms during notexin-induced regeneration of rat soleus muscles. *Dev. Biol.*, **141**, 24–40.

Williams, W.V., Boyer, J.D., Merva, M. *et al.* (1993) Genetic infection induces protective *in vivo* immune responses. *DNA Cell Biol.*, **12**, 675–83.

Wolff, J.A., Malone, R.W., Williams, P. *et al.* (1990) Direct gene transfer into mouse muscle *in vivo*. *Science*, **247**, 1465–8.

Wolff, J.A., Williams, P., Acsadi, G. *et al.* (1991) Conditions affecting direct gene transfer into rodent muscle *in vivo*. *BioTechniques*, **11**, 474–85.

World Health Organization Programme for Vaccine Development (1990) Potential use of live viral and bacterial vectors for vaccines. *Vaccine*, **8**, 425–37.

Xiang, Z.Q., Spitalnik, S., Tran, M. *et al.* (1994) Vaccination with a plasmid vector carrying the rabies virus glycoprotein gene induces protective immunity against rabies virus. *Virology*, **199**, 132–40.

Yankauckas, M.A., Morrow, J.E., Parker, S.E. *et al.* (1993) Long-term anti-nucleoprotein cellular and humoral immunity is induced by intramuscular injection of plasmid DNA containing NP gene. *DNA Cell Biol.*, **12**, 771–6.

Index